46 Topics in Current Chemistry
Fortschritte der chemischen Forschung

Photochemistry

Springer-Verlag Berlin Heidelberg GmbH 1974

This series presents critical reviews of the present position and future trends in modern chemical research. It is addressed to all research and industrial chemists who wish to keep abreast of advances in their subject.

As a rule, contributions are specially commissioned. The editors and publishers will, however, always be pleased to receive suggestions and supplementary information. Papers are accepted for "Topics in Current Chemistry" in either German or English.

Any volume of the series may be purchased separately.

ISBN 978-3-662-15871-5   ISBN 978-3-540-37908-9 (eBook)
DOI 10.1007/978-3-540-37908-9

Contents

Physical Basis of Qualitative MO Arguments in Organic Photochemistry

Prof. Josef Michl*

Department of Chemistry, University of Utah, Salt Lake City, Utah, USA

Contents

---

* A. P. Sloan Foundation Fellow, 1971—1973.

1

J. Michl

Introduction

Although far from complete, present-day understanding of ground-state reactivity of organic molecules is relatively well advanced. The underlying physical model is based on the Born-Oppenheimer approximation [1] which leads to the concept of a ground-state energy hypersurface on which reactions occur. The organic chemist has a fair feeling for the shape of this hypersurface, based on his "intuition" and on the valence theory. The latter can be simply the basic theory of directional valence and knowledge of bond angles, lengths, force constants, steric repulsion and substituent and delocalization effects, but in recent years quantum mechanical arguments, such as those leading to the Woodward-Hoffmann rules [2-12] and even numerical quantum mechanical calculations of the shape of parts of the hypersurfaces [13] have found their way into this arsenal.

On the contrary, understanding of photochemical reactivity is lagging far behind. This appears to be due to the much more limited amount of empirical knowledge that has been acquired to date, as well as to the intrinsically more complicated nature of the processes involved. The familiar ground of the Born-Oppenheimer approximation often has to be abandoned at some phase of the reaction process, which typically starts with a species in an excited state and ends with another species in the ground state. The excited state Born-Oppenheimer hypersurfaces undoubtedly still play a role, but various authors disagree as to what it is, and besides, less "intuition" is available as to the shape of these hypersurfaces. Singlet and triplet behavior clearly must be distinguished, yet the naive MO arguments at the Hückel level, which were so useful for ground state processes, fail to do so. It is thus hardly surprising that the treatment of photochemical reactivity in existing reviews covering topics such as Woodward-Hoffmann rules is rather vague and emphasizes the uncertainty which prevails in the area. Some authors believe that the Woodward-Hoffmann interpretation of correlation diagrams is completely irrelevant [14,15], even for reactions believed to be concerted. Texts on organic photochemistry generally reflect similar ambiguous attitudes toward the use of simple MO arguments for the interpretation of photochemical reactivity: the Woodward-Hoffmann rules work sometimes but must be applied with caution and their physical basis is debatable.

Considering the existing lack of consensus, this may seem to be an odd time to choose to write a review on the use of simple MO arguments in photochemistry. Yet, in the author's opinion, this should be done. Whether with or without a reasonable or properly understood physical basis, such arguments are and undoubtedly will be used extensively by the great number of practicing organic photochemists, simply because they often work and nothing better is easily available. By outlining a reasonable (even if far from unique) view of such a physical basis and pointing out some of the complexities, limitations, and pitfalls, one can hope to contribute to a better understanding of the nature of the problem[a].

The organic photochemist without great interest in theory may want to read the present review to find out about the kind of physical background that simple MO arguments can be given. On the other hand, each mechanistic and physical photochemist probably already has a set of views of his own and may be bored by those parts of the review which agree with his opinions and irritated by those which do not. Better future reviews may result. Those doing purely theoretical work on photochemical problems rarely are interested to actually use simple MO arguments of the nature discussed here, but may be intrigued by them and perhaps decide to provide physically sounder rationalizations.

All of the above groups of potential readers will agree that the description provided here is oversimplified and overgeneralized. It is, first, because the author concentrated on what appears to be typical behavior, rather than exceptional, and second, because of the complexity of the subject. If it provokes thought, discussion, experiments or calculations which will help the development of a less naive, more universally acceptable, and yet relatively simple and practically useful theory for photochemical behavior of large organic molecules, particularly in solutions, the author's goal will be achieved, even if it turns out that the attempt to present a unified view of organic photochemistry, incorporating most or all of the essential features and useful as a basis of future elaboration, was premature.

The text assumes a certain familiarity with the derivation and use of correlation diagrams for ground state reactions and with fundamentals of their derivation for excited states, say on the level of Woodward and

---

[a] The author originally became interested in finding a simplified unified physical view of organic photochemical reactivity in connection with a discovery of some unusual two-photon electrocyclic reactions.[16] Attempts to understand the phenomena in terms of Woodward-Hoffmann rules looked promising [17], but a need was felt for a relatively well-defined physical basis for their application. This was not found ready-made in existing literature, but could be synthesized [18] by choosing from the great number of views already expressed by various authors, tested on a variety of known photochemical processes [19], and then applied to the problem at hand.[20]

Hoffmann's review.[2] The number of specific examples mentioned in the text is severely limited in order to save space; they can be easily found elsewhere.[2] Instead, space is devoted to detailed discussion of topics likely to be less familiar to the organic chemist, such as some of the properties of potential energy hypersurfaces in multidimensional nuclear configuration space, etc. When in doubt, the author erred on the side of sounding too explicit and trivial at the risk of offending the reader with good physical background.

Most of the arguments are illustrated on hydrocarbons, for the sake of simplicity. In many instances, generalization to more complex molecules is straightforward, and much of the discussion will be applicable directly.

It is appropriate to single out at least a few of the previous reviews.[15,21-23] The most important points in which the present review differs are the emphasis on the combination of MO arguments with those concerning the physical fundamentals of photochemical processes, and the introduction of a simple physical basis for distinguishing between singlet and triplet reactivity.

Of course, the literature abounds with discussions of the use of qualitative MO and VB arguments in photochemistry.[24] However, these are often not very suitable for introduction to the field. Usually, they suffer from the lack of a well-defined physical model for photochemical reactions, which is either assumed to be well known to the reader, or simply ignored. In its absence, however, it is not really clear what the critical feature of the reactions is that needs to be estimated. Some authors took this to be the presence or absence of a symmetry-imposed barrier in the lowest excited state hypersurface[2,25], others claimed that this was irrelevant and what mattered was the location of minima in this surface.[14,15] It seems preferable to first adopt a physical model for the processes involved and to proceed to the use of numerical calculations or qualitative MO arguments only after it is determined from the model what is to be estimated (of course, one will then run the risk of adopting a poor model). The present review is divided into three parts accordingly. The model adopted is based on current thinking of molecular spectroscopists and physical photochemists and only the future will show whether it is realistic. Combined with simple MO arguments, it seems to account well for many known tendencies in organic photochemistry and has even allowed some predictions.[19,20] Its solution for the existing controversy is simple and obvious: both minima and barriers are important.

Right at the beginning, the author would like to disclaim originality although he does not wish to disclaim responsibility for selecting ideas in a mutually compatible way from the vast reservoir of views already expressed in the literature, which are of course often contradictory. In other words, while it is hoped that the overall picture is not exactly like any of those drawn before, the individual building blocks from which the mosaic was assembled already existed. A generalization [17,19,26] of the treatment of singlet-triplet reactivity differences very briefly mentioned by Fukui and Kita [5,27], and recognition of the general importance of "abnormal" orbital crossovers [17,20,26], previously mentioned by Chu and Kearns [28], are the only important parts not taken over from other authors.

However, since this is not a historical review, and since the literature on the topic is quite vast, with many important ideas buried in discussions of experimental results, no claim is made of having correctly assigned priorities for the various contributions. Rather, the sources given are the ones which were actually used to develop the author's present views, those which are particularly illustrative, which are of review nature, and the most recent ones, and further references can be found there.

I. A Qualitative Physical Model for Photochemical Processes

The physical basis for the present model is taken from the work of molecular spectroscopists, physical photochemists, and theoreticians such as Born [1], Teller [29], Condon [30], Kasha [31,32], Herzberg and Longuet-Higgins [33], Coulson [34], Porter [35], Siebrand [36–39], Noyes [40], Jortner, Rice, Hochstrasser, Englman, and Freed [41–50], Becker [51], Förster [23], Ross [52], Simpson [53], Lippert [54], Berry [55] and their respective collaborators, and summarized in books dealing with spectroscopy and physical aspects of photochemistry. [56–62] For qualitative applications to a real system, it is necessary to know at least the shapes of the ground (S_0) and first excited (S_1) singlet hypersurfaces and the lowest triplet hypersurface (T_1), in particular the location of minima, valleys and barriers. Such simple kind of thinking, which ignores problems of molecular dynamics, has served as a starting point for most of the recent discussions of experimental results in mechanistic organic photochemistry, two voluminous examples being provided by the long series of articles by Hammond, Zimmerman, and their respective collaborators (for references see Refs. [63,64]). More sophisticated and more quantitative applications such as those outlined in Ref. [41] would require much more information an additional excited states, on densities of vibrational levels, on vibronic coupling matrix elements, etc., and presently appear out of reach.

A. Born-Oppenheimer Hypersurfaces[b]

It will be remembered that these potential energy hypersurfaces are obtained in principle by solving the molecular Schrödinger equation for a great number of nuclear geometries (points in the nuclear configuration space), assuming in each calculation that the nuclei are stationary. At each geometry, an infinite number of solutions are obtained, each associated with a total energy. The many-dimensional plot of the lowest energy against the nuclear geometry for which the energy was calculated provides a multidimensional hypersurface (sheet) for the ground state. The molecular wavefunction associated with a point on the hypersurface does not appear anywhere in the plot, but there is one for every such point. All isomers which can be formed from a given collection of atoms and electrons have a common ground state hypersurface and correspond to minima in it. Barriers in between correspond to unstable geometries.

The infinite number of higher energy solutions of the Schrödinger equation obtained at each geometry will combine similarly to give rise to an infinite number of excited state hypersurfaces, of which usually only the lowest few are of interest. While in principle the various hypersurfaces can cross as the nuclear configuration coordinate is varied along various paths [33], this is a relatively uncommon occurrence and along most paths such crossings, even if "intended", are more or less strongly avoided[c]. Also states of molecules differing only in a number of electrons, e.g., ionized states, can be plotted in the same picture.

---

[b] Throughout the article, we will have frequent occasion to refer to sketches of cross-sections through potential energy hypersurfaces such as shown in Fig. 1. It is important to keep in mind the limitations of such two-dimensional representations. For example, we indicate the initial excitation by an arrow starting at the S_0 surface and ending at the excited surface, in keeping with the Franck-Condon principle (the electronic excitation itself does not change the energy of the nuclear motion). However, in polyatomic molecules, one could draw the arrow so that it would start and end above the curves representing the surfaces, and still obey the Franck-Condon principle by acknowledging that some vibration in a direction perpendicular to that represented in our two-dimensional drawing is excited. In other words, the excitation occurs outside of the particular cross-section chosen. This presentation is perhaps a little confusing and is not used here, but it allows one to represent many related cross-sections in a single two-dimensional drawing, effectively eliminating what one considers to be unimportant directions of nuclear motion, such as perhaps some stretching or twisting in a region unrelated to the actual reaction center. It is thus helpful for consideration of the effect of the energy of the absorbed photon on the subsequent motion on the excited hypersurface. The efficiency with which motion in "unimportant" directions is transformed into motion along the chosen reaction coordinate determines how much of the "extra" energy of the photon can be utilized for the reaction, say to move over some barriers in the way.

[c] Even in highly symmetrical molecules, at most points in the nuclear configuration space the molecule has very low or no symmetry.

Fig. 1 emphasizes the convention used in this review for labelling of Born-Oppenheimer hypersurfaces: at any given geometry, Born-Oppenheimer states are labelled sequentially in the order of their increasing energies. This differs from the convention commonly used by spectroscopists and physical photochemists, which would follow a state through a crossing keeping the same label, say S_0 or S_1. The present convention is advantageous when one considers large portions of the nuclear configuration space simultaneously, as is often required in photochemistry, but not in electronic spectroscopy. As usual, spin-orbit coupling is not included in the Born-Oppenheimer Hamiltonian. As a result, singlet and triplet states can be distinguished, and their surfaces then cross freely. Such crossings do not disturb the labelling S_0, S_1, S_2, and T_1, T_2, T_3.... The existence of spin-orbit coupling is added as an afterthought in the usual manner [32,39,41,43,50]; this procedure is justified in the absence of heavy atoms. Probability of leakage between S and T states (intersystem crossing) is then always relatively small even if they cross: the nuclear motions are basically governed by either an S surface or a T surface. In the weak coupling limit (no state crossing), the rate of intersystem crossing in a series of related molecules can be, in general, expected to increase exponentially with decreasing energy gap between the two states and this agrees with experiments on aromatic hydrocarbons. [37,39,42, 43,49,50]

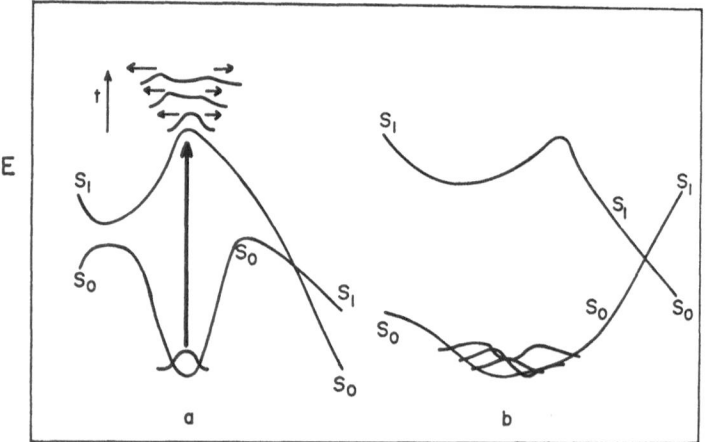

Fig. 1. The lowest singlet (S_0) and first excited singlet (S_1) surfaces of two hypothetical molecules. Vibrational wavefunctions for one and three vibrational levels, respectively, are indicated. Top part of *a* indicates schematically the time development of the nuclear geometry probability distribution after initial excitation

On the other hand, if states of the same multiplicity cross, they exchange their labels as defined by our convention (Fig. 1), but not with the spectroscopist's convention. Here, at any nuclear geometry, the lowest singlet state is S_0, the lowest triplet state T_1, etc. This may appear awkward to the spectroscopist but is useful for our present purposes. The convenience gained is that all points throughout the nuclear configuration space are treated on the same footing and it is not necessary to distinguish between S_0 of one isomer and S_0 of another isomer — the S_0 surface is common to all isomers. In spectroscopist's convention, S_0 of one isomer could simultaneously be say S_2 of another isomer, and S_1 of a molecule at one geometry could simultaneously be S_2 of an isomeric or even the same molecule at another (or even the same) geometry if it happened to cross another state in the appropriate manner. Using the convention adopted here, one can make general statements such as that molecules always eventually return to the surface labelled S_0 (or T_1), no matter which part of the nuclear space they happen to end up in. All this is typically of little or no consequence for molecular spectroscopists interested in a small area in the neighborhood of the initial geometry.

The inconvenience introduced by the labelling system used here is that the physical nature of any given state, say S_1, will often change, possibly rather abruptly, as the nuclei wander throughout the nuclear configuration space (Fig. 1). It may be $n\pi^*$ at some geometries, $\pi\pi^*$ at others, "singly excited" at some, "doubly excited" at others.

In areas without state crossings both conventions are identical. A state, say S_2, governs the nuclear motion, terms neglected in the Born-Oppenheimer approximation may induce a slow transition to S_1, whose rate again should depend exponentially on the energy gap [50], etc. In areas with state crossings, or weakly avoided state crossings, the two labelling schemes differ but both describe the same physical situation: after going through an area of avoided crossing, a molecule may find its nuclear motions governed by the hypersurface which brought it in, or by the one with which an avoided crossing occurred. This is illustrated in Fig. 2, which shows a weakly avoided crossing between two singlet states. The physical nature of S_0 changes abruptly in the region of an avoided crossing, and so does that of S_1. Instead of following the same Born-Oppenheimer hypersurface after reaching the area of the avoided crossing, some molecules will then emerge from the area with their motions governed by a hypersurface with a new label (they "jump" from one surface to the other). The probability of this happening will increase with increased velocity of the nuclear motion and also if the crossing becomes less avoided, and approach unity if the crossing is not avoided at all. As the crossing becomes more strongly avoided, the probability will become very small [64a]. If one were to judge the suitability of a

9

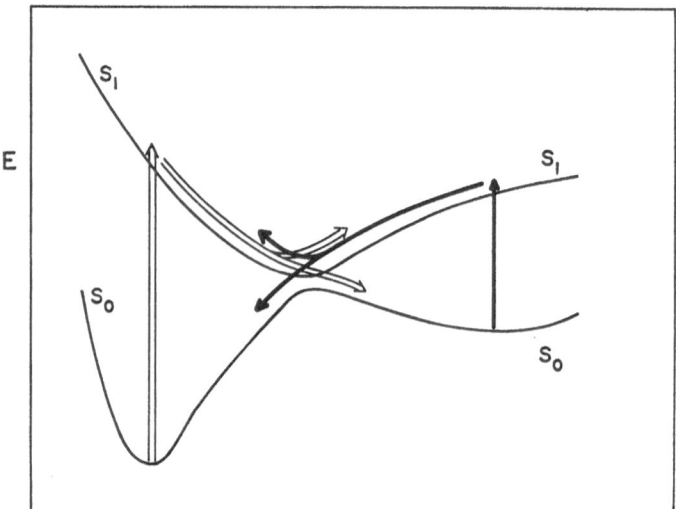

nuclear configuration coordinate

Fig. 2. Avoided crossing of the S_0 and S_1 hypersurfaces. Vertical arrows indicate initial excitation; additional arrows show how the surfaces govern nuclear motion afterwards

labelling system on the basis of how often it allows the nuclear motions of a molecule to be governed by a hypersurface with the same label after it goes through an area of avoided crossing, neither convention is perfect since some "jumps" will always occur. In the limit of a weakly avoided crossing, the "spectroscopic" convention is better (fewer molecules change state labels); in the limit of strongly avoided crossing, the convention adopted here is better. In the language of perturbation theory, one can say that Born-Oppenheimer states are a good zero-order approximation for description of nuclear motion (our convention) if non-Born-Oppenheimer corrections are small, since the large terms which cause the crossing to be avoided are already included in the zero-order Hamiltonian (*e.g.*, electron repulsion), and diabatic states ("spectroscopic" convention) are a good zero-order approximation in the opposite case. The choice of a labelling convention is purely a matter of convenience. To a spectroscopist investigating a predissociation phenomenon, the "spectroscopic" labelling is much more natural; to a quantum chemist calculating potential energy hypersurfaces using the Born-Oppenheimer Hamiltonian, the one used here will appear more natural.

It is now clear why the abrupt changes in the physical nature of a state, encountered in our convention whenever there is a weakly avoided crossing, are inconvenient: they will induce jumps from one surface to another (change of label) with relatively high probability. Moreover, note that the fate of a molecule which finds itself in an area of a weakly avoided crossing will, in general, depend on where it came from, so that the minimum in S_1 in Fig. 2 can be expected to have some peculiar properties, as will be mentioned later. Also, as indicated in Fig. 3, the shape of any given surface, such as S_1, may be rather bumpy due to a variety of avoided crossings. Motion over the bumps will be unusually difficult if they result from weakly avoided crossings, since "transmission coefficients" of the barriers will be low: probability of a jump to a higher surface, following the "diabatic" or "spectroscopic" curve, will be high, particularly if the nuclei are moving fast, and when a turning point is reached, the molecule will return from above the barrier although it had enough energy to pass it. Considerations such as this show that predictive calculations of the dynamics of molecular motion on the S_1 (T_1) surface are exceedingly difficult.

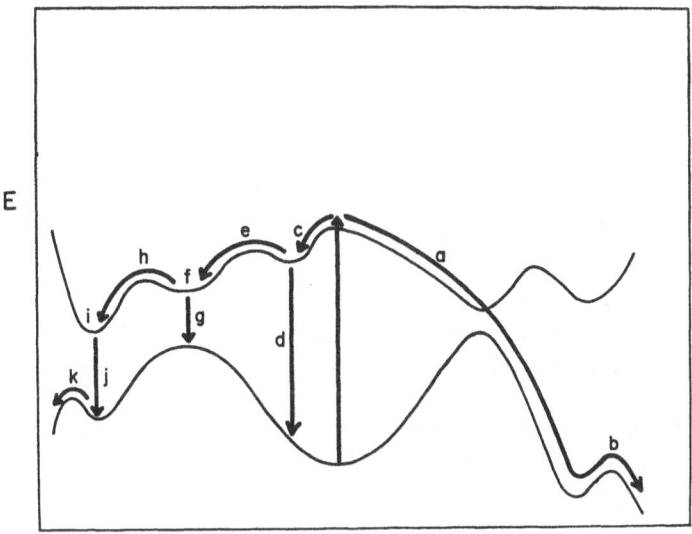

nuclear configuration coordinate

Fig. 3. The S_0 (lower) and S_1 (upper) hypersurface and various processes following initial excitation

B. Initial Excitation

In the Born-Oppenheimer approximation, the total wavefunction of a molecule is approximated as a product of parts which describe the translational motion, the rotational motion, the vibrational motion, the electronic motion, etc. According to the (approximate) Franck-Condon principle, light absorption is a "vertical" process. Assuming that the electronic transition probability is the same for all nuclear geometries accessible to the ground state molecule, the vibrational (as well as rotational, translational, nuclear, etc.) part of the total wavefunction of the molecule remains unchanged while the electronic part changes: $\psi_{el} \cdot \psi_{vib} \xrightarrow{h\nu} \psi'_{el} \cdot \psi_{vib}$. Before irradiation, the molecule was in a stationary state: ψ_{vib} was an eigenfunction of the vibrational Hamiltonian containing kinetic energy of the nuclei and their potential energy defined by the hypersurface corresponding to ψ_{el}. At relatively low temperatures, and in molecules lacking low frequency vibrations, ψ_{vib} is the lowest-energy (zero-point energy) vibrational eigenfunction with respect to all normal coordinates, as indicated schematically in Fig. 1a for a hypothetical molecule (S_0 surface). At higher temperatures, or if low-frequency normal modes are available, a considerable fraction of molecules may start with ψ_{vib} in which one or more of the low-frequency modes are excited by one or more quanta of vibration. Then their ψ_{vib} typically have relatively large amplitude throughout a larger region of nuclear geometries (Fig. 1b, S_0 surface). If the electron transition probability changes significantly as a function of geometry, the picture becomes somewhat more complicated and the initial distribution of probabilities of various nuclear geometries in the excited state then differs from that given by $|\psi_{vib}|^2$.

C. Changes in Nuclear Geometry after Excitation

As long as the molecule was in ground electronic state and its stationary vibrational wavefunction ψ_{vib} was correctly "adapted" to the ground state potential energy hypersurface, the probability that a measurement would find one or another nuclear geometry did not change in time and was largest at the equilibrium geometry (only the complex phase of ψ_{vib} changed in time). However, after excitation to another electronic state, the motions of the nuclei are suddenly governed by a new potential energy hypersurface, to which ψ_{vib} is not an appropriate eigenfunction, so that its time development becomes more complicated[d]. Crudely viewed, the wavepacket "rolls downhill", spreads out, possibly separates

---

[d] The wavefunction is complex. Fig. 1a indicates the square of its absolute value as a function of time in a very schematic way. It is more usual, but less instructive for our purposes, to expand ψ_{vib} in a complete set of vibrational functions such as the eigenfunctions of the excited hypersurface. The probability that a measurement of the total energy after excitation will give a value corresponding to one or another of these stationary vibrational levels is then related to the familiar Franck-Condon overlap integrals.

into two or more packets, etc. (Fig. 1a, S_1 surface). It is easy to imagine that the overall direction of motion of the bulk of the wavepacket can be a sensitive function of just what the starting vibrational wavefunction was, *i.e.*, results such as quantum yields of photochemical reactions can depend on the starting conformation and temperature already for this reason alone. [65] While one can hope to estimate the general direction in which the bulk of the wavepacket travels from calculations of the shape of the potential energy hypersurface at the starting geometry, this is unfortunately not the whole story. If other excited state hypersurfaces lie nearby, as is common in large molecules, the motion of the nuclei need not be simply dictated by a single hypersurface, because the Born-Oppenheimer approximation will be rather poor, and vibronic mixing need not be negligible. The motion will then be affected by several surfaces simultaneously, particularly in regions in which these undergo a (generally avoided) crossing as already discussed, and even crude calculations become much more difficult.

D. Effect of Medium

So far, the description has been limited to the case of an isolated molecule. In practice, however, the organic chemist typically deals with molecules in solution, or in gas phase at relatively high pressures. The medium then acts as a heat sink and efficiently removes excess vibrational energy. These are the conditions to which we shall limit our attention in the following. The simplest description would be that the overall motion of the wavepacket is slowed down by friction so that the nuclei never acquire very much kinetic energy in spite of the acceleration they receive from the hypersurface corresponding to the excited state. Attempts at calculations, even crude, become even more complicated. More realistic pictures of the effect of the heat bath presently appear to be hopelessly complex for detailed calculations.

E. Role of Minima in S_1 and T_1

After a very short period of time, 10^{-11}—10^{-12} sec, crossing from higher to lower excited states [36] (internal conversion) and thermal equilibration [67] will bring the molecule to one or another of the numerous minima in its first excited state hypersurface. If the initial excitation was into the triplet manifold, this will typically be T_1; if it was into the singlet manifold, it will typically be S_1 (Kasha's rule [31]), exceptionally S_2 if the internal conversion to S_1 is slow (azulene [48,68]), or T_1 if intersystem crossing into the triplet manifold proceeds unusually fast and is able to compete with the relaxation to S_1 (particularly in the presence of heavy atoms). [69] It is important to realize that there are usually more minima in the S_1 and T_1 surfaces than in the S_0 surface; in other words, many of the minima in S_1 and T_1 surfaces do not correspond to stable ground state molecules. The reason for this is discussed in the next section. It is common to associate minima in the S_0 surface with names of conformers of a molecule, and to call those minima in S_1 and T_1 which lie in their neighborhood, and are accessible directly by light absorption, (spectroscopic) excited state of that same molecule (conformer). However, no good names are available for those minima which have no ground state counterparts (cf. "phantom triplets" of olefins [70]). We will run into this problem in our later discussion of MO arguments, but it will be helpful to provide examples now. The situation is still relatively easy in the case of minima reached by stretching one bond. For instance, the minimum in T_1 hypersurface of formaldehyde which exists close to the ground state equilibrium geometry and is well known from spectroscopic studies (pyramidal $^3CH_2O$) is simply triplet formaldehyde, and the minimum in its T_1 which corresponds to large separation of H and CHO fragments can be called a triplet radical pair. An example of a worse problem is the minimum calculated [14] to lie in S_1 hypersurface halfway between butadiene and cyclobutene geometries along the disrotatory pathway, near a high ridge in the S_0 surface. No good name appears to be available for minima of this kind (the author used the terms Van der Lugt — Osterhof minimum and pericyclic biradicaloid minimum in previous articles [17,19]).

After a shorter or longer sojourn in a minimum in S_1 the molecule goes to S_0 (or T_1), either by emission of light or by radiationless conversion. Natural fluorescence lifetimes are usually quite long, say 10^{-8}—10^{-9} sec, and emission is able to compete successfully with return to S_0 only if radiationless conversion is also slow. This is generally the case if the S_0—S_1 gap is relatively large. If the S_1 hypersurface lies close above the S_0 surface, fast nonradiative return will prevail. As already discussed, particularly fast return to S_0 can be expected if the minimum in S_1 is a result of a crossing, or more usually avoided crossing[e], between S_0 and S_1 states (Fig. 2). This is so since the probability that motions of the nuclei will start to be governed by a new potential energy hypersurface is related to the rate at which the electronic wavefunction changes as the nuclei

[e] This is a common designation of the situation depicted in Fig. 2, derived from the "spectroscopic" state labelling, and we shall use it for lack of a better expression. Adopting our labelling convention, the S_0 and S_1 states do not cross or attempt to cross, but instead attempt to touch and exchange their physical nature. The problem is purely one of semantics.

14

move, and this rate of change is large in the region of avoided crossings (infinite at the point where the two hypersurfaces cross if the crossing is not avoided at all). In a pictorial way, one could say that the molecule dislikes sudden changes in its electronic wavefunction as its nuclei move, and will "jump" from one surface to another in order to avoid them.

In the limit, the probability that a $S_1 \rightarrow S_0$ jump will occur when a molecular geometry of the avoided crossing is reached from a suitable direction and fast enough could reach unity, $i.e.$, a single vibration which takes the molecule through this area of the S_1 hypersurface will induce return to S_0. Under such conditions, the molecule would clearly have no chance to reach thermal equilibrium in the minimum in S_1, which would be more properly described as a funnel rather than a well. It is interesting to note that different valleys in S_0 are reached through the same "funnel" depending on which direction the molecule first came from, so that the sum of quantum yields of all processes proceeding from the same excited state, namely that of the funnel, could exceed one (Fig. 2).

An (avoided) crossing of S_0 and S_1 need not even always correspond to a minimum in the excited state surface [64a]. Fig. 1 shows that this will depend on the slopes of the surfaces. Even such crossings which do not actually lead to a minimum in the excited state surface will be called funnels since they can provide efficient return to S_0. Further, if S_2 lies very close to S_1, vibronic coupling may make it impossible to describe the nuclear motion as specifically due to only one or the other hypersurface and both will have to be considered. Then, whether S_1 or S_2 undergoes an avoided crossing with S_0 may be immaterial — a funnel is still present and return of the molecule to S_0 probable.

Another interesting situation occurs if S_0 has a maximum and becomes almost degenerate with S_1 in an area where S_1 does not have a minimum nor a funnel. If a molecule happens to pass through such an area, vibronic mixing may again be strong and the molecule may emerge with its motions governed by either S_1 or S_0. Then, we would again have to label such an area of the S_1 surface a "funnel". It is difficult to assess the importance of this case without a more detailed analysis.

Typically, both of these conditions will occur together: if S_2 undergoes a weakly avoided crossing with S_0, and S_1 lies between them, S_2 will usually have a minimum and S_0 a maximum in the same area, and both will be close to S_1. Arguments given below indicate that S_1 and S_2 will be typically much closer to each other than they will be to S_0. It will still be very difficult to determine the relative importance of the processes outlined above and it may be necessary to consider vibronic mixing of all three states. For further practical pruposes, we shall simply remember that a crossing, or avoided crossing, of S_1 or S_2 with S_0 is likely to effectively provide S_1 with a funnel taking molecules down to S_0.

We can then amend our previous statement by saying that after a very short period of time, say 10^{-11}—10^{-12} sec, the molecule will have reached one of the numerous minima in its S_1 or T_1 hypersurface or have already reached the S_0 surface through one of the funnels leading to it.

Minima in T_1 are usually above the S_0 hypersurface, but in some cases, below it (ground state triplet species). In the latter case, the photochemical process proper is over once relaxation into the minimum occurs, although under most conditions further ground-state chemistry is bound to follow, *e.g.*, intermolecular reactions of triplet carbene. On the other hand, if the molecule ends up in a minimum in T_1 which lies above S_0, radiative or non-radiative return to S_0 occurs similarly as from a minimum in S_1. However, both of these modes of return are slowed down considerably in the $T_1 \rightarrow S_0$ process, because of its spin-forbidden nature, at least in molecules containing light atoms, and there will usually be time for vibrational motions to reach thermal equilibrium. One can therefore not expect "funnels" in the T_1 surface, at least not in light-atom molecules.

F. Two Kinds of Photochemical Reactions

The above-mentioned two possibilities — return to S_0 through a funnel or through a minimum in the lowest excited state — correspond to the two kinds of photochemical reactions sometimes distinguished in texts on photochemistry [57]: the unquenchable essentially instantaneous ones such as photodissociation, and those in which the molecule first equilibrates thermally in a minimum in S_1 or T_1. In our opinion, it is unlikely that return to S_0 could occur efficiently in an area where the excited state has neither a minimum nor a funnel (the whole region populated when thermal equilibrium is reached in a given excited state minimum is counted as one minimum). The molecule is not likely to pass through sloping areas of the excited S_1 or T_1 hypersurface more than a relatively small number of times after initial excitation since it loses excess vibrational energy fast, and if the area contains no funnel, there is no particular reason to expect a jump to S_0 to be highly probable and to occur during the very limited amount of time spent by the molecule in such vicinity.

1. "Instantaneous" Reactions

In these processes, the time development of the initial excited state leads a fraction of the molecules to a funnel in S_1, which is subsequently abandoned for S_0 essentially as fast as it was reached from the state reached by original excitation. The first time thermal equilibration of the vibrational motion in these molecules is achieved after initial excitation is at the very end of the photochemical process proper, *i.e.*, in the first minimum reached in S_0, and the photochemical reaction proper has no intermediates (Fig. 3, path a). Of course, the primary photoproduct may undergo further ground state reactions, and in that sense be an intermediate in the overall complex process (Fig. 3, path k). It should be often possible to prevent such complications by work at very low temperatures. It is worth repeating that in photochemical processes without a vibrationally equilibrated intermediate there is no physical basis to expect the sum of quantum yields of all processes which proceed from the same funnel in the excited hypersurface to be unity. In other words, a common funnel is not quite the same as a common intermediate. The reason for this is that a molecule in a funnel is not sufficiently characterized by giving the positions of the nuclei — the directions and velocities of their motion are needed as well.

2. Reactions with Intermediates

In these processes, return to S_0 from the minimum in S_1 or T_1 originally reached by the molecule is slow enough that vibrational equilibration in the minimum occurs first and the reaction can be said to have an excited state intermediate. Sum of the quantum yields of all processes which proceed from a given minimum (intermediate) then cannot exceed one.

(a) If the minimum in S_1 or T_1 which is originally reached occurs near the ground-state equilibrium geometry of the starting molecule (Fig. 3, path c), it corresponds to one of its spectroscopic excited states. If return to S_0 then occurs, whether radiative or non-radiative (Fig. 3, path d), the molecule ends up in the same minimum in S_0 where it started and the whole process is considered photophysical.

However, often the minimum in S_1 or T_1 which is reached at first is shallow and thermal energy will allow escape into other areas on the S_1 or T_1 surface before return to S_0 occurs (Fig. 3, path e). This is particularly true in the T_1 state which has longer lifetimes due to the spin-forbidden nature of both its radiative and non-radiative modes of return to S_0. The rate of the escape should depend on temperature and is determined in the simplest case by the height and shape of the wall around the minimum, similarly as in ground state reactions (concepts such as activation energy and entropy should be applicable). In cases of intermediate complexity, non-unity transmission coefficients may become important, as discussed above. Finally, in unfavorable cases, vibronic coupling between two or more states has to be considered at all times and simple concepts familiar from ground-state chemistry are not applicable. Presently available experimental data are too scarce to estimate the relative frequency with which the three cases will occur. The first, simplest situation appears to exist for some hydrogen abstraction reactions. [38,71]

Of course, in the case of intermolecular reactions, the rate also depends on the frequency with which diffusion brings in the reaction partner. In the presence of a reaction partner or quencher, new directions in the nuclear configuration space open and these may provide accessible valleys with low or no barriers along the way leading to new minima (exciplexes) or funnels, in S_1 or T_1. Return from these to the S_0 surface then may lead back to the starting point (quenching) or to a new minimum in S_0 (photochemical reaction). It appears that at least small activation energies (a few kcal/mole) due to the presence of walls around the minimum S_1 or T_1 should be expected in general for monomolecular photochemical reactions with intermediates. In bimolecular cases, this need not be so: there may be enough time for thermal equilibration in a "minimum" which does not have walls all around but instead slopes downhill in some directions, if motion in those directions is slow due to finite diffusion rates.

If thermal motion on the T_1 (or S_1) surface leads to a quasi-equilibrium distribution of molecules between several minima, some of them are likely to provide a faster return to S_0 than others and they will then "drain" the excited state population and determine which products will be formed. This is a straight-forward kinetic problem and it is clear that the process need not be dominated by the position of the lowest-energy accessible minimum in the excited hypersurface. Such minima may correspond to conformers, valence isomers, etc. Of course, it is well known that ground-state conformers may correspond to excited-state isomers, which are not in fast equilibrium. [65,72] Also, there is no reason why several separate minima in S_1 or T_1 could not correspond to one minimum in S_0, and there is some evidence that this situation indeed occurs in certain polycyclic cyclohexenones. [73,74]

Another process which allows the molecule to escape from an originally reached well in S_1 is intersystem crossing to the triplet manifold. This process can again be temperature dependent since the rate of the crossing into the triplet hypersurface can, in general, depend on the amount of vibrational energy available to the molecule. [45,75,76] Other possibilities are return to the S_0 hypersurface by classical energy transfer to another molecule, excitation to higher excited states by absorption of another photon or by triplet-triplet annihilation, etc. The result of all of these processes is that the molecule eventually again ends up temporarily in some minimum in S_1 or T_1, or returns back to S_0 through a funnel.

(b) If a molecule ends up in a minimum in S_1 or T_1 which is further away from the nuclear geometry of the starting species, it may correspond to a "non-spectroscopic" excited state (Fig. 3, minimum f), or even to a "spectroscopic" excited state of another molecule (or conformer, Fig. 3, minimum i). Reactions of the latter kind can sometimes be detected by product emission (Fig. 3, path j), comparison of final products with those obtained by direct excitation of the new molecule from its ground state to the S_1 or T_1 minimum, etc. However, only a few such reactions have been detected (e. g., Refs.[23, 77-80]).

On the other hand, reactions in which the return to S_0 occurs from a "non-spectroscopic" minimum (Fig. 3, path g) are probably the most common kind. The return is virtually always non-radiative[f]. This may be the very first minimum in S_1 (T_1) reached, e.g., the twisted triplet ethylene, or the molecule may have already landed in and again escaped out of a series of minima (Fig. 3, sequence c, e). For instance, triplet excitation of trans-stilbene [70,81-83] gives a relatively long-lived trans-stilbene triplet corresponding to a first "spectroscopic" minimum in T_1. This is followed by escape to the "non-spectroscopic", short-lived "phantom" twisted stilbene triplet, corresponding to a second and last minimum in T_1. This escape is responsible for the still relatively short lifetime of the planar $\pi\pi^*$ triplet compared to $\pi\pi^*$ triplet of, say, naphthalene. A jump to nearby S_0 and return to S_0 minima at cis- and trans-stilbene geometries complete the photochemical process[g].

Escape from the minimum described in this paragraph is again possible in a variety of ways (e.g., path h in Fig. 3), and the process may go on for awhile before return to S_0 from one of the minima or funnels finally occurs.

3. The Term "Intermediate"

Many authors may object to our usage of the word "intermediate" for the "spectroscopic" excited state at the original geometry, such as trans-stilbene triplet, and there may even be some opposition to the usage of this term for the "non-spectroscopic" excited state, such as twisted stilbene triplet, and a tendency to reserve the term "intermediate" only for those species which have a minimum in the S_0 hypersurface.

---

[f] A very interesting case of a radiative return from such a minimum in T_1 by phosphorescence has been reported recently.[84]

[g] Other authors prefer to draw the shape of the potential energy surface for triplet stilbene differently [85], but the existence of controversies does not detract from the illustrative value needed here.

We prefer the present usage because it emphasizes the analogy between the role of the S_1 and T_1 hypersurfaces in photochemical reactions and the role of the S_0 hypersurfaces in thermal reactions, and because it emphasizes the physical similarity of "spectroscopic" and "non-spectroscopic" minima in S_1 and T_1. Indeed, the distinction between them does not depend on what the shape of S_1 or T_1 is, but what the shape of S_0 is! In reality, S_0 is only of secondary importance — it co-determines the rate of return from various minima and funnels in S_1 and T_1, and comes into play after the photochemical process proper is over. In this sense, the sequence of initial excitation, then nuclear motions dictated by the various excited states reached originally and along the way, and finally by S_1 or T_1, and at the end the return to S_0, acts like a conveyor belt which lifts molecules from their original well in S_0 and delivers them at some new and often unusual points at S_0, from which they roll back down to some possibly different well in S_0. Further ground state chemistry may then occur, and although we do not consider it a part of the photochemical process proper, it is clear that it is very important for the observable overall chemical change. In some instances, valuable information on the course of the overall reaction course can be obtained by studies of ground-state chemical reactions occurring in the same region of the S_0 hypersurface (*e.g.*, Ref. [63]). It seems best to call the minima in S_1 and T_1, in which the molecule reached thermal equilibrium, "excited intermediates", whether they are "spectroscopic" or "non-spectroscopic", and those in S_0 in which it equilibrates after return to S_0 "ground-state intermediates". Those minima in S_1 in which thermal equilibration cannot be achieved for lack of time are classified as "funnels" (and it is necessary to remember that some "funnels" in S_1 may correspond to inflex points rather than minima).

G. Back in the Ground State

Once the nuclear motions of a molecule are again governed by the familiar S_0 surface, the organic chemist is on firmer ground. If the return was non-radiative, as is commonly the case, the molecule will usually have a large excess of vibrational energy. In dense media, àt ordinary temperatures, this will be lost [67] in 10^{-11}—10^{-12} sec, the molecule ending up in one of the minima in S_0. One can expect to be able to estimate which minima in S_0 will be reached from approximate knowledge of the shape of the S_0 hypersurfaces and of the location of the area in which return to S_0 occurred.

Nevertheless, the situation differs in an important respect from the familiar thermal reactivity. The molecule will have "landed" in an area of S_0 which typically slopes in some direction, possibly quite steeply.

As it "rolls down the hill", its nuclei may acquire considerable kinetic energy already during the first vibration. Even in dense media, where collisions are frequent and efficient, removal of this excess energy may not be fast enough to prevent the molecule from going over nearby barriers which normally would be prohibitively high. This is likely to occur if the original impulse which the nuclei received is aimed towards the barrier. The products formed will then be classified as due to a "hot ground state" reaction (Fig. 3, path b). If the original motion is in another of the numerous directions available, it is quite likely that enough energy will have been dissipated by collisions as heat before the barrier is approached, and this is the usual argument against the occurrence of hot ground state reactions in dense media. Of course, in gas phase at low pressure and possibly at very low temperatures in rare gas matrices [86] (where heat conductivity is low), hot ground state reactions can be expected to play a role in the final steps of many photochemical processes. However, even in dense media at ordinary temperatures, evidence for such reactions is slowly accumulating. [87]

In general, the nuclear kinetic energy acquired after the nuclear motions are already governed by the S_0 surface is only a part of such energy available. The $S_1 \rightarrow S_0$ or $T_1 \rightarrow S_0$ jump itself generates nuclear kinetic energy at the expense of electronic energy, and finally, the molecule need not even have been thermally equilibrated at the time of the jump if it proceeded through a funnel rather than a minimum in S_1 (Fig. 3, path a, b).

H. Effect of Reaction Variables on the Photochemical Process Proper

The four variables which can be relatively easily controlled are the reaction medium and temperature, and wavelength and intensity of the exciting light.

Effects of the medium can be divided into those which directly modify the potential energy hypersurfaces of a molecule, such as polarity or hydrogen-bonding capacity which affect energies by strong solvation, in particular $n\pi^*$ as opposed to $\pi\pi^*$ state energies, and those which operate by more subtle means, such as microscopic heat conductivity determining the rate of removal of excess vibrational energy, presence of heavy atoms, enhancing rates of spin-forbidden processes, or viscosity, which affects diffusion rates and thus influences the frequencies of bimolecular encounters, through which it may also control triplet lifetimes. At present, many of these variables are investigated routinely in serious mechanistic work.

Very high solvent viscosity also effectively alters shapes of potential energy hypersurfaces by making large changes in shape difficult or impossible. [83,88] In the first approximation, it could then be expected to increase the probability that the excited molecule will not change its initial geometry much and will eventually emit light rather than react. However, even in rigid media at cryogenic temperatures local heating may occur in the neighborhood of the excited molecule as much of the photon energy is converted into heat, particularly if the heat conductivity of the medium is low.

Changes in temperature generally affect ratios of conformer concentration of the starting ground state molecules and also their distribution among individual vibrational levels, and thus co-determine the nuclear configuration the average molecule will have just after excitation (Fig. 1b). Further, they will affect the rates at which molecules escape from minima in S_1 or T_1 (excited state intermediates). At very low temperatures, even barriers of a few kcal/mole may be sufficient to slow down escape from a minimum to such a degree that processes which were too slow at room temperature can compete, *e.g.*, fluorescence. It is, of course, generally known that many compounds with weak or undetectable fluorescence at room temperature emit strongly and react less or not at all at low temperatures and this cannot be always ascribed to increased viscosity. However, systematic efforts to investigate the effect of temperature on photochemical reactivity in solution are rare. [89]

Changes in light wavelength determine the initial total amounts of energy available to the excited species. As pointed out above, in dense media part of this energy is lost very fast and in 10^{-11}—10^{-12} sec the molecule reaches one or another of the minima or funnels in the S_1 (or T_1) surface. However, the probability that one or another of the latter is reached may change drastically as the energy of the starting point changes (internal conversion need not be "vertical"). In general, one can imagine that higher initial energy will allow the molecule to move above barriers which were previously forbidding, and that additional minima will become available. Whether the additional energy is actually used for motion towards and above such barriers, or whether it is used for motion in other "unproductive" directions and eventually lost as heat could be a sensitive function of the electronic state and vibrational level reached by the initial excitation. For our purposes, the present state of knowledge does not warrant an attempt to differentiate between say a "hot" S_1 state and an isoenergetic "less hot" S_2 state in a large molecule, where the density of vibronic states is high, the Born-Oppenheimer approximation poor, and mixing likely to be extremely efficient. [47,52,90] Indeed, experimentally, changes in quantum yields or even nature of products sometimes occur half-way through a given electronic absorption band. [51,91,92] As the sophistication of photochemical models increases, even such subtle distinctions will probably be called for. The number of known wavelength-dependent photochemical reactions in solution is already considerable, although a search has not yet really started in earnest (for leading references, see Refs. [18,51,87,93]).

Light intensity at the usual levels seldom has an effect on the primary photochemical step if all other variables are kept constant, although it may affect overall results considerably since it may control the concentrations of reactive intermediates. However, it will affect the outcome of a competition between primary one-photon and two-photon processes. The latter are still somewhat of a rarity but may be more important than is commonly realized, namely in rigid media where triplets have long lifetimes and quite a few of them are likely to absorb a second photon. The additional available energy may permit motion to new minima in T_1 and thus give new products.

It is clear that the reaction variables can play a decisive role in photochemical reactivity. It is likely that a concentrated effort in this direction will greatly improve an understanding of the nature of photochemical processes.

II. Quantum Mechanical Calculations

The above discussion shows the great variety of basic processes which the photoexcited molecules undergo. It is probably safe to assume that each of these processes has some finite probability in most photochemical reactions and at this point, specific predictions of any nature for real molecules might appear hopeless. Yet quite a few calculations have been reported and virtually all of them claimed at least partial success. For the purpose of our discussion they will be divided into two classes according to the nature of the questions asked. The relatively simple case of certain proton-transfer reactions will be mentioned separately afterwards.

A. Calculations of Regions of Excited Potential Surface

In the first kind of calculations, the authors wish to obtain a relatively detailed description of all that happens during the photochemical process and ask how well one can account for some reaction paths being followed and certain products formed and others not, starting in an excited state of a given multiplicity. A good example is Van der Lugt and Oosterhoff's calculation [14] investigating if it is reasonable that excited singlet butadiene should give cyclobutene in a disrotatory, but not conrotatory fashion, whereas triplet butadiene does not cyclize at all. In such calculations, the shape of the potential surface is typically computed in a region of interest, using a more or less sophisticated theoretical method. For examples of recent use of semiempirical methods, see Refs. [14,17,27,94–104]; for representative ab initio calculations, see Refs. [73,105–110].

It is not always clear how meaningful the calculated shapes of excited hypersurfaces are. Recent ab initio calculations with large basis sets and extensive configuration interaction, or equivalent, are probably quite reliable. On the other hand, the common semiempirical methods suffer from serious defects. Zero differential overlap methods such as PPP, CNDO, and INDO mistreat overlap badly and should not be trusted in the case of twisted unsaturated systems. [111] Hückel-type methods such as Extended Hückel or its iterative versions mistreat electron repulsion and do not permit interaction of configurations. This is perhaps often permissible for T_1, but quite likely fatal in the case of S_1, particularly near biradical-like geometries. [19,112] π-Electron methods treating the σ-framework classically are limited to a small region of the nuclear configuration space in which the lengths of σ-bonds are close to normal and $n\pi^*$, $\sigma\pi^*$, $\pi\sigma^*$, and similar excited states are unimportant. Therefore, they may work well for certain types of reactions, but are inapplicable to others.

Unless there is some obvious flaw such as those mentioned above, semiempirical methods still give a more reliable answer than qualitative MO arguments which we are about to discuss, and it certainly makes sense to use them if a program and computer time are available, provided that limitations such as the above are recognized. However, what is really needed most is a cheap semiempirical method free of the above disadvantages.

A basic problem of all calculations of photochemical reactivity, whether semiempirical or ab initio, is the need for repeated calculation of energies at a huge number of geometries if anything like a decent mapping of the important portion of the excited surface is to be obtained. Indeed, just determining which portion is important may be very costly. Here, a theoretical breakthrough providing a short-cut would be of fundamental importance.

A calculation of the slopes of excited state hypersurfaces usually gives a fairly good idea of paths which are impossible or unlikely, but little if any information is obtained on the probabilities with which the likely paths are actually followed, particularly if the original excitation is into a higher excited electronic state.

Such calculations attempt to answer two questions. First, where are the minima and funnels in the S_1 and T_1 surfaces, assuming that knowledge of ground state chemistry or a calculation will allow us to estimate the fate of a molecule once it reaches the S_0 surface, as long as we know just where it occurred. Second, where are the barriers in S_1 or T_1 which may prevent access to otherwise perfectly reasonable minima, given a starting geometry and excitation energy, and how high are they. As outlined above, the model gives no information about which minima will actually be reached except for suggesting to follow the paths of steepest descent. These may clearly depend on the conformation of the starting species as well as the wavelength of light used, which determine the direction of the first push felt by the nuclei immediately after excitation and the likelihood that the molecule will move toward any given barrier and above it when it gets to it. A detailed understanding requires a considerably better knowledge of details of the excited state hypersurfaces and of vibronic coupling than is presently available for large molecules. [41,46,47,113,114] Hopefully, detailed experimental studies on relatively small molecules [115] will help in further development of the necessary theory. Information about the direction of the very first push can, of course, be obtained from calculations of the potential energy gradient at the starting geometry, as discussed below. However, once the molecule departs considerably from the starting geometry, the problem becomes more difficult. Also consideration of the effect of other nearby surfaces (vibronic mixing) would complicate matters considerably.

B. Calculations of Potential Energy Gradient at or Near the Starting Geometry

This brings us to the other and less ambitious class of calculations, in which it is taken for empirical fact that certain kinds of products can be formed in a given set of reactions, and one inquires how the yields of the various products depend on molecular structure in a series of related molecules, or how successful will various possible closely related paths be in competing against each other. This information is sought from a calculation for the excited state at the starting geometry and in its vicinity, typically from the expected amount of stabilization energy due to formation of new bonds along each path using perturbation theory in one form or another. This is sometimes formulated as looking for the direction of the easiest motion [2] or of the steepest slope. [99,116] The differences in slope of the excited hypersurfaces between different paths in the same molecule or pair of molecules, *e.g.*, conrotatory versus disrotatory ring opening, or head-to-head versus head-to-tail dimerization, or between similar paths in different but related molecules (substituent effects) is typically estimated by use of perturbation theory. [5,117–122] Numerous authors use excited state "static" reactivity indices and molecular diagrams (*e.g.*, Refs. [123,130]), a dynamic reactivity index (refs. [126, 131–133]), or PMO approximation (*e.g.*, Refs. [15,25,120,121,134–136]). Use of the PMO approximation and interaction diagrams on simple molecules whose molecular orbitals can be guessed (*e.g.*, Refs. [25,120,121]) already could be classified along with qualitative MO arguments (next section), since no calculations are required, but is not elaborated here in any detail because recent reviews are available. [25,120,121] Sometimes, orbital energies are taken from experiment, although this presents difficulties since differences between singlets and triplets are ignored (see Refs. [137,138] for discussions of the relation between orbital and state energies in the absence of configuration interaction).

The usual reactivity indices, such as elements of the first-order density matrix, are also incapable of distinguishing properly between singlet and triplet behavior. Recently, French authors [139,140] have discussed the problem and shown how electron repulsion terms can be introduced to obtain meaningful results. The particular case of interest to them was excited state basicity, but their arguments have general applicability. In particular, the PMO approach, which loses much of its potential appeal because of its inability to distinguish between singlet and triplet behavior [25,121], could profit considerably from an extension in this direction. [119,122]

This second approach to calculations is attractive in its simplicity, but unappealing in its exclusive concentration on one step of the photochemical process, namely the initial push at the nuclei, which may but need not be product-determining. For example, it appears quite possible that the nature and slope of the S_1 (or T_1) hypersurface change soon after the molecule departs from the starting geometry (*e.g.*, S_1 becomes $\pi\pi^*$ instead of $n\pi^*$), so that the initial direction of push is soon changed. Also, the reaction may proceed via an excited intermediate at a geometry far away from the initial one and the critical factor may be some property of this new minimum in S_1 (T_1), such as the height of walls around it, rather than any property of S_1 (T_1) at the initial geometry. The difficulties already mentioned with the first approach are further exacerbated here. For instance, if the photochemical excitation is into S_1 (or T_1), the application appears straightforward, since this is the state for which computations are done. However, if the initial excitation is into one of the higher states, one should do a calculation for that particular state, or assume that the molecule still has the original geometry when its motions start to be governed by the S_1 (T_1) surface after it has lost its excess energy, *i.e.*, that internal conversion was vertical. This may be often unrealistic, since the higher excited states also do their part of "pushing the nuclei around". Further, if S_1 (T_1) lies close to other excited states, it appears unlikely that its slope alone will determine the direction of nuclear motions, owing to vibronic perturbations.

In the discussion of the physical meaning of the calculations of the gradient of the potential energy in the S_1 (or T_1) state at the initial geometry, we have so far assumed along with most investigators that the slope is negative. This would imply no minimum in S_1 (T_1) at the starting geometry, and thus no fluorescence (phosphorescence) of the starting material if the reaction is monomolecular, since emission can hardly be expected to compete successfully with vibrational relaxation to a lower-lying minimum. In bimolecular reactions, of course, fluorescence can occur even if the slope along the reaction coordinate is negative, since the rate of the motion of the "supermolecule" in that direction is limited by the relatively slow rate of diffusion.

Yet some of the calculations of this type were performed for monomolecular reactions known to involve a slight barrier (experimental activation energy) and for cases in which the photochemical process is believed to compete with fluorescence. [99] They still may make physical sense if they are interpreted somewhat differently. First, such calculations will work if S_1 (T_1), for which the perturbation calculation is being done, has a minimum at the starting geometry since only relative slopes are being computed anyway. Lower calculated slope would then imply a lower barrier and easier escape from the minimum towards the product geometry in complete analogy to predictions of ground state reactivity. Second, there is yet another physical situation in which this kind of calculation might provide reasonable results. The calculation is often done for the HOMO $\rightarrow$ LFMO excited state, which need not be the lowest excited state at the starting geometry. Even if it is not, the magnitude of its negative slope along the reaction coordinate may be important if the slope for the lowest state is positive. For instance, in benzenoid compounds, the lowest state could be of ^1L_b type. Then, an (avoided) crossing will occur sooner or later along the reaction coordinate and the resulting barrier in S_1 will be lower if the calculated slope was strongly negative. The three physical situations corresponding to a possibly meaningful calculation of the kind limited to initial geometry alone are summarized in Fig. 4.

Finally, in many of the perturbation calculations of the effect of substituents and other structural changes, an important tacit assumption is made and it is far from obvious that it is always fulfilled. As already discussed, the physical argument on which the calculation is based is that the value of the initial slope, or the height of a small barrier along the way, determine the rate at which the photochemical reaction occurs. However, the experimental value with which comparison is made usually is not the reaction rate but the quantum yield, which of course also depends on rates of other competing processes and these may be affected by substitution as well. For instance, the rate at which fluorescence occurs is related to the absorption intensity of the first transition, the rate of intersystem crossing may be affected by introduction of heavy atoms

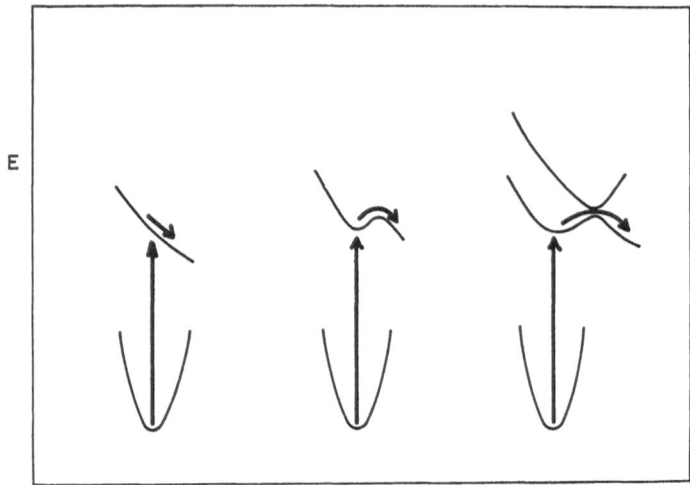

nuclear configuration coordinate

Fig. 4. Physical significance of calculations of the potential energy gradient at the starting nuclear geometry. From left to right, negative slope, positive slope, and a calculation for the second excited state

or low-lying $n\pi^*$ states, the rates of radiationless conversion to S_1 as well as intersystem crossing are, in general, affected by energy differences between the respective states, and all these factors may change when substituents are introduced. Moreover, the substituent may introduce entirely new minima in the S_1 or T_1 surface and thus new possibilities for quenching and new photochemical processes (say, cleavage of a carbon-halogen bond or of a nitrogen-oxygen bond in halogenated or nitrated derivatives); and finally, it may change the dynamics of the molecular motion and thus the probabilities with which the various already existing minima are reached during electronic and vibrational relaxation of the excited state reached by light absorption.

Unfortunately, determination of rates of photochemical reactions is often very laborious and besides, for many practical purposes, quantum yields are of great interest. The best that can be done at present probably is to be aware of the existence of the various pitfalls, to exclude as many of them as possible on the basis of available experimental evidence, and to try a correlation with quantum yields anyway.

C. Calculations for Proton-Transfer Reactions in Excited States

Many proton-transfer reactions can be handled very straightforwardly. In some instances, a fast proton-transfer equilibrium is established in the excited state and the dependence of the ratio of reactant to product concentration on pH defines a pK* in the usual way. [79,141] Calculations of total energies are then all that is needed for rationalization of results, provided that reasonable assumptions can be made about the entropy terms (e.g., Refs. [126,139,142]). In some instances of intramolecular proton transfers, the equilibrium is displaced completely to one side. If this happens to be the opposite side than that favored in the ground state, return to S_0 produces a phototautomer. Arguments based on calculated charge distributions also appear to work reasonably well except that direct comparisons of singlet with triplets are impossible [139, 140] as already mentioned above.

III. Use of Qualitative MO Arguments

After this brief description of the present status of quantum mechanical calculations of organic photochemical reactivities, we shall turn our attention to the second principal ingredient of the simplified model of organic photochemistry, namely the use of simple MO arguments. Again, we shall start by listing some leading references which were found most useful.

The simple MO arguments provide a wealth of qualitative information about the crucial S_0, S_1, and T_1 hypersurfaces. Already Hückel-type arguments can be used to estimate the nuclear geometries at which S_0 and S_1 undergo (generally avoided) crossing and at which motion from one to the other should be particularly easy. [8,14,15,143] Such arguments can also be used to estimate positions of barriers in the hypersurface and distinguish between "allowed" and "forbidden" processes. [2,5,144,145] The occurrence of "abnormal" orbital crossovers [17,20,28] accounts for barriers encountered even along some of the reaction paths which at first sight appear "allowed". Purely qualitative perturbation theory can be used to estimate shapes or slopes of certain regions of the hypersurfaces, as discussed in great detail elsewhere. [25,120,121] Further, simple MO arguments lead to the recognition of the special role played by biradicaloid geometries, and introduction of electron-repulsion terms into the theory permits an estimate of the general differences to be expected in the location of biradicaloid minima in the S_1 and T_1 surfaces. [5,19]

A. Scope and Limitations of Qualitative MO Arguments

Qualitative arguments at the Hückel level can be used to predict positions of minima (funnels) and barriers in the hypersurfaces, but do not distin-

guish properly between states of different multiplicities. [2,15,25,143)]
In the MO description, minima and barriers appear as a result of nodal
properties of molecular orbitals. Since in highly symmetrical molecules
these are determined by symmetry, the minima and barriers are usually
referred to as "imposed by symmetry" although this really is a mis-
nomer. In addition to the minima (funnels) and barriers imposed by the
nodal characteristics of molecular orbitals, there are undoubtedly others
whose existence does not follow from qualitative MO arguments but
would be revealed by a good calculation. In this respect, the situation
is as in ground state chemistry where "symmetry-imposed" barriers are
usually higher than others, but by no means the only ones that exist.

It is to be noted that there is nothing fundamentally sacred in nodal
properties of molecular orbitals. They only appear at the heart of the
argument since it is couched in MO terms. Other equivalent approaches
are possible, *e.g.*, in VB terms [9,11,14)] and in those other concepts will
appear to be of primary importance. The choice of the MO theory is
merely a matter of convenience, which is particularly pronounced in
photochemical applications.

Addition of electron-repulsion terms to the naive Hückel arguments
permits a differentiation between singlet and triplet hypersurfaces.
Most of the previous state crossings [8,15,143)] will now be avoided [14)],
and biradicaloid minima will occur at different geometries depending
on multiplicity. [5,19)] It is important to emphasize at this point that the
commonly encountered tendency to assume that a singlet and triplet
of the same MO configuration, and thus of similar charge-bond order
matrix, will have similar reactivity (for example, Refs. [121,146,147)])
is incorrect. This is most clearly seen from the analysis published by
Constanciel [140)] and also from the arguments given in Ref. [19)] and
summarized briefly further below.

Since the qualitative MO arguments, even those including electron
repulsion explicitly, provide us at best with positions of minima (funnels)
and barriers, and not all of them at that, it is clear they are only capable
of providing very limited answers to photochemical problems discussed
in the preceding section. Specifically, we can expect them to predict
the nature of many or most possible products, but hardly their relative
quantum yields or relative rates of their formation. Further, they should
be able to predict whether there is particular potential for wavelength
dependence in the formation of any of the products, and if we are lucky
and can somehow estimate the energy of the barrier in the excited state
which stands in the way, it will be possible to specify the minimum
amount of "extra" energy needed to form the product, but it will still
be impossible to predict if the product will indeed be formed when such
energy is made available.

This may appear as a rather poor achievement since one doesn't really care much about the products which are possible in principle, but rather about the products that are experimentally detectable. Some consolation for the advocates of the qualitative MO arguments, such as the author, is found in the observation that even the most elaborate and expensive of numerical calculations on organic molecules mentioned above fare no better in this respect, and a breakthrough does not yet appear in sight, at least not for large molecules. Of course, the part of the job that both qualitative MO arguments and elaborate calculations can do is done much more reliably by the latter, but it is often still nice to have a "back of the envelope" approach available. The following discussion of qualitative MO arguments should be approached with this in mind.

B. Location of Minima and Funnels in S_1 and T_1 Hypersurfaces

The position of minima in the S_1 and T_1 surfaces is one of the two most important features of the model of photochemical reactivity described earlier and it is therefore lucky that approximate positions of many of the minima appear to be predictable from very simple MO arguments.

The use of MO theory to find deep minima in the S_0 surface, or geometries of stable molecules, is well known. A simplified rule would be to choose the geometry so as to allow efficient overlap of valence orbitals of the constituent atoms in a way giving bonding orbitals for all available electrons from pairs or larger sets of suitably hybridized atomic orbitals. No atomic orbitals occupied by one electron should be left over dangling free and unable to interact with others, since that would give radicals, biradicals, etc. "Chemical intuition" allows one to proceed almost automatically in cases of molecules of familiar types.

The process is much more difficult when minima are sought in the S_1 or T_1 surface, but two likely candidates come to mind right away. First of all, there are often likely to be minima in S_1 and T_1 at geometries close to those of minima in S_0. These occur whenever the electronic excitation does not radically change the overall picture of bonding in the molecule, for instance in singlet or triplet $n\pi^*$ excited ketones or in $\pi\pi^*$ excited aromatic molecules. Although there are typically some differences in bond lengths and angles between the equilibrium geometry in S_0 and S_1 or T_1, there is usually no difficulty in assigning these minima as spectroscopic excited states.

Now in addition to all these S_1 and T_1 minima which have counterparts in the ground state surface, minima (or funnels) in S_1 and T_1 can also be expected at biradicaloid geometries. These are geometries at which the molecule, in the MO description, has two approximately

31

non-bonding orbitals occupied by a total of two electrons[h]). These geometries are clearly unfavorable in the S_0 state, and rarely correspond to significant minima, since the two electrons are not contributing to bonding, whereas a change in nuclear geometry, which would make the two orbitals interact and combine into one bonding and doubly occupied orbital, and one antibonding and empty orbital, will lead to stabilization. Expressed differently, at the biradicaloid geometry, there is one less bonding pair of electrons and thus effectively one less bond than at "ordinary" geometries of the molecule. In many cases, this agrees well with the usual way of writing chemical formulas for biradicals; for instance, the trimethylene biradical has one bond less than cyclopropane. On the other hand, it is not obvious from the usual valence formula that two of the four π-electrons of square planar cyclobutadiene do not contribute to bonding, while the other two contribute more than they would in an ordinary double bond, but it is easily seen from simple MO arguments.

In the S_1 and T_1 states, on the other hand, one can expect the biradicaloid geometry to be relatively favorable[i]). This is perhaps not obvious from the usual valence formulas, but is again easily derived from the simple MO picture. If the excited molecule is distorted from a biradicaloid toward an "ordinary" geometry, and the pair of non-bonding orbitals splits into a bonding and an anti-bonding combination, one of the two electrons is assigned into the bonding, the other into the antibonding orbital and there is no net gain of bonding. To the contrary, often there will be a loss since the energy of the antibonding orbital typically increases faster than that of the bonding one decreases (Fig. 5). This argument is only valid in the vicinity of the biradicaloid geometry, where the S_1 and T_1 states can be expected to arise from S_0 by electron redistribution between the two still approximately non-bonding levels, rather than from some other type of electron excitation which perhaps does not involve the non-bonding levels at all. However, this is all that is needed for the argument in favor of the likelihood of a local minimum in S_1 and T_1.

C. Biradicaloid Minima

Many types of biradicaloid geometries can be envisaged for a given molecule.[19] They can be derived from its ordinary geometry (minimum

---

[h] The noun *biradicaloid* is sometimes used to describe singlet ground state molecules with a thermally accessible low-lying triplet state (p. 183 in Ref. [148]). It will be noticed that we use the adjective *biradicaloid* in a related but more general sense.

[i] A direct observation of a triplet state of substituted trimethylene biradical has recently been claimed.[84] This is in good agreement with the prediction [104,105] that the lowest excited state of cyclopropane will have an energy minimum at a geometry in which one C—C bond is broken.

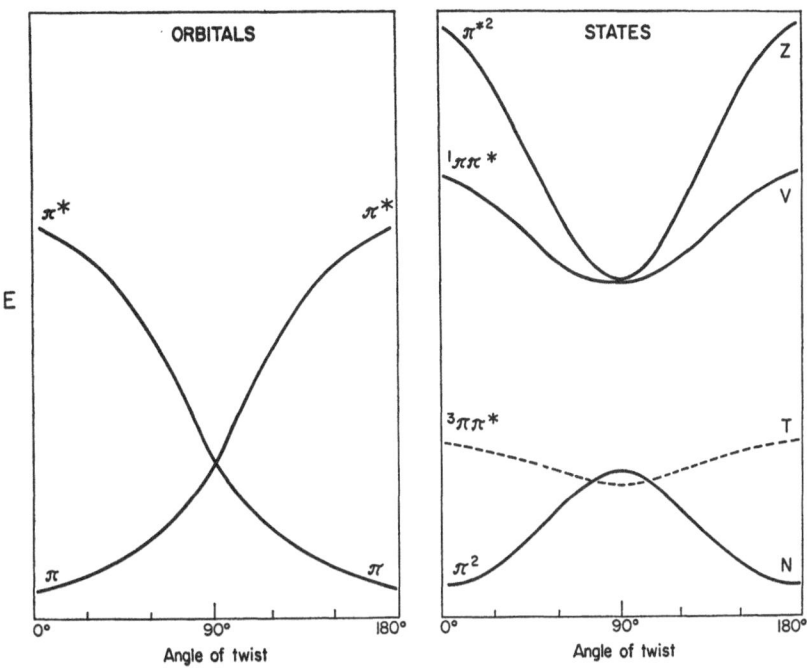

Fig. 5. Energies of π-electron orbitals and states of ethylene as a function of twist angle

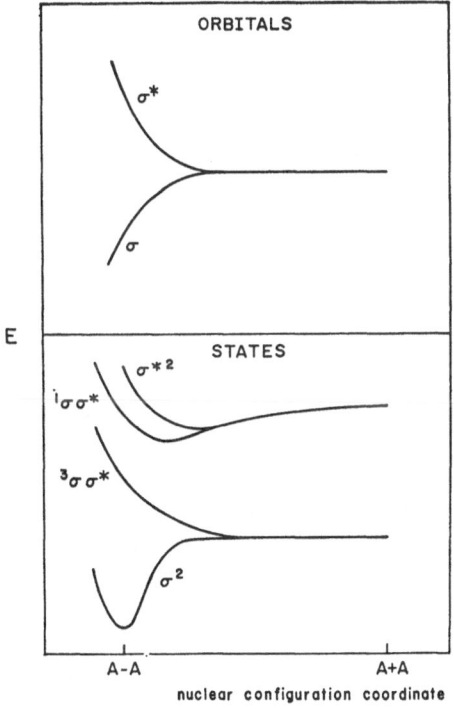

Fig. 6. Energies of orbitals and states of a single-bonded A_2 molecule as a function of the bond length

in S_0) by stretching a single bond (Fig. 6), by twisting a double bond (Fig. 5), by rearranging atoms to make a carbene, by going part way along a ground-state symmetry-forbidden pericyclic reaction path (Fig. 7), etc. Each of these geometries is likely to correspond to a local minimum in S_1 and T_1, but to find the bottom of the minimum, one needs to investigate all kinds of conformations and decide which one is most favorable. For example, one way to derive a biradicaloid geometry from cyclobutane is to break one of the C—H bonds and this gives the biradicaloid system "radical pair of H atom and cyclobutyl radical" (1). Another is to break one of the C—C bonds, and this gives the tetramethylene biradicaloid geometry (2). To find a minimum in the S_1 or T_1 surface that corresponds to it, one needs to estimate relative energies of conformations such as 2 and 3. To do this, one would investigate steric repulsions, bond angles and lengths, delocalization effects, etc., similarly as in the

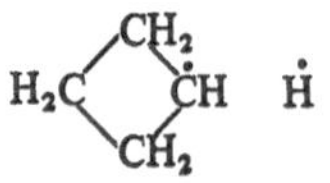

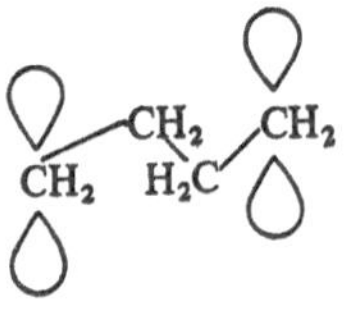

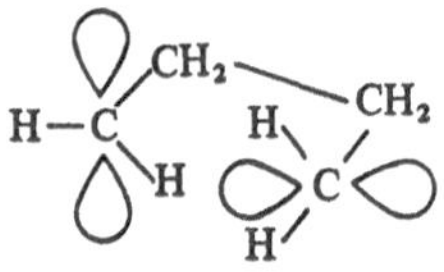

usual ground state conformational analysis, except for the condition that two of the orbitals must be kept at least approximately non-bonding in order to stay near the minimum in S_1 or T_1. For instance, one of the conformations which would have to be taken into consideration is 4, obtained by stretching two C—C bonds rather than breaking one, and corresponding on the S_0 surface to the "antiaromatic" transition state found along the concerted $2s + 2s$ cycloaddition of two ethylene molecules at the point of orbital crossing in the well-known correlation diagram (Fig. 7). 4 is isoconjugate with cyclobutadiene and has two non-bonding orbitals, just like 2 or 3.

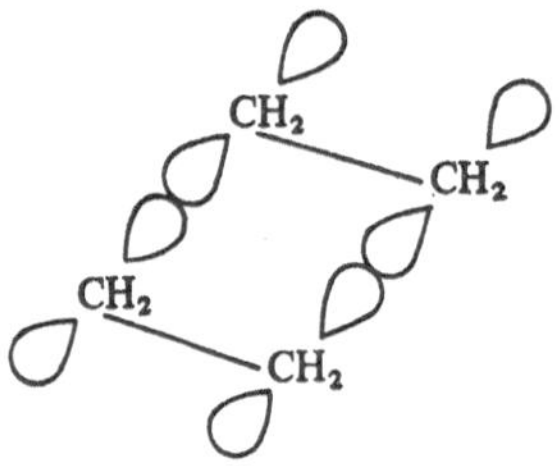

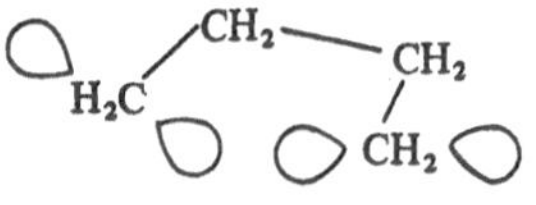

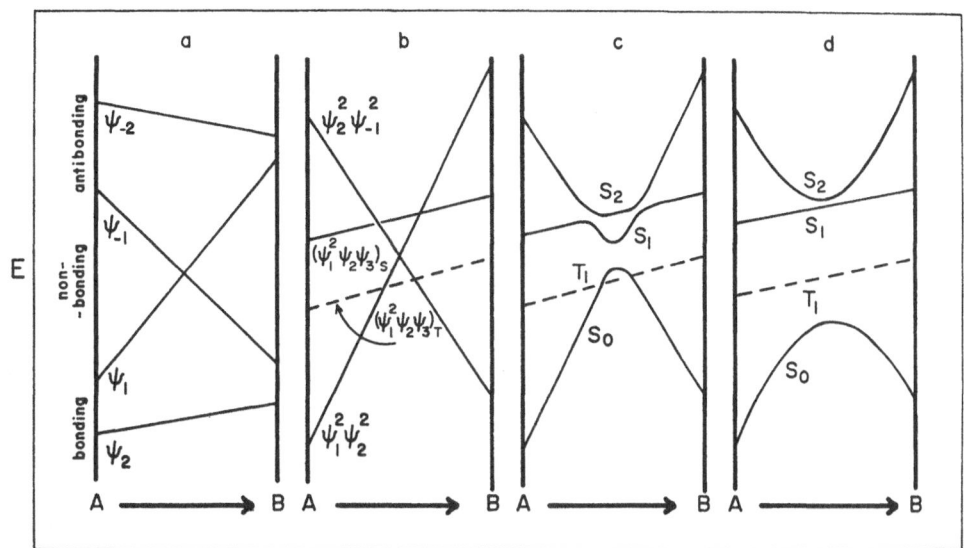

Fig. 7. Orbital (a), configuration (b), and state (c, d) correlation diagrams for a typical ground-state symmetry-forbidden pericyclic reaction

In the presently discussed naive MO approximation, the S_0, S_1, and T_1 states all have the same energy at a biradicaloid geometry such as 2–4, since they differ by assignment of electrons into orbitals which are non-bonding and which contribute to the total energy equally. Moreover, the energy of S_0 in any of these confirmations is above the cyclobutane minimum in S_0 by approximately the energy of one C–C bond. There will be slight differences between 2, 3 and 4, due to effects such as the presence of some cyclic delocalization in 4, but in a reasonable approximation, the heat of formation is given in all three cases by the fact that there are effectively three C–C bonds and eight C–H bonds. Consequently, at the present level of approximation, 2–4 are also equally favorable in S_1 and T_1.

D. Singlet-Triplet Differences

On a somewhat higher level of qualitative MO argumentation, one can allow for the fact that the total energy of the molecule is not related only to the sum of energies of occupied orbitals, but also to certain electron repulsion terms. This leads to a better understanding of the nature of electronic states of molecules at biradicaloid geometries [19,112] and, in particular, of the difference between S_1 and T_1 hypersurfaces.[19] We

35

continue to assume that the contribution to the total energies of *2—4* provided by the bonding electrons is approximately the same, since in each species they provide effectively three C—C bonds and eight C—H bonds,[)] but look at the role of the two non-bonding electrons in more detail. We shall specify the choice of the two orthogonal non-bonding orbitals more explicitly: they will be as localized as possible, *i.e.*, atomic rather than molecular orbitals (see Ref. [19] for details; the argument really is in VB rather than MO terms).

It is well known [19,112] that two electrons can be assigned to the two non-bonding orbitals in six ways leading to three singlet states and the three components of one triplet state. The sum of the energies of occupied orbitals is the same in all cases, but if electron repulsion is allowed for, the four resulting states nevertheless have different energies. In the triplet state electron motion turns out to be well correlated — when one of the electrons is in one of the non-bonding orbitals, the other electron is in the other orbital. This follows from the Pauli principle. The lowest of the three singlets can also be labelled as well-correlated; when one of its electrons is in one of the non-bonding orbitals, the other electron is most likely (but usually not certain) to be in the other. This singlet will form the S_0 state. On the other hand, in the other two singlet states, which form the S_1 and S_2 states, the electron motions are very poorly correlated — when one electron is in the first non-bonding orbital, and thus in the vicinity of one atom, the other most likely is there as well. At any moment in time, there is excess negative charge in one of the non-bonding orbitals (in one region of the molecule) and excess of uncompensated positive charge in the other. For this reason, these states are often referred to as "ionic" although there is no net polarity (the electrons travel together, but they spend an equal amount of time in each of the non-bonding orbitals, provided that the latter are really of the same energy). It is hardly surprising that the two "ionic" singlets, S_1 and S_2, which suffer from constant separation of charge, are higher in energy than the well correlated singlet and the triplet state.

Now, the energy differences between the four states, T_1, S_0, S_1, and S_2, will depend on the conformation of the biradicaloid molecule.[19] If the two localized non-bonding orbitals are far apart in space, the cost

---

[)] One could argue that *4* should be favored because of cyclic delocalization, although this is usually not important in σ-bonded systems. The same argument, however, applies to the S_0 state and if this effect were significant, the thermal cycloaddition of olefins would prefer to proceed along the symmetry-forbidden concerted path in a *4*, whereas in reality it is non-concerted and geometries such as *2* are more accessible than the transition state which probably [149] resembles *5* or *4*.

of charge separation which has to be paid in the two ionic singlets S_1 and S_2 will be particularly high. These singlets will clearly prefer as tight a geometry as possible, in which both non-bonding orbitals are in the same region of space, *i.e.*, *4*, not *2*. On the other hand, the well-correlated states S_0 and T_1 have nothing to gain from pushing the two non-bonding orbitals together. To the contrary, they stand to lose, provided that the orbitals really remain non-bonding, since all that is achieved is that the non-bonding electrons as well as their respective nuclei now repel more. This latter fact has, of course, been long recognized in the rule ,,triplet molecules prefer geometries at which the two unpaired electrons stay apart".[5] As a result, among biradicaloid geometries, S_0 and T_1 energetically[k] prefer "loose" ones such as *2*, S_1 (and S_2) "tight" ones such as *4*. Consequently, the location of minima in T_1 and S_1 is not the same, although it appeared to be until electron repulsion effects were taken into account.

To obtain an estimate of the magnitude of the effects, some very simple calculations have been performed.[150] In these, the Extended Hückel Method was used to obtain the form of non-bonding orbitals of various molecules at biradicaloid geometries, and the Mulliken and Mataga approximations for electron repulsion integrals were then used to calculate the separation of the S_0, S_1, and S_2 states obtained by a configuration interaction calculation. The method is not considered reliable enough for total energies and for mapping of the S_0 surface, but results for the relative separation of the states derived from the various assignments of electrons in the non-bonding orbitals are considered qualitatively correct.

It was found that at biradicaloid geometries the S_1 and S_2 states are generally within 10 kcal/mole of each other. At tight pericyclic biradicaloid geometries such as *4*, they were calculated to be typically 35—40 kcal/mole above S_0, whereas at loose open-chain biradicaloid geometries with the two radical centers separated in space, they were well over 100 kcal/mole above S_0, and there might possibly be other excited singlets below them (not considered in the calculations). Bond rotation which allowed hyperconjugation and delocalization of the non-bonding orbitals decreased this huge difference considerably, in spite of the fact that the orbital energy difference involved in the excitation increased. This indicates that bond rotation is strongly restricted even in the open-chain form of a species such as tetramethylene biradical in its excited singlet state. All the calculated results were easily understandable in the qualitative terms outlined above.

---

[k] The entropy term is generally better for the loose geometry.

Similar results were obtained for C—C bond dissociations in ethane. The S_1 and S_2 states were degenerate at infinite separation, decreased in energy as the bond length was shortened, went through a minimum at about twice the calculated ground state bond length, and then increased steeply as ground state bond length was approached. Again, other states such as Rydberg states were not considered.

The expected general tendency of T_1 hypersurfaces to have minima at loose biradicaloid geometries and of S_1 hypersurfaces to have similar minima at tight biradicaloid geometries is also in good agreement with experimental data [151] as well as calculations [152, 153] for the simplest prototype of a singly-bonded molecule, namely H_2. When the H—H distance is increased, the geometry rapidly acquires biradicaloid character (Fig. 6). As expected, the T_1 surface has a minimum at the loosest possible geometry (the two H atoms at infinite separation, the $^3\sigma\sigma^*$ T_1 state is purely repulsive). On the other hand, the $^1\sigma\sigma^*$ singlet surface, with the same orbital occupancy, has a minimum at $R = 1.3$ Å, although at this distance the molecule is just barely biradicaloid (ground state bond length is $R = 0.74$ Å). Evidently, in the "ionic" excited singlet, it even pays to have one of the two "non-bonding" orbitals slightly bonding, the other slightly antibonding, and thus incur a penalty in that a non-zero orbital energy difference is involved in the excitation, just so the electron repulsion term can be kept at minimum. Even the σ^{*2} state, which on the basis of orbital energies alone would be expected to be purely repulsive since it only has electrons in the antibonding orbital, actually has a shallow minimum at $R = 2.3$ Å, and this can again be understood using the above argument in favor of tight biradicaloid geometries in S_1 and S_2 states.

If one accepts the previously described model for the course of photochemical reactions, the electron repulsion argument provides a natural rationalization of the differences in singlet and triplet reactivities. In the case of reactions such as bond dissociation or hydrogen abstraction, return to S_0 from T_1 should occur through "broken σ-bond" minima at loose geometries, and thus result in completely broken bonds. On the other hand, return to S_0 from S_1 through "broken σ-bond" biradicaloid minima would occur generally at tight geometries, resulting in stretched bonds with reasonable probability for bond restoration (amounting to overall quenching). This is discussed in more detail in Ref. [19] where specific examples are given. The experimental evidence is far from complete and a more detailed investigation seems warranted. It should be pointed out that the argument applies to efficiencies rather than rates of reactions. Relative rates are most easily rationalized by assuming different barriers in the S_1 and T_1 surfaces on the way to the respective "broken σ-bond" minima, as in Ref. [154]. It is not uncommon to visualize the reaction from the singlet state as involving intersystem crossing to the purely repulsive T_1 surface [154-156], and this, of course, is a distinct possibility. On the other hand, at least in some cases [157], CIDNP evidence shows that the dissociation occurs from an excited singlet state since the resultant pair of radicals is born in its singlet state. Our interpretation would then be that return to S_0 occurs from the tight biradicaloid minimum in S_1, generates a ground state molecule with excess vibrational energy and one considerably stretched bond and that there is a certain probability that the excess vibrational energy will be used to break the bond completely before it is lost as heat, leading to a singlet radical pair. This fits in nicely with recent results for dissociation of benzylic bonds [64] from S_1 state which indicate that crossing to T_1 is not involved. The increased rate of formation of allyl as opposed to methyl radical in this case would then be simply related to a lower barrier in S_1 on the way from the initial geometry to the "broken σ-bond" minimum, assuming other factors remain constant. This also accounts for the increased rate of radiationless conversion to S_0 via the same minimum.

The evidence is clearer in the case of the well-known tendency of singlet excited species to give products of concerted pericyclic reactions and of triplet excited species to give their non-concerted counterparts (for references see Refs. [121,158]). There, the rationalization based on the electron-repulsion argument is on firmer ground and is discussed in more detail below. Still, presently unexplained exceptions to the general tendency appear to exist.[159]

It should be emphasized that there are other reasons for differences in singlet and triplet reactivities. The shapes of the S_1 and T_1 hypersurfaces can be affected differently by configuration interaction, the various minima in S_1 and T_1 will in general be reached at different rates after initial excitation owing to differences in vibronic couplings, more vibrational energy will typically be liberated during a final $S_1 \rightarrow S_0$ jump as compared to final $T_1 \rightarrow S_0$ jump and this may be important in hot ground state reactions, etc.

Finally, it should be stressed that the rationalization of singlet-triplet differences based on electron repulsion effects as outlined here is relatively new and not generally accepted. For alternative rationalizations in the case of β,γ-unsaturated ketones, see Refs. [147, 160].

E. "Stretched σ-bond", "Twisted π-bond" and Carbene Minima

It is easy to see how simple MO arguments lead one to expect that two non-bonding orbitals will result if a single bond is stretched sufficiently (Fig. 6), as a double bond is twisted by 90 degrees (Fig. 5), or as atoms are rearranged to make a carbene. Return through such biradicaloid minima in S_1 or T_1 then accounts for bond dissociation and atom abstraction reactions, carbene formations, olefin *cis-trans* isomerizations, etc.[19] It is worth noting that the fact that S_1 of planar olefins is not of $\pi\pi^*$ HOMO $\rightarrow$ LFMO nature but rather $\pi\sigma^*$ or Rydberg [161, 162] which is not even shown in Fig. 5, would only matter if this state increased in energy as the twist angle is increased and thus presented a barrier in the way towards the "twisted π-bond" minimum at 90°. However, its energy is believed to change very little with the twist angle.[161]

The present model accounts well for the known features of the *cis-trans* isomerization of polyenes. The observation that methyl-labelled butadiene and hexatriene isomerize two double bonds upon triplet excitation but only one upon singlet excitation [163,164] is in good agreement with the expected much longer lifetime of the twisted triplet such as allyl methylene, which would allow rotation at both ends [97] while the excited singlet should return to S_0 extremely rapidly after twist at one end[1]. However, the recent reports [66, 164] that triplet hexatrienes probably isomerize faster at the end bonds than at the central bond is out of line with simple arguments as well as semiempirical calculations [97,165] of the energy surface and other explanations are under investigation. [164]

F. Pericyclic Minima (Funnels) and Their Open-Chain Counterparts

A still different class of biradicaloid geometries, called here pericyclic minima [19], has already been mentioned. Pericyclic reactions are defined by Woodward and Hoffmann as those in which all first-order changes in bonding relationships take place in concert on a closed curve.[2] Best known examples are concerted electrocyclic, sigmatropic, and cheletropic reactions, cycloadditions, and cycloreversions. In pericyclic biradicaloid minima, the two non-bonding orbitals can be imagined to be formed from one bonding and one anti-bonding orbital of an "ordinary" molecule or pair of molecules (minimum in S_0) by following a ground-state symmetry-forbidden path to the antiaromatic [4] geometry. Rules for distinguishing ground-state symmetry-allowed and symmetry-forbidden paths have been discussed by many authors from various points of view.[2–12] It will be assumed in the following that the reader is familiar with the derivation and use of correlation diagrams in ground-state reactions as outlined by Woodward and Hoffmann.[2] A typical orbital correlation diagram for a ground-state forbidden path is given in Fig. 7a. Half-way through the

---

[1] However, it is possible [166] that *cis-trans* isomerization occurs, also or exclusively, via 1,3-bonded biradicals which have been postulated as intermediates in the formation of cyclopropenes.[167]

reaction, the species has two approximately non-bonding orbitals and its geometry is biradicaloid. Both non-bonding orbitals are located in the same region of space throughout the cyclic array of orbitals (*e. g., 4*) typical of a pericyclic reaction (tight biradicaloid geometry).

1. Singlets

The corresponding singlet configuration correlation diagrams are shown in Fig. 7b, and finally, the singlet state diagrams resulting after introduction of configuration interaction are shown in Figs. 7c,d. This kind of orbital crossover results in a barrier in the S_0 surface and a funnel (or a minimum) in one of the excited singlet surfaces. Whether this will be S_1 (Fig. 7c) or S_2 (Fig. 7d) depends on the degree to which the crossing is avoided. In some early diagrams [144], this splitting was estimated to be very large. Some calculations [14] on the butadiene-cyclobutene system indicate that the chances are good that the funnel will actually be in S_1, whereas others [10] indicate that it will be in S_2. At any rate, previous arguments and approximate calculations suggest that for biradicaloid geometries such as these the S_1 and S_2 states are generally very close in energy. In the more distantly related case of twisted ethylene, ab initio calculations [109] showed S_1 and S_2 to be virtually degenerate (as indicated in Fig. 5). While it is impossible to claim at present with certainty that an orbital crossover of the pericyclic type will always lead to a funnel or minimum in the S_1 surface, it seems safe to claim it will lead to a funnel (or minimum) in either S_1 or S_2, and moreover, that these two states are of quite comparable energies in the region of the funnel while S_0 is only somewhat farther below. If the funnel occurs in S_1, return to S_0 at this biradicaloid geometry should be particularly effective. If it occurs in S_2, return to S_0 at this particular geometry appears perhaps less likely, but still quite feasible, as already mentioned on p. 15. Indeed, in order to accommodate experimental data within the framework of the present simple model, one is forced to conclude that funnels in S_1, or in nearby S_2, due to orbital crossover in pericyclic reactions, are extremely efficient traps for excited molecules and take them down to S_0 practically each time nuclear motion brings the molecule to the funnel. Otherwise, whenever energetically feasible, it should be possible for some of the molecules to reach minima in S_1 on the other side of the funnel, which typically correspond to spectroscopic excited states of products (Figs. 7c,d).[78] This is sometimes impossible because of the energetics [168], but even if the product is relatively stable and has a low-lying excited state known to fluoresce strongly, emission from it is not observed. Actually, the first observation of any product fluorescence in a ground-state-forbidden excited singlet pericyclic reaction has only been reported very recently and the quantum yield of the emission is extremely small.[78] Other similar searches have been fruitless.[169] On the other hand, product fluorescence is well known in several reactions of other types, in which no funnel occurs between the initial and product geometry, such as dissociations of diatomic or other very small molecules. Of greater importance in organic chemistry is the case of intramolecular and intermolecular proton transfer reactions. [79,141]

It is interesting to note that most recent calculations [72,171–173] and some experimental data [174] indicate strongly that the lowest excited state of butadiene, and probably polyenes in general, does not correspond to simple HOMO→LFMO excitation, but to a complicated mixture of configurations including the doubly excited HOMO, HOMO→LFMO, LFMO configuration which correlates with the ground state of electrocyclic products along the ground-state-forbidden pathway (for butadiene, for disrotatory ring closure), and that the HOMO→LFMO excited state is a little higher in energy[m]. As a result, the minimum encountered halfway along the reaction coordinate as a result of attempted crossing must almost certainly be in the S_1 surface and moreover, there should be no correlation-imposed barrier along the way from the starting planar geometry.

Some ground-state allowed pericyclic reactions have also been reported to proceed in a concerted manner upon singlet excitation (e.g., Ref. [170]). These could be hot ground state reactions. However, if they are not, the minimum in S_1 through which they proceed could lie at product (or exiplex) geometry and product (or exciplex) emission might be detectable. Of course, they could also proceed through a minimum whose existence does not follow from simple MO arguments of the kind discussed here.

We have only discussed the biradicaloid geometry reached from the initial (or final) geometry by concerted motion, forming and breaking all bonds along a closed curve simultaneously, although not necessarily all to the same degree at all times (e.g., cyclobutane→4, or two ethylenes →4). This resulted in a cyclic array of interacting orbitals (tight geometry), and the antiaromatic nature of the cyclic conjugated system, guaranteed by the fact that we only chose to investigate ground-state-forbidden pathways, gave us the two non-bonding orbitals of a biradicaloid. Now there are many non-concerted open-chain counterparts to this pericyclic minimum, in which the two approximately non-bonding orbitals can be localized in different regions of space (loose geometry)[n].

---

[m] In what appears to be the best presently available calculation [175], the HOMO → LFMO state comes out lowest, but the energy difference is minimal.

[n] These are essentially just "broken σ-bond" minima.

They can be obtained from the pericyclic geometry by disrupting the cyclic conjugation by suitable bond stretching and twisting, for instance *2, 3, 5* from *4*. We have already used this particular example to show that, roughly speaking, all these biradicaloids have comparable energies since they all effectively have the same number of bonds, but that the loose geometries are favored in S_0 and T_1 states while the tight geometry is favored in S_1 state as soon as one goes beyond the most primitive approximation. Because of the highly unfavorable nature of the loose geometries in the S_1 state, we can expect the S_1 hypersurface to slope downhill towards the tight geometry and the minimum can be expected there. During its sojourn in this minimum, the molecule has little freedom for motion such as bond rotation, since that would destroy the cyclic array of orbitals, lead to a spatial separation of the two non-bonding orbitals (loose geometry), and therefore require considerable energy. Besides, the duration of this sojourn is undoubtedly very short, and in the limit, the molecule could just pass through the funnel during a single vibrational period[o]. A molecule returning through this minimum (funnel) will find itself high on the S_0 surface, (since it still is at a biradicaloid geometry, which is unfavorable in S_0) and the steepest slope is likely to lead to a minimum in S_0 with one or the other set of partially formed bonds broken and the other formed fully, *i.e.*, to starting material or to products. This is in good agreement with the well known tendency of excited singlet molecules to give products of concerted pericyclic reactions along pathways forbidden in the ground state. In molecules other than hydrocarbons, this argument may need more careful examination if significant substituent effects are present, but even then, biradicaloid minimum (funnel) can be expected to have a general tendency to be located at concerted pericyclic geometries.

2. Triplets

The preceding argument works also for the T_1 hypersurface, but exactly in the opposite sense. Now loose geometries are more favorable and this is where the biradicaloid minima should be sought (for instance, *2*, not *4*). These minima in T_1 will typically allow considerable freedom of motion such as bond rotation, since there now is no rigid geometrical requirement such as a need for a cyclic array of orbitals was in the singlet case. Also, return to S_0 is spin-forbidden and may be relatively slow,

---

[o] This apparently happens in the excited singlet dedimerization of thymine photodimer in a rigid matrix as well as subsequent excited singlet dimerization of the pair of thymine molecules formed, since quantum yields of both processes are unity.[176]

possibly permitting detection by direct observation (observation of a related broken σ-bond triplet has been claimed recently [84]). It is then probable that much stereochemical information originally contained in the molecule will be lost. The ground state is subsequently reached at a geometry which is, of course, still biradicaloid, but this time because one bond is broken, not because several are partially broken and several others partially formed, as was the case in S_0 at the tight geometry. Some additional stereochemical information may then possibly be lost while the two loose ends are groping for each other to form a bond[p]) (cyclobutane from 2) or perhaps even while the molecule is coming apart (two ethylenes from 2). Again, this is in good agreement with the well known tendency of triplet excited molecules to give products of non-concerted analogs of pericyclic reactions (for example, $2+2$ dimerization of olefins, and $2+4$ photo-Diels-Alder reaction).

According to our approximate calculations, the preference of triplets for the loose geometry is less pronounced than the preference of excited singlets for the tight geometry. If suitable structural features make loose geometries unfavorable, the molecule may be coaxed into staying on the concerted path. However, there is no particular reason to expect a minimum in T_1 near the concerted pericyclic biradicaloid geometry.[14] This is clear from Fig. 7b which shows the triplet configuration correlation diagram for the pericyclic reaction path. The triplet state correlation diagram (Figs. 7c, d) is identical with it to a good approximation, since there is no first-order configuration interaction. There is no apparent reason to expect a minimum anywhere else than near the starting or final geometries. If that at the initial geometry is lower in energy, the molecule will most likely end up there after excitation into the triplet manifold, and return to S_0 by emission of light or by non-radiative decay. In either case, the process will be labelled photophysical. If the minimum at the product geometry is lower, the molecule will end up there so that we expect formation of a triplet excited product, detectable in principle by triplet-triplet absorption spectroscopy. If its phosphorescence is able to successfully compete with radiationelss decay, product emission should be observable.

As long as the loose biradicaloid geometries are more favorable in T_1 than the typically non-biradicaloid geometries of the starting material or the products, as is usually the case, the above-described process does not occur. For example, although the T_1 hypersurface goes downhill from cyclobutene geometry to butadiene geometry, it goes even further downhill to the loose biradicaloid geometry of twisted butadiene. Triplet

---

p) See Refs. [102,149,177] and literature cited there for current views on the behavior of 1,4-biradicals in their S_0 state.

excited cyclobutene should reach the S_0 surface at the geometry of twisted butadiene and finally relax to ordinary ground-state butadiene in a non-stereospecific manner.

One way to endow the starting or product geometry with the additional stabilization needed to make the process feasible is to make it planar and highly conjugated. Then, motion towards the non-planar loose biradicaloid geometry, favorable as far as the two unpaired electrons are concerned, will destroy the delocalization stabilization of a large number of other electrons and will be unfavorable[r]. For example, going on the T_1 surface from the geometry of Dewar naphthalene 6 or naphthvalene 7 to the T_1 minimum at planar naphthalene geometry is undoubtedly downhill, but in contrast to the case of butadiene, this time motion toward a loose biradicaloid geometry 8 (one C—H bond out of plane) is uphill, because much of the overall aromatic stabilization

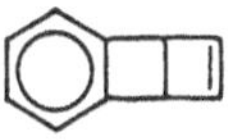

6    7 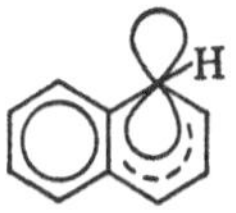   8

would be lost. One should therefore expect [19] triplet excitation of 6 and 7 to give excited triplet naphthalene, and this has been recently found experimentally in the case of 7.[80] Also the formation of a triplet excited 11,12-dihydrocarbazole derivative upon triplet excitation of triphenylamine [77] can be easily understood in these terms and predictions of other such cases can be made. If the initial excited triplet has a choice of paths toward the product excited triplet minimum, the presence of barriers in the T_1 surface should determine which one will be taken. As is discussed in more detail below, one can generally expect the ground-state forbidden path to be more favorable in the excited T_1 state (the Woodward-Hoffmann rules apply).

G. Energy Barriers in S_1 and T_1 Hypersurfaces

These are the other essential feature of in the simple model of photochemical reactivity described at the outset. Their importance is in limiting access to certain parts of the potential hypersurfaces and thus making some otherwise quite reasonable minima inaccessible under given conditions (temperature, photon energy). Again, we are lucky in that the

---

[r] An illuminating discussion of the geometrical preferences of $\pi\pi^*$ triplets of unsaturated compounds has been given by Baird.[82]

locations of at least some of the barriers can be derived from simple MO arguments, namely from correlation diagrams. Unfortunately, it is hard to say much about the heights of these barriers. Also, the information can only be used in a negative sense: if a barrier is in the way, a minimum will not be reached. However, if there is no such barrier, or if enough energy is available to overcome it, the minimum still need not be reached by any significant fraction of molecules. The existence of the barriers is therefore hard to prove experimentally in a totally unequivocal manner.

The use of correlation diagrams for the derivation of the location of barriers in S_1 and T_1 is entirely analogous to their use for ground-state reactions. Fig. 8 shows that a barrier in S_1 (T_1) will result if, going along

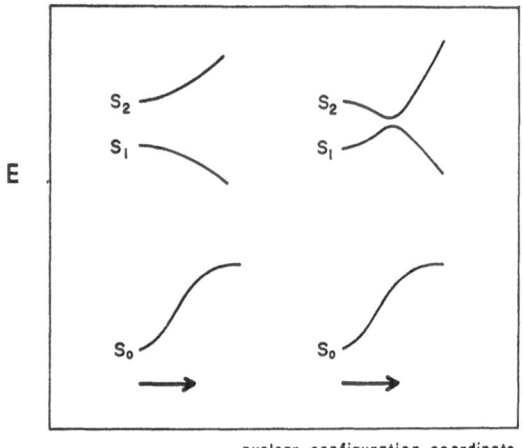

nuclear configuration coordinate

Fig. 8. Origin of correlation-imposed barriers in the S_1 hypersurface

the reaction coordinate towards the desired minimum, the lowest excited state of the reactant geometry does not correlate with one of the lowest excited states, but instead with a highly excited state at the final geometry. Correlation diagrams have been in use for this purpose essentially independently of related developments in ground-state chemistry (for recent examples, see Refs. [154,156,178]). However, they are perhaps easiest to derive and interpret in the case of concerted pericyclic reactions [2,144] and this will be taken up first. We have already seen that this is the case of practical importance for reactions in excited singlet states, and only relatively infrequently for triplet state reactions.

1. Pericyclic Reaction Paths

Along concerted pericyclic pathways which are ground-state allowed, correlation-imposed barriers are generally encountered in S_1 as well as

T_1, since the $1 \to -1$ excited configuration of the reactant correlates with a higher singly excited configuration of the product at whose geometry the next minimum in S_1 (or T_1) would be expected [2,144] (there is no special reason to expect any minima or funnels along the way [14]). These barriers are apparently part of the reason why singlet excited products are so rarely seen in photochemistry. They block off minima in S_1 which are not separated from the starting geometry by funnels and which would thus at first sight have a good chance of being reached. As already pointed out, the few photochemical singlet reactions which are believed to follow a concerted pericyclic path allowed in the ground state should have a reasonable chance to give excited products (unless of course they really are hot ground state reactions). Some additional known puzzling cases await further experimental work (*e.g.*, Ref. [179]).

Along ground-state forbidden concerted pericyclic paths, we expect a funnel in S_1, and therefore a potential source of singlet photochemical reactivity. Correlation diagrams need to be constructed for the part of the reaction coordinate between the starting material and the geometry of the funnel. This might seem difficult, since virtually nothing is known experimentally about electronic states of the molecule at the latter geometry. Fortunately, the information is easily available from a correlation diagram extending all the way between the starting material and product geometries, from which, of course, even the existence of the funnel itself was deduced. In the case of those triplet pericyclic reactions which proceed through a minimum in T_1 at product geometry rather than through a loose biradicaloid minimum as would be usual, the same kind of a correlation diagram will provide the answers, although the point at which S_0 is finally reached is quite different (it lies at the end of the reaction coordinate rather than half-way through).

The barrier in S_0 encountered along the ground-state symmetry-forbidden concerted pathway is due to the loss of one occupied bonding orbital, and an empty antibonding orbital, and their replacement by a pair of approximately non-bonding orbitals half-way through the reaction, which amounts to a net loss of one bond (Figs. 7a, 9). Excited configurations of the reactant in which the former orbital is occupied and the latter orbital empty (Fig. 9, case a) will correlate with triply excited states of the product if the ground-state configuration correlates with a doubly excited state, and will thus go uphill as much as the ground configuration itself. Those excited configurations of the reactant in which the former orbital has only one electron and the latter is empty, as well as those in which the former is full and the latter has one electron (Fig. 9, case b), should correlate with doubly excited states of the product and thus still go uphill. Only the configuration in which an electron has been removed from the former orbital and placed in the latter ("charac-

47

teristic configuration") will correlate with a singly excited configuration of the product and as a result, there is no particular reason for it to go uphill (Fig. 9).

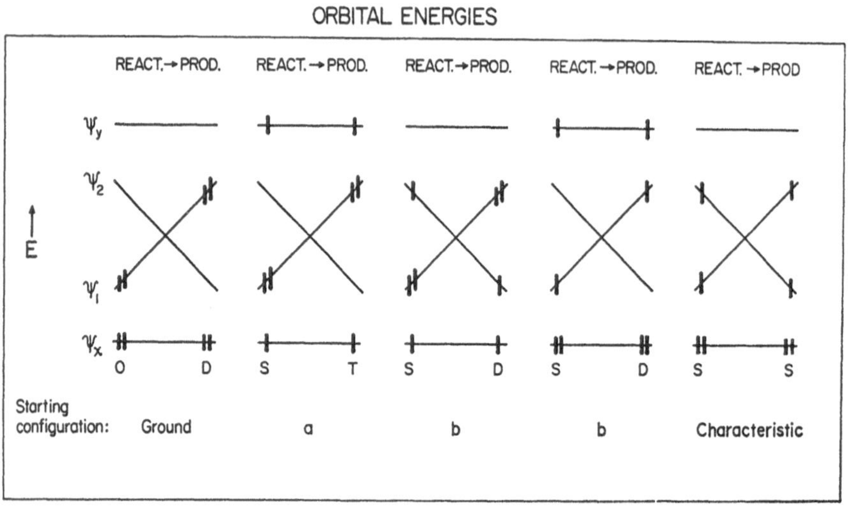

ORBITAL ENERGIES

Fig. 9. Fate of various starting configurations during a ground-state symmetry-forbidden pericyclic reaction. Schematic

Electronic states of many common unsaturated molecules can be reasonably well approximated by one or only a few configurations constructed from delocalized molecular orbitals. Whenever this happens, it is easy to go from a configuration correlation diagram to a state correlation diagram. If the "characteristic configuration" happens to represent the lowest excited state of the reactant (S_1 or T_1), the photochemical reaction from S_1 or T_1 will have no symmetry-imposed barrier. If it happens to represent one of the higher excited states, a barrier can be expected. In such a case, some other configuration or configurations describe the lowest excited state, and since they necessarily correlate with higher excited states of the product, they will go uphill sooner or later, while the state represented by the "characteristic configuration" does not change its energy much. A crossing, or avoided crossing, and a barrier in the S_1 (or T_1) hypersurface will result (Fig. 8). The height of the barrier depends on a variety of factors, such as how strongly will the crossing be avoided, how far above $S_1(T_1)$ the "characteristic state" was at starting geometry, how fast its energy changes along the reaction coordinate, etc.

The techniques which can be used for the construction of orbital correlation diagrams are well known [2,5,8,20] and need not be reviewed here. It will only be pointed out that it is necessary to actually identify the originally bonding orbital which becomes antibonding in the product (i) and the originally antibonding orbital which becomes bonding in the product (j) in order to completely specify the "characteristic configuration" ($i \rightarrow j$). This can be done by inspection.[20] The second step in the argument consists of an assignment of states of the starting material and product to specific configurations, and may require an appeal to spectroscopic data, such as polarization directions, and to some CI calculations.

Very often, the orbital i is the highest occupied bonding orbital (HOMO) of the starting molecule, and j its lowest free antibonding orbital (LFMO), and as they cross along the reaction coordinate i becomes the LFMO and j the HOMO of product. Of course, the crossing may be avoided along paths of low symmetry but this does not change the arguments. This kind of crossing will be referred to as "normal orbital crossover"[20]; the HOMO $\rightarrow$ LFMO configuration of the reactant correlates with the HOMO $\rightarrow$ LFMO configuration of the product both in singlet and triplet. This case is easily handled by the frontier orbital theory.[5] Often these configurations predominate in the lowest excited states of both the reactant and product, both S_1 and T_1, and then no barriers due to orbital correlation are imposed on either side. Sometimes, particularly in the case of aromatic chromophores such as benzene, naphthalene, etc., the singlet state represented by the HOMO $\rightarrow$ LFMO configuration is not lowest but only second in energy. In these instances, the lowest singlet state is represented by a mixture of $1 \rightarrow -2$ and $2 \rightarrow -1$ configurations[s], and a barrier is expected. Since the separation between the two states is usually very small, the resulting barrier can be expected to be relatively small as well (Fig. 8). On the other hand, the triplet state represented by the HOMO $\rightarrow$ LFMO excitation typically is the lowest triplet even in these instances and no symmetry-imposed barriers are expected in the T_1 surface. Experimentally, predissociative line broadening, and photochemistry and emission with wavelength-dependent quantum yields have been observed in several such singlets, for instance on molecules with benzene [51,92,180-182] and biphenyl [183] chromophores, and although they cannot be ascribed to the presence of the predicted [181] barriers with absolute certainty, they are compatible with the MO argument.

Much larger barriers can be expected in at least some of the "abnormal orbital crossovers"[20], $i.e.$, those in which either the orbital i or the orbital j (or both) is not a frontier orbital of the starting material (HOMO or LFMO), but some other orbital. In these instances, a straight-

---

[s] Bonding orbitals are labelled with positive integers, anti-bonding ones with negative integers, starting with HOMO (1) and LFMO (-1).

forward application of the frontier orbital theory would be misleading. Sometimes, the "characteristic configuration" $i \to j$ may be very high in energy and represent a highly excited state. On the other hand, it may be relatively low, or it may interact strongly with low-lying configurations, so that the resulting barriers may be small. Examples of high barriers have been claimed in certain electrocyclic reactions [16,184,185] (the barriers in T_1 can be overcome when the molecule absorbs another photon of light); an example of a low barrier which can be overcome using thermal energy has been reported for a cycloreversion reaction. [28] While some of the other instances of experimentally determined small barriers (e.g., Refs. [95,186–188]) may turn out to be amenable to discussion in similar terms, there are likely to be many others for which such attempts will be unprofitable. It is the author's belief that small barriers will prove to be a common occurrence in photochemical reactions and that they have escaped detection in many cases because detailed temperature and wavelength dependence are rarely studied.

It is perhaps worth pointing out that the initial excitation need not be into the "characteristic state" in order for the barrier to be overcome. All that is needed is to supply enough energy for the motion of the nuclei toward the barrier to enable them to pass above it. Changes in reactivity as a function of wavelength of exciting light can thus easily occur halfway through an electronic absorption band rather than only when a new electronic band is reached and this is well known experimentally.[51,91,92,186].

2. Other Reaction Paths

Derivation of correlation diagrams along other kinds of reaction paths is often more difficult because of lack of symmetry and since it is often harder to estimate the effects of mutual interaction of orbitals and of configuration interaction along the way. If they are strong, such interactions can, of course, completely wipe out barriers expected from simpler arguments. We shall limit the discussion to one well known example, namely photolytic dissociation of the benzylic C—H bond in toluene. When light of sufficiently short wavelengths is used, this proceeds upon absorption of a single photon. With light of longer wavelengths, the reaction no longer proceeds from the singlet state, nor the triplet state, reached by internal conversion (at 77 °K), but does proceed when the triplet absorbs another photon.[189] Thus, it appears very likely that the minima in T_1 and S_1 at the starting geometry are separated from the product minima by barriers which are insurmountable at 77 °K. The origin of the barriers is rationalized on the basis of the following argument, illustrated in Fig. 10 for the triplet reaction. In the first approximation, the "characteristic configuration" in this case is the triplet $^3\sigma\sigma^*$ configuration where σ and σ^* are the bonding and antibonding localized bond orbitals of the C—H bond to be broken. Along the reaction coordinate, they become the non-bonding orbitals in the hydrogen atom and the benzyl radical. If the phenyl group were completely isolated from the methyl group, this process would not be at all affected by triplet $^3\pi\pi^*$ excitations in the phenyl chromophore and $^3\pi\pi^*$ configurations would go up in energy along the reaction coordinate just as much as the ground configuration does (total increase is 85 kcal/mole, *i.e.*, the dissociation energy of the C—H bond).[190] Only the high-energy triplet $^3\sigma\sigma^*$ characteristic configuration would be coming down in energy and at some point it would cross the rising curve of the lowest triplet $\pi\pi^*$ configuration. This would be the top of a rather high barrier, as shown in Fig. 10 by dotted lines. Now, when interaction between the σ and σ^* orbitals and the π orbitals of the phenyl is allowed for, the crossing will be avoided (solid lines in Fig. 10) and the barrier will be lower. It is, however, very difficult to predict without a detailed calculation whether there will be any barrier left and if so, how high it will be. The experimental results suggest that some barrier is indeed left[t]. It is interesting to note that in the process of reaching the triplet state by intersystem crossing from relaxed S_1 state, some 23 kcal/mole are dumped into energy of vibrational motion, mostly into C—H stretching vibrations. This amount of energy might be sufficient to overcome the barrier if it were all concentrated in the direction of our reaction coordinate. Apparently, the probability of this happening before the energy is lost to the surrounding heat bath is vanishingly small. A figure similar to Fig. 10 can be drawn for the excited singlet reaction where experiments also indicate the presence of a barrier, in spite of greater overall exothermicity. This is understandable since both the $^1\pi\pi^*$ and $^1\sigma\sigma^*$ states will be displaced to higher energies compared to their triplet counterparts. It does not seem likely that the photo-dissociation from the $^1\pi\pi^*$ state proceeds via crossing to the $^3\sigma\sigma^*$ triplet [64], as was proposed originally.[155]

It is also possible to make similar arguments for other cases, say photo-decomposition of phenyldiazonium salts [178] or phenyl halides.

---

t) The minimum in the lowest $^3\pi\pi^*$ state of toluene is about 83 kcal/mole above the minimum in S_0, [59] so that the overall motion on the T_1 surface is approximately thermoneutral and there is not much driving force for the reaction. In the absence of any barrier altogether, it should however be observed, since there are no efficiently competing other processes (relatively long triplet lifetime), the endothermicity is very low, and entropy change favorable. Cases in which a minimum in T_1 is inaccessible simply because it is too high in energy and becomes accessible as the energy of the initial excitation increases are also known (*e.g.*, ketene [191]).

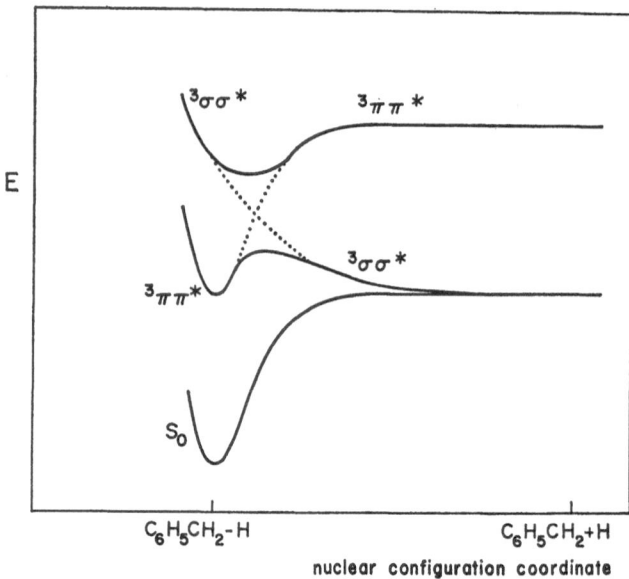

Fig. 10. Energies of selected states during dissociation of a benzylic C—H bond in toluene

For methyl derivatives of an aromatic hydrocarbon, one could expect a rough correlation of the height of the barrier with the size of the expansion coefficients of the HOMO and LFMO orbitals of the aromatic residue at the atom where the methyl group is attached. Also, since the origin of the barrier is related to that of the ground state barrier, substitution which decreases the C—H bond energy should decrease the barrier in T_1 as well. Thus, it should be lower in diphenylmethane and triphenylmethane. Experimental data on barrier heights are presently not available. Relative quantum yields of benzyl radical formation from a series of compounds have been determined and do not correlate with bond energies. As the authors pointed out [155], the interpretation of this result is difficult since cage effects will come into play (a hydrogen atom can diffuse through a rigid matrix much better than larger radicals can). One could also extrapolate to photolysis of other bonds, say benzylic C—C bond in ethylbenzene in comparison to the strained benzylic C—C bond in benzocyclobutene, since in the latter the ring opening can be expected to be semi-rotatory[u] in the triplet state and interaction with

---

[u] One methylene rotating before the other. Such paths are rarely considered in ground state reactions [192,193] although they are the non-concerted analogs of the concerted conrotatory and disrotatory paths and as such are probably followed when the allowed concerted mode is out of reach, say due to steric constraints.[19] They should be quite common in the T_1 state.[19]

the benzene π orbitals should only occur on one of the two benzylic carbon atoms (the minimum in T_1 would be expected at the loose biradicaloid geometry of o-xylylene with one methylene group twisted out of plane [82]). Again, the barrier should be lower in benzocyclobutenes phenyl-substituted in benzylic positions, and this is indeed observed[v]. This kind of consideration removes an apparent paradox: photochemical reactions seem to give some of the least stable products imaginable, as if they did not care whether a little energy is wasted here or there, and yet, are often facilitated considerably by substitution which makes the products a little more stable. Within the framework of the present model, the reason is that the nature of the products is determined by the location of minima in S_1 or T_1, and they tend to occur at biradicaloid geometries where S_0 energies are high, so that even minima located very high in S_0 can be reached after return to ground state. On the other hand, in order to reach a given minimum in S_1 or T_1, the excited molecule may have to pass over barriers and having these a few kcal/mole higher or lower may mean the difference between passing and being stopped.

IV. Summary

Much of organic photochemistry in solution can be understood qualitatively within the Born-Oppenheimer framework by appeal to calculations or to simple arguments directly concerning the energy of the S_1, T_1, and S_0 states and its variation with changes in nuclear geometry, particularly the location of minima and barriers. Other arguments, such as distribution of electron spin, need for continuous electron redistribution, for conservation of orbital symmetry, etc., are secondary, but can be useful if they provide a simple guide for the crux of the matter, namely for the shapes of the energy hypersurfaces.

A higher level of understanding would require a knowledge of molecular dynamics and presently represents a rather distant goal. In addition to reliable knowledge of the shapes of potential energy hypersurfaces, it would also require information such as vibronic coupling elements, densities of vibrational states, detailed mechanism of the action of the heat bath, etc.

---

v) It has been reported [194] that under identical conditions of low-temperature irradiation benzocyclobutene is inert whereas its phenyl derivatives open to substituted o-xylylenes.

J. Michl

To summarize, it is perhaps best to repeat a sentence from the introduction: Simple MO arguments work sometimes but must be applied with caution.

Acknowledgment. The author is indebted to Professor Pete D. Gardner for encouragement and discussions during many a brown-bag lunch hour. Acknowledgment is made to donors of the Petroleum Research Fund, administered by the American Chemical Society, for support of work of which this article is an outgrowth.

References

[1] Born, M., Oppenheimer, R.: Ann. Physik *84*, 457 (1927). — Born, M., Huang, K.: Dynamical theory of crystal lattices. New York: Oxford University Press 1954. For recent discussions see Kołos, W.: Advan. Quantum Chem. *5*, 99 (1970). — Nikitin, E. E.: Advan. Quantum Chem. *5*, 135 (1970).

[2] Woodward, R. B., Hoffmann, R.: Angew. Chem. Intern. Ed. Engl. *8*, 781 (1969).

[3] Trindle, C.: J. Am. Chem. Soc. *92*, 3521, 3255 (1970).

[4] Dewar, M. J. S.: Angew. Chem. Intern. Ed. Engl. *10*, 761 (1971).

[5] Fukui, K.: Accounts Chem. Res. *4*, 57 (1971); Topics Curr. Chem. *15*, 1 (1970).

[6] George, T. F., Ross, J.: J. Chem. Phys. *55*, 3851 (1971).

[7] Pearson, R. G.: Accounts Chem. Res. *4*, 152 (1971); J. Am. Chem. Soc. *94*, 8287 (1972).

[8] Zimmerman, H. E.: Accounts Chem. Res. *4*, 272 (1971).

[9] Goddard, W. A., III: J. Am. Chem. Soc. *94*, 793 (1972).

[10] Langlet, J., Malrieu, J.-P.: J. Am. Chem. Soc. *94*, 7254 (1972).

[11] van der Hart, W. J., Mulder, J. J. C., Oosterhoff, L. J.: J. Am. Chem. Soc. *94*, 5724 (1972).

[12] Zimmerman, H. E., Sousa, L. R.: J. Am. Chem. Soc. *94*, 834 (1972).

[13] For a recent list of references to calculations of parts of ground state potential energy hypersurface see Salem, L.: Accounts Chem. Res. *4*, 322 (1971).

[14] van der Lugt, W. Th. A. M., Oosterhoff, L. J.: J. Am. Chem. Soc. *91*, 6042 (1969).

[15] Dougherty, R. C.: J. Am. Chem. Soc. *93*, 7187 (1971).

[16] Michl, J., Kolc, J.: J. Am. Chem. Soc. *92*, 4148 (1970).

[17] Michl, J.: J. Am. Chem. Soc. *93*, 523 (1971).

[18] Michl, J.: Mol. Photochem. *4*, 243 (1972).

[19] Michl, J.: Mol. Photochem. *4*, 257 (1972).

[20] Michl, J.: Mol. Photochem. *4*, 287 (1972). In: Chemical Reactivity and Chemical Paths (ed. G. Klopman). New York: John Wiley and Sons, Inc. (in the press).

[21] Hammond, G. S.: Advan. Photochem. *7*, 373 (1969).

[22] Zimmerman, H. E.: Angew. Chem. Intern. Ed. Engl. *8*, 1 (1969).

[23] Förster, Th.: Pure Appl. Chem. *24*, 443 (1970).

[24] For an elementary introduction, see Katz, H.: J. Chem. Educ. *48*, 84 (1971).

[25] Epiotis, N. D.: J. Am. Chem. Soc. *94*, 1941, 1946 (1972).

[26] Michl, J.: presented at the 6th International Conference on Photochemistry in Bordeaux, France, September 1971.

[27] Kita, S., Fukui, K.: Bull. Chem. Soc. Japan *42*, 66 (1969).

[28] Chu, N. Y. C., Kearns, D. R.: J. Phys. Chem. 74, 1255 (1970).

[29] Teller, E.: J. Phys. Chem. 41, 109 (1937).

[30] Condon, E. U.: Am. J. Phys. 15, 365 (1947).

[31] Kasha, M.: Discussions Faraday Soc. 9, 14 (1950).

[32] Henry, B. R., Kasha, M.: Ann. Rev. Phys. Chem. 19, 161 (1968); see also Schlag, E. W., Schneider, S., Fischer, S. F.: Ann. Rev. Phys. Chem. 22, 465 (1971).

[33] Herzberg, G., Longuet-Higgins, H. C.: Discussions Faraday Soc. 35, 77 (1963).

[34] Coulson, C. A.: In: Reactivity of the photoexcited organic molecule, p. 1. New York: Wiley-Interscience 1967.

[35] Porter, G.: In: Reactivity of the photoexcited organic molecule, p. 79. New York: Wiley-Interscience 1967.

[36] Siebrand, W., Williams, D. F.: J. Chem. Phys. 49, 1860 (1968).

[37] Siebrand, W.: In: The triplet state (ed. A. B. Zahlan), p. 31. Cambridge, Great Britain: Cambridge University Press 1967; J. Chem. Phys. 46, 440 (1967); 47, 2411 (1967).

[38] Yip, R. W., Siebrand, W.: Chem. Phys. Letters 13, 209 (1972).

[39] Henry, B. R., Siebrand, W.: J. Chem. Phys. 54, 1072 (1971). — Lawetz, V., Orlandi, G., Siebrand, W.: J. Chem. Phys. 56, 4058 (1972).

[40] Phillips, D., Lemaire, J., Burton, C. S., Noyes, W. A., Jr.: Advan. Photochem. 5, 329 (1968).

[41] Jortner, J., Rice, S. A., Hochstrasser, R. M.: Advan. Photochem. 7, 149 (1969). — Rice, S. A., Gelbart, W. M.: Pure Appl. Chem. 27, 361 (1971). — Jortner, J.: Pure Appl. Chem. 27, 389 (1971).

[42] Englman, R., Jortner, J.: Mol. Phys. 18, 145 (1970); see also Voltz, R.: Mol. Phys. 19, 881 (1970).

[43] Freed, K. F.: Topics Curr. Chem. 25, 105 (1972).

[44] Hochstrasser, R. M.: Accounts Chem. Res. 1, 266 (1968).

[45] For dependence of the rate of radiationless transitions on individual vibrational levels, see Heller, D. F., Freed, K. F., Gelbart, W. M.: J. Chem. Phys. 56, 2309 (1972). — Nitzan, A., Jortner, J.: J. Chem. Phys. 56, 2079 (1972). — Lin, S. H.: J. Chem. Phys. 56, 4155 (1972), and references therein.

[46] Rice, S. A., McLaughlin, I., Jortner, J.: J. Chem. Phys. 49, 2756 (1968).

[47] Gelbart, W. M., Rice, S. A., Freed, K. F.: J. Chem. Phys. 52, 5718 (1970).

[48] Huppert, D., Jortner, J., Rentzepis, P. M.: J. Chem. Phys. 56, 4826 (1972).

[49] Freed, K. F., Jortner, J.: J. Chem. Phys. 52, 6272 (1970).

[50] Nitzan, A., Jortner, J.: J. Chem. Phys. 56, 3360 (1972).

[51] Becker, R. S., Dolan, E., Balke, D. E.: J. Chem. Phys. 50, 239 (1969).

[52] Byrne, J. P., Ross, I. G.: Australian J. Chem. 24, 1107 (1971).

[53] Simpson, W. T.: Theoret. Chim. Acta 23, 295 (1972).

[54] Lippert, E.: Accounts Chem. Res. 3, 74 (1970).

[55] For a non-mathematical introduction, see Berry, R. S.: Rec. Chem. Progr. 31, 9 (1970).

[56] Herzberg, G.: Molecular spectra and molecular structure. III. Electronic spectra and electronic structure of polyatomic molecules, Chapt. IV. Princeton, N. J.: D. Van Nostrand 1966.

[57] Wayne, R. P.: Photochemistry. London: Butterworths 1970.

[58] Simons, J. P.: Photochemistry and spectroscopy, New York: Wiley-Interscience 1971.

[59] Birks, J. B.: Photophysics of aromatic molecules, New York: Wiley-Interscience 1970.

60) Calvert, J. G., Pitts, N. J., Jr.: Photochemistry, New York: Wiley-Interscience 1966.

61) McGlynn, S. P., Azumi, T., Kinoshita, M.: Molecular spectroscopy of the triplet state. Englewood Cliffs, N. J.: Prentice-Hall 1969.

62) Becker, R. S.: Theory and interpretation of fluorescence and phosphorescence. New York: Wiley-Interscience 1969.

63) Zimmerman, H. E., Epling, G. A.: J. Am. Chem. Soc. 94, 7806 (1972); for original formulation see Zimmerman, H. E., Schuster, D. I.: J. Am. Chem. Soc. 84, 4527 (1962).

64) Carroll, F. A., Hammond, G. S.: J. Am. Chem. Soc. 94, 7151 (1972).

64a) For more detail see, e.g., Preston, R. K., Tully, J. C.: J. Chem. Phys. 54, 4297 (1971).

65) Dependence on the starting conformation is now well established: Hammond, G. S.: In: Reactivity of the photoexcited organic molecule, p. 119. New York: Wiley. Interscience 1967. — Dauben, W. G.: Plenary lecture presented at the IUPAC Symposium on Photochemistry, Baden-Baden, Germany, July 1972. — Havinga, E.: Plenary lecture presented at the International Symposium on Organic Chemistry of the Excited State, Reading, Great Britain, July 1972; Ref. 66) and references therein.

66) Padwa, A., Brodsky, L., Clough, S.: J. Am. Chem. Soc. 94, 6767 (1972).

67) Rentzepis, P. M.: Chem. Phys. Letters 2, 117 (1968).

68) Huppert, D., Jortner, J., Rentzepis, P. M.: Chem. Phys. Letters 13, 225 (1972). — Eaton, D. F., Evans, T. R., Leermakers, P.A.: Mol. Photochem. 1, 347 (1969).

69) Li, Y. H., Lim, E. C.: J. Chem. Phys. 56, 1004 (1972).

70) Hammond, G. S., Saltiel, J., Lamola, A. A., Turro, N. J., Bradshaw, J. S., Cowan, D. O., Counsell, R. C., Vogt, V., Dalton, C.: J. Am. Chem. Soc. 86, 3197 (1964).

71) Wagner, P. J., Kemppainen, A. E.: J. Am. Chem. Soc. 94, 7495 (1972). — Previtali, C. M., Scaiano, J. C.: J. Chem. Soc., Perkin II 1972, 1667, 1672.

72) Dumbacher, B.: Theoret. Chim. Acta 23, 346 (1972).

73) Devaquet, A.: J. Am. Chem. Soc. 94, 5160 (1972).

74) Marsh, G., Kearns, D. R., Schaffner, K.: Helv. Chim. Acta 51, 1890 (1968); J. Am. Chem. Soc. 93, 3129 (1971).

75) Loper, G. L., Lee, E. K. C.: Chem. Phys. Letters 13, 140 (1972).

76) Spears, K. G., Rice, S. A.: J. Chem. Phys. 55, 5561 (1971).

77) Förster, E. W., Grellmann, K. H.: Chem. Phys. Letters 14, 536 (1972).

78) Menter, J., Förster, Th.: Photochem. Photobiol. 15, 289 (1972).

79) Vander Donckt, E.: Progr. Reaction Kinetics 5, 273 (1970).

80) Turro, N. J.: Plenary lecture presented at the 4th IUPAC Symposium on Photochemistry, Baden-Baden, Germany, July 1972.

81) Herkstroeter, W. G., McClure, D. S.: J. Am. Chem. Soc. 90, 4522 (1968).

82) Baird, N. C.: J. Am. Chem. Soc. 94, 4941 (1972).

83) Saltiel, J., D'Agostino, J. T.: J. Am. Chem. Soc. 94, 6445 (1972). — Saltiel, J., Rousseau, A. D., Sykes, A.: J. Am. Chem. Soc. 94, 5903 (1972).

84) Becker, R. S., Edwards, L., Bost, R., Elam, M., Griffin, G.: J. Am. Chem. Soc. 94, 6584 (1972).

85) Lamola, A. A.: In: Energy transfer and organic photochemistry (ed. P. A. Leermakers and A. Weissberger), p. 68. New York: Interscience 1969.

86) Chapman, O. L.: Plenary lecture presented at the International Symposium on Organic Chemistry of the Excited State, Reading, Great Britain, July 1972.

87) Ullman, E. F.: Accounts Chem. Res. 1, 353 (1968).

88) Sharafy, S., Muszkat, K. A.: J. Am. Chem. Soc. *93*, 4119 (1971) and references therein.

89) A classical example is provided by the work of Malkin and Fischer: Malkin, S., Fischer, E.: J. Phys. Chem. *66*, 2482 (1962); *68*, 1153 (1964). For a recent study, see Coyle, J. D.: J. Chem. Soc., Perkin II *1972*, 683; and Ref. [83, 84].

90) Hochstrasser, R. M., Marzzacco, C.: J. Chem. Phys. *49*, 971 (1968).

91) Simonaitis, R., Pitts, J. N., Jr.: J. Am. Chem. Soc. *91*, 108 (1969).

92) Parmenter, C. S.: Advan. Chem. Phys. *22*, 365 (1972).

93) Singh, B., Zweig, A., Gallivan, J. B.: J. Am. Chem. Soc. *94*, 1199 (1972).

94) Feler, G.: Theoret. Chim. Acta *12*, 412 (1968).

95) Becker, R. S., Inuzuka, K., King, J., Balke, D. E.: J. Am. Chem. Soc. *93*, 43 (1971).

96) Borrell, P., Greenwood, H. H.: Proc. Roy. Soc. (London), Sec. A. *298*, 453 (1967).

97) Baird, N. C., West, R. M.: J. Am. Chem. Soc. *93*, 4427 (1971). — Baird, N. C.: Mol. Photochem. *2*, 53 (1970). — Hoffmann, R.: Tetrahedron *22*, 521 (1966).

98) Lorquet, A. J.: J. Phys. Chem. *74*, 895 (1970).

99) Muszkat, K. A., Schmidt, W.: Helv. Chim. Acta *54*, 1195 (1971). — Schmidt, W.: Helv. Chim. Acta *54*, 862 (1971).

100) Zimmerman, H. E., Binkley, R. W., Givens, R. S., Grunewald, G. L., Sherwin, M. A.: J. Am. Chem. Soc. *91*, 3316 (1969).

101) Becker, R. S., Inuzuka, K., King, J.: J. Chem. Phys. *52*, 5164 (1970). — Inuzuka, K., Becker, R. S.: Bull. Chem. Soc. Japan *44*, 3323 (1971).

102) Stephenson, L. M., Gibson, T. A.: J. Am. Chem. Soc. *94*, 4599 (1972).

103) White, G. M., Yarwood, A. J., Santry, D. P.: Chem. Phys. Letters *13*, 501 (1972).

104) Hoffmann, R.: Pure Appl. Chem. *24*, 567 (1970); J. Am. Chem. Soc. *90*, 1475 (1968).

105) Buenker, R. J., Peyerimhoff, S. D.: J. Phys. Chem. *73*, 1299 (1969).

106) Karplus, S., Bersohn, R.: J. Chem. Phys. *51*, 2040 (1969).

107) Hayes, D. M., Morokuma, K.: Chem. Phys. Letters *12*, 539 (1972).

108) Fink, W. H.: J. Am. Chem. Soc. *94*, 1073, 1078 (1972).

109) Kaldor, U., Shavitt, I.: J. Chem. Phys. *48*, 191 (1968); see also Buenker, R. J., Peyerimhoff, S. D., Kammer, W. E.: J. Chem. Phys. *55*, 814 (1971).

110) Buenker, R. J., Peyerimhoff, S. D.: J. Chem. Phys. *53*, 1368 (1970).

111) Baird, N. C.: Mol. Phys. *18*, 39 (1970); Chem. Commun. *1970*, 199.

112) Salem, L., Rowland, C.: Angew. Chem. Intern. Ed. Engl. *11*, 92 (1972).

113) Fischer, S.: J. Chem. Phys. *56*, 5199 (1972). — Nitzan, A., Jortner, J.: J. Chem. Phys. *56*, 5200 (1972), and references therein.

114) Caplan, C. E., Child, M. S.: Mol. Phys. *23*, 249 (1972).

115) Evans, K., Rice, S. A.: Chem. Phys. Letters *14*, 8 (1972).

116) Devaquet, A.: J. Am. Chem. Soc. *94*, 5626 (1972).

117) Salem, L.: J. Am. Chem. Soc. *90*, 543 (1968).

118) Salem, L.: J. Am. Chem. Soc. *90*, 553 (1968).

119) Devaquet, A., Salem, L.: J. Am. Chem. Soc. *91*, 3793 (1969).

120) Herndon, W. C., Giles, W. B.: Mol. Photochem. *2*, 277 (1970).

121) Herndon, W. C.: Topics Curr. Chem., in press.

122) Devaquet, A.: Mol. Phys. *18*, 233 (1970).

123) Zimmerman, H. E.: Advan. Photochem. *1*, 183 (1963).

124) Zimmerman, H. E., Swenton, J. S.: J. Am. Chem. Soc. *89*, 906 (1967).

125) Daudel, R.: In: Reactivity of the photoexcited organic molecule, p. 51. New York: Wiley-Interscience 1967.

126) Daudel, R.: Advanc. Quantum Chem. *5*, 1 (1970).
127) Seiler, P., Wirz, J.: Tetrahedron Letters *1971*, 1683.
128) Havinga, E.: In: Reactivity of the photoexcited organic molecule, p. 201. New York: Wiley-Interscience 1967. — Havinga, E., Kronenberg, M. E.: Pure Appl. Chem. *16*, 137 (1968).
129) Shizuka, H., Ono, S., Morita, T., Tanaka, I.: Mol. Photochem. *3*, 203 (1971). — Tinland, B., Décoret, C.: Tetrahedron Letters *1971*, 2467. — Tinland, B.: Tetrahedron *26*, 4795 (1970). — Feler, G.: Theoret. Chim. Acta *10*, 33 (1968). — Tyutyulkov, N., Fratev, F., Petkov, D.: Theoret. Chim. Acta. *8*, 236 (1967). — Song, P.-S., Kurtin, W. E.: Mol. Photochem. *1*, 1 (1969).
130) Sommer, U., Kramer, H. E. A.: Photochem. Photobiol. *13*, 387 (1971). — Laarhoven, W. H., Cuppen, Th. J. H. M., Nivard, R. J. F.: Tetrahedron *26*, 4865 (1970).
131) Chalvet, O., Daudel, R., Schmid, G. H., Rigaudy, J.: Tetrahedron *26*, 365 (1970).
132) Bertran, J., Schmid, G. H.: Tetrahedron *27*, 5191 (1971).
133) Godfrey, M.: J. Chem. Soc., Perkin II, *1972*, 1690.
134) Higuchi, J., Ito, T.: Theoret. Chim. Acta *22*, 61 (1971).
135) Fukui, K., Morokuma, K., Yonezawa, T.: Bull. Chem. Soc. Japan *34*, 1178 (1961).
136) Herndon, W. C.: Chem. Rev. *72*, 167 (1972).
137) Michl, J., Becker, R. S.: J. Chem. Phys. *46*, 3889 (1967). — Michl, J.: J. Mol. Spectroscopy *30*, 66 (1969).
138) Haselbach, E., Schmelzer, A.: Helv.Chim.Acta *54*, 1575 (1971); *55*, 1745 (1972).
139) Bertrán, J., Chalvet, O., Daudel, R.: Theoret. Chim. Acta *14*, 1 (1969).
140) Constanciel, R.: Theoret. Chim. Acta *26*, 249 (1972); the importance of exchange terms for differentiation between singlet and triplet reactivity is also recognized in ref. 133).
141) Weller, A.: Progr. Reaction Kinetics *1*, 187 (1961).
142) Pechenaya, V. I., Danilov, V. I.: Chem. Phys. Letters *11*, 539 (1971).
143) Zimmerman, H. E.: J. Am. Chem. Soc *88*, 1566 (1966).
144) Longuet-Higgins, H. C., Abrahamson, E.W.: J. Am. Chem. Soc. *87*, 2045 (1965).
145) Bryce-Smith, D.: Chem. Commun. *1969*, 806.
146) Wagner, P. J.: J. Am. Chem. Soc. *89*, 2503 (1967).
147) Houk, K. N., Northington, D. J., Duke, R. E., Jr.: J. Am. Chem. Soc. *94*, 6233 (1972).
148) Daudel, R.: Advan. Quantum Chem. *3*, 161 (1967).
149) Hoffmann, R., Swaminathan, S., Odell, B. G., Gleiter, R.: J. Am. Chem. Soc. *92*, 7091 (1970).
150) Harris, F. E., Michl, J.: unpublished results.
151) Herzberg, G.: Molecular spectra and molecular structure. I. Spectra of diatomic molecules, pp. 373, 532. Princeton, N. J.: D. Van Nostrand 1950 (2nd edit.). — Dieke, G. H., Cunningham, S. P.,: J. Mol. Spectry. *18*, 288 (1965).
152) Kołos, W., Roothaan, C. C. J.: Rev. Mod. Phys. *32*, 219 (1960).
153) Dewar, M. J. S., Kelemen, J.: J. Chem. Educ. *48*, 494 (1971).
154) Dalton, J. C., Dawes, K., Turro, N. J., Weiss, D. S., Barltrop, J. A., Coyle, J. D.: J. Am. Chem. Soc. *93*, 7213 (1971).
155) Porter, G., Strachan, E.: Trans. Faraday Soc. *54*, 1595 (1958).
156) Plotnikov, V. G.: Opt. Spectry. (USSR) English Transl. *27*, 322 (1969).
157) Ward, H. R.: Accounts Chem. Res. *5*, 18 (1972).
158) Bartlett, P. D., Helgeson, R., Wersel, O. A.: Pure Appl. Chem. *16*, 187 (1968).
159) Givens, R. S., Oettle, W. F.: J. Am. Chem. Soc. *93*, 3963 (1971).

[160] Schuster, D. I., Underwood, G. R., Knudsen, T. P.: J. Am. Chem. Soc. 93, 4304 (1971).

[161] Merer, A. J., Mulliken, R. S.: Chem. Rev. 69, 639 (1969).

[162] Watson, F. H., Jr., McGlynn, S. P.: Theoret. Chim. Acta 21, 309 (1971).

[163] Saltiel, J., Rousseau, A. D., Sykes, A.: J. Am. Chem. Soc. 94, 5903 (1972), and references therein.

[164] Liu, R. S. H., Butt, Y.: J. Am. Chem. Soc. 93, 1532 (1971).

[165] Simmons, H. E.: Progr. Phys. Org. Chem. 7, 1 (1970).

[166] Saltiel, J., Metts, L., Wrighton, M.: J. Am. Chem. Soc. 92, 3227 (1970).

[167] Boué, S., Srinivasan, R.: J. Am. Chem. Soc. 92, 3226 (1970).

[168] Dauben, W. G.: In: Reactivity of the photoexcited organic molecule, p. 171. New York: Wiley-Interscience 1967.

[169] Zweig, A.: Plenary lecture presented at the 4th IUPAC Symposium on Photo chemistry, Baden-Baden, Germany, July 1972.

[170] Houk, K. N., Northington, D. J.: J. Am. Chem. Soc. 93, 6693 (1971).

[171] Koutecký, J.: J. Chem. Phys. 47, 1501 (1967).

[172] Schulten, K., Karplus, M.: Chem. Phys. Letters 14, 305 (1972).

[173] Clark, P. A., Csizmadia, I. G.: J. Chem. Phys. 56, 2755 (1972).

[174] Hudson, B. S., Kohler, B. E.: Chem. Phys. Letters 14, 299 (1972).

[175] Shih, S., Buenker, R. J., Peyerimhoff, S. D.: Chem. Phys. Letters 16, 244 (1972).

[176] Lamola, A. A.: Photochem. Photobiol. 7, 619 (1968).

[177] Casey, C. P., Boggs, R. A.: J. Am. Chem. Soc. 94, 6457 (1972). — Bartlett, P. D.: Quart. Rev. 24, 473 (1970).

[178] Cox, R. J., Bushnell, P., Evleth, E. M.: Tetrahedron Letters 1970, 207.

[179] Berson, J. A., Olin, S. S.: J. Am. Chem. Soc. 92, 1086 (1970).

[180] Callomon, J. H., Parkin, J. E., Lopez-Delgado, R.: Chem. Phys. Letters 13, 125 (1972).

[181] Haller, I.: J. Chem. Phys. 47, 1117 (1967).

[182] Bryce-Smith, D.: Pure Appl. Chem. 16, 47 (1968).

[183] Kolc, J., Michl, J.: Presented at the 4th IUPAC Symposium on Photochemistry, Baden-Baden, Germany, July 1972.

[184] Kolc, J., Michl, J.: Abstracts, First Rocky Mountain Regional Meeting, A. C. S., Fort Collins, Colo., June 1972, p. 32; J. Am. Chem. Soc., 95, 7391 (1973).

[185] Meinwald, J., Samuelson, G. E., Ikeda, M.: J. Am. Chem. Soc. 92, 7604 (1970).

[186] Hemminger, J. C., Rusbult, C. F., Lee, E. K. C.: J. Am. Chem. Soc. 93, 1867 (1971). — Hemminger, J. C., Lee, E. K. C.: J. Chem. Phys. 56, 5284 (1972).

[187] Muszkat, K. A., Fischer, E.: J. Chem. Soc. (B) 1967, 662.

[188] Srinivasan, R., Boué, S.: J. Am. Chem. Soc. 93, 550 (1971).

[189] Schwartz, F. P., Albrecht, A. C.: Chem. Phys. Letters 9, 163 (1971).

[190] Golden, D. M., Benson, S. W.: Chem. Rev. 69, 125 (1969).

[191] McIntosh, J. S. E., Porter, G. B.: Can. J. Chem. 50, 2313 (1972).

[192] Bauld, N. L., Farr, F. R., Chang, C.-S.: Tetrahedron Letters 1972, 2443.

[193] Jones, W. M., Krause, D. L.: J. Am. Chem. Soc. 93, 551 (1971).

[194] Quinkert, G., Finke, M., Palmowski, J., Wiersdorff, W.-W.: Mol. Photochem. 1, 433 (1969).

Recent Advances in Research on the Chemiluminescence of Organic Compounds

Prof. Dr. Karl-Dietrich Gundermann

Organisch-Chemisches Institut der Technischen Universität, Clausthal-Zellerfeld

Contents

I. Introduction

Research on the chemiluminescence of organic compounds is making very rapid progress. Since the appearance in 1968 of the first monograph dedicated exclusively to the field [1], many new results have been published, ranging from chemiluminescent autoxidation reactions of hydrocarbons in which singlet oxygen is involved to complicated bioluminescent, *i.e.* enzyme-catalyzed, reactions. It becomes more and more impossible to draw a complete picture of the whole field. Therefore this review concentrates on some special aspects which have been and still are in the center of interest. These are chemiluminescence reactions involving 1. cyclic and non-cyclic hydrazides and related compounds; 2. dioxetane derivatives; 3. radical ions.

Special review articles published since 1968 on these topics are: one by E. H. White and D. F. Roswell [2] on hydrazide chemiluminescence; M. M. Rauhut [3] on the chemiluminescence of concerted peroxide-decomposition reactions; and D. M. Hercules [4,5] on chemiluminescence from electron-transfer reactions. The rapid development in these special fields justifies a further attempt to depict the current status. Results of bioluminescence research will not be included in this article except for a few special cases, *e.g.* enzyme-catalyzed chemiluminescence of luminol, and firefly bioluminescence [6].

II. General Concepts of the Mechanism of Chemiluminescence

Chemiluminescence is defined as the production of light by chemical reactions. This light is "cold", which means that it is not caused by vibrations of atoms and/or molecules involved in the reaction but by direct transformation of chemical into electronic energy. For earlier discussions of this problem, see [7-9]. Recent approaches towards a general theory of chemiluminescence are based on the relatively simple electron-transfer reactions occurring in aromatic radical-ion chemiluminescence reactions [10] and on considerations of molecular orbital symmetry as applied to 1.2-dioxetane derivatives, which very probably play a key role in a large number of organic chemiluminescence reactions [11].

A. Physical Basis of Chemiluminescence

D. M. Hercules [4,166] gives the following main criteria for the production of light. To be chemiluminescent, a reaction must provide

a) sufficient excitation energy,

b) at least one species capable of transfer into an electronically excited state,

c) a chemical reaction proceeding at a sufficiently high rate to provide the excitation energy,

d) a system of reaction coordinates favoring the production of excited states rather than ground states.

For the emission of visible light, the necessary energy ranges from 40 kcal/mole (red light) to 70 kcal (violet light). Thus, only rather exergonic reactions can be chemiluminescent in the visible range of the spectrum.

The chemiluminescence quantum efficiency of a reaction

$$A + B \longrightarrow C^x + D$$

(C^x: electronically excited product) depends on the efficiency ϕ_{es} of the production of excited product molecules, and on the efficiency of the excited product molecules (or other molecules present in the reaction mixture) in transforming excitation energy into light. In most of the chemiluminescence reactions investigated so far this efficiency is identical with the fluorescence efficiency of the molecules concerned, so that

$$\phi_{CL} = \frac{\text{Einsteins of } h\nu}{\text{reacted moles A and/or B}} = \phi_{es} \times \phi_{fl}$$

ϕ_{CL}: chemiluminescence quantum efficiency, ϕ_{es}: efficiency for production of excited-state molecules, ϕ_{fl}: fluorescence efficiency

Brundrett, Roswell, and White [12] subdivide the efficiency ϕ_{es}, the "chemical efficiency" of a chemiluminescent reaction, into the efficiency ϕ_r (fraction of molecules following the "correct" chemistry) and the efficiency ϕ_{es} (fraction of molecules crossing over to the excited state after having taken the correct chemical path).

Therefore a low chemiluminescence quantum yield can be due to the fact that the fluorescence efficiency of the product molecules is high but the chemical efficiency of the reaction producing excited molecules is low, or the reverse, or that ϕ_{es} and ϕ_{fl} are both low.

Recent investigations (see e.g. [12,13]) therefore make a special point of differentiating as far as possible between the chemical and physical efficiencies of chemiluminescence reactions.

If the emitting species is not a reaction-product molecule directly formed by the exergonic reaction (as is the case in the luminol reaction, for example), chemiluminescence can occur via energy transfer processes:

$$A + B + \ldots \longrightarrow P^x$$
$$P^x + X \longrightarrow X^x + P$$
$$X^x \longrightarrow X + h\nu$$

A, B....: reactants, P^x: electronically excited reaction product, P: reaction product in the ground state, X^x: energy acceptor in electronically excited, X: energy acceptor in the ground state

These energy-transfer processes are especially interesting in those chemiluminescence reactions where the primary electronically excited product is formed in its triplet state (autoxidation reactions, radical-ion recombination reactions; see Sections III and VIII), although some reactions have been reported to involve direct emission from the excited triplet state [14].

In the crossing over of product molecules to the excited state the geometry of the excited-state product as compared with that of the reactants is of decisive importance. This is demonstrated by a theory of

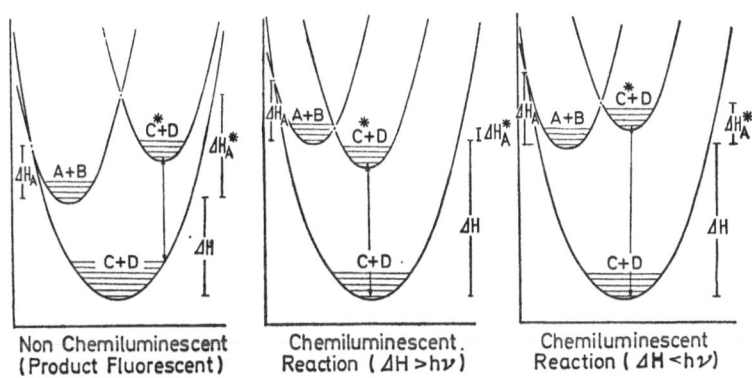

| Non Chemiluminescent (Product Fluorescent) | Chemiluminescent. Reaction ($\Delta H > h\nu$) | Chemiluminescent Reaction ($\Delta H < h\nu$) |

A + B = Reactants
C + D = Products formed in ground states
C* + D = Products formed with C in an excited state and D in the ground state
ΔH = Energy available from the reaction according to the usual thermo-dynamic criteria
ΔH_A = Activation energy for formation of products in the ground state
ΔH_A^* Activation energy for formation of one product in an excited state
$h\nu$ Energy necessary for the excitation C $\longrightarrow$ C*

Fig. 1. Reaction coordinate diagrams for chemical excitation processes [4]

R. A. Marcus for simple electron-transfer reactions [10] which, in appropriate modification, appears to be theoretically valid for more complex chemiluminescence reactions, too (e.g. [13]).

According to this theory, the activation energy of a reaction is high when the enthalpy of a reaction is high, as demonstrated in Fig. 1.

If a reaction can yield products in the ground state or in an electronically excited state, the activation energy for the formation of the latter will therefore be less than that required for the formation of the products in the ground state — provided that there is no significant change in the configuration of the excited-state molecules as compared with the reactant molecules.

In this case the intersection point of the potential curve for the reactant molecules $A + B$ and of that of the excited products $(C^x + D)$ will be only a little above the respective minima of the curves; this means a relatively small activation energy.

If, however, there is a significant configurational change, the intersection point moves upward in respect of the energy minima of the curves $A + B$ and $C^x + D$, corresponding to an increase in activation energy.

Chemiluminescent reactions must proceed at a sufficiently fast rate to provide the minimum number of quanta per time unit, as determined by the sensitivity of the detector used. According to Hercules [4] "a chemiluminescence reaction with 100% efficiency emitting only one photon per fortnight would not be detected".

B. Radical-Ion Reaction Mechanisms

In complex organic molecules calculations of the geometry of excited states and hence predictions of chemiluminescent reactions are very difficult; however, as is well known, in polycyclic aromatic hydrocarbons there are relatively small differences in the configurations of the ground state and the excited state. Moreover, the chemiluminescence produced by the reaction of aromatic hydrocarbon radical anions and radical cations is due to simple one-electron transfer reactions, especially in cases where both radical ions are derived from the same aromatic hydrocarbon, as in the reaction between 9.10-diphenyl anthracene radical cation and anion. More complex are radical ion chemiluminescence reactions involving radical ions of different parent compounds, such as the couple naphthalene radical anion/Wurster's blue (see Section VIII. B.).

But even in the case of electron-transfer reactions between radical ions of the same hydrocarbon, the mechanism leading to emission is simple only in the case of direct formation of the respective hydrocarbon in its first excited singlet state. R. E. Visco and E. A. Chandross [15] suggested the following scheme in terms of simplified HMO treatment:

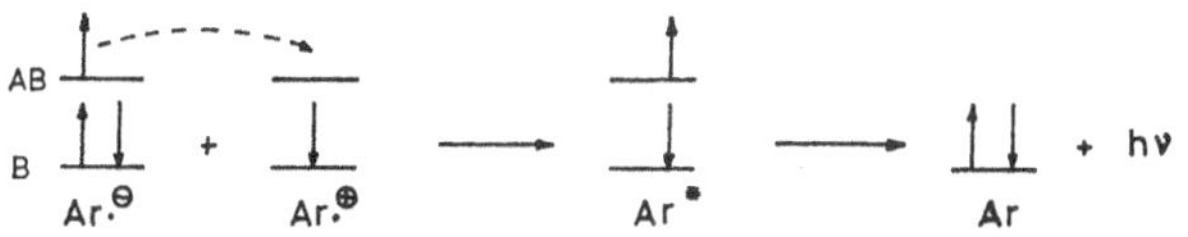

Fig. 2

In the radical anion (produced either by chemical or electrochemical treatment of the hydrocarbon) the highest bonding π orbital is filled and an additional electron is placed on the lowest antibonding π orbital. The radical cation has only one electron in the highest bonding π orbital. The direct formation of the first excited singlet state occurs when the electron on the lowest antibonding π orbital of the radical anion is transferred to the lowest antibonding orbital of the radical cation. When the spin orientation of the reacting radical ions is as depicted in Fig. 2, *i.e.* when the transferred electron is of opposite spin from the lone electron on the highest bonding π orbital of the radical cation, the hydrocarbon is actually formed in its first excited singlet state. The chemiluminescence occurring on return of the "excited" electron to the highest bonding orbital is fluorescence emission. The formation of an excited state in this electron-transfer reaction instead of the direct formation of the ground state of the hydrocarbon is explained by the above-mentioned theory of Marcus (see also [16]): if there are only small differences between the geometry of the reactant radical ions and the excited singlet state of the hydrocarbon, it will require a smaller activation energy to form the excited state of the hydrocarbon than the ground state.

It has been pointed out, however, that the primary requirement of a chemiluminescent reaction is that it should be energy-sufficient. In a considerable number of radical-ion reactions the energy is not sufficient to produce the first excited singlet state (the term "energy-deficient systems" has been introduced for these cases, see Section VIII.). The first excited triplet state is commonly situated at a lower energy level. Therefore triplet states should be involved in radical-ion chemiluminescence, especially in energy-deficient systems, but also in those cases involving quick loss of energy of the primarily formed singlet states. (This does not, of course, mean totally radiationfree deactivation).

However, the chemiluminescence emission of radical-ion reactions is in most cases identical with the fluorescence emission of the electron-accepting species. Therefore the following other possibilities have been suggested:

$$Ar^+_{\cdot} + Ar^-_{\cdot} \longrightarrow {}^3Ar^* + Ar \qquad \text{(formation of excited triplet states)}$$

$${}^3Ar^* + {}^3Ar^* \longrightarrow {}^1Ar^* + A \qquad \text{(formation of excited singlet states by triplet-triplet annihilation)}$$

$$Ar^+_{\cdot} + Ar^-_{\cdot} \longrightarrow {}^1Ar^*_2 \qquad \text{(excimer formation)}$$

$${}^1Ar^*_2 \longrightarrow 2\,Ar + h\nu \qquad \text{(excimer fluorescence)}$$

$$Ar^-_{\cdot} + P^+_{\cdot} \longrightarrow {}^3Ar^* + P \qquad \text{(reaction of radical anion with an electron acceptor other than its corresponding radical cation; formation of excited triplet states)}$$

$$Ar^+_{\cdot} + P^-_{\cdot} \longrightarrow {}^3Ar^* + P \qquad \text{(reaction of radical cation with electron donor forming excited triplet states)}$$

$$Ar^-_{\cdot} + P^+_{\cdot} \longrightarrow {}^1(ArP)^* \qquad \text{(reaction of radical anion with electron acceptor forming a hetero excimer [17])}$$

There is experimental evidence that triplet states indeed play an important role in radical ion reactions. The formation of excimers has been suggested on the basis of chemiluminescence emission spectra, e.g. in the case of N-phenylcarbazole [15] and in some other experiments [18]. Other authors [19,20] have observed that the "excimer fluorescence" reported is probably produced by decomposition products of the radical ions or other impurities, as is very probably the so-called preannihilation chemiluminescence which occurs in electrogenerated chemiluminescence (see [21]).

C. Dioxetane Intermediates

The application of the Woodward-Hoffmann theory [22] of electrocyclic reactions to chemiluminescence has proved a very useful and productive approach, suggested independently by E. H. White and M. J. C. Harding [23], F. McCapra, D. G. Richardson, and Y. C. Chang [11, 24], and M. M. Rauhut and coworkers [25].

If a four-membered ring peroxide (1.2-dioxetane) is involved in a reaction, its concerted bond cleavage into two carbonyl moieties should yield one of these in its excited electronic state on the basis of the orbital symmetry conservation rules of R. B. Woodward and R. Hoffmann:

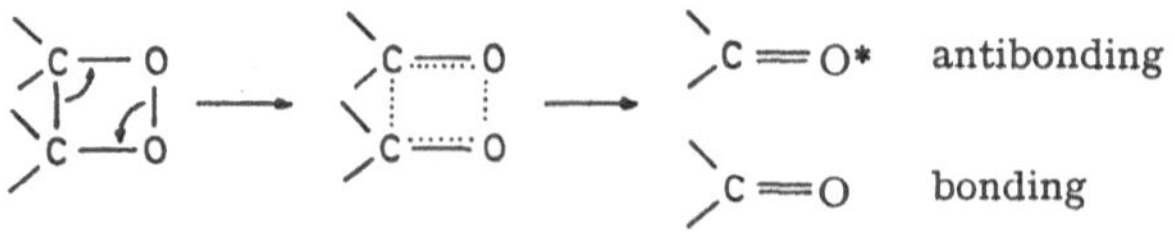

McCapra in particular proposed [11] that the chemiluminescence reactions of a large number of organic compounds had this concerted dioxetane decomposition step as key reaction in the production of electronically excited products, namely: acridinium salts [25,26,27], indolylperoxides [28], activated oxalic esters [29], diphenyl carbene [30], tetrakis-dimethylamino-ethylene [31,32], lucigenin [33], and substituted imidazoles [23].

The recently discovered preparative methods for the synthesis of 1.2-dioxetane derivatives (see Section V.) have made these compounds and their chemiluminescent decomposition the subject of especially intensive study.

As to the nature of the electronically excited state, the investigation of the thermolysis of tetramethyl-1.2-dioxetane revealed a high yield (about 50%) of excited triplet acetone [34]:

$$\begin{array}{c} (H_3C)_2C-O \\ |\quad| \\ (H_3C)_2C-O \end{array} \longrightarrow \quad ^3(H_3C-CO-CH_3)* \; + \; ^1(H_3C-CO-CH_3)*$$
$$50\% \qquad\qquad 1\%$$

III. Autoxidation Reactions

Numerous autoxidation reactions of aliphatic and araliphatic hydrocarbons, ketones, and esters have been found to be accompanied by chemiluminescence (for reviews see [1], p. 19; [14]) generally of low intensity and quantum yield. This weak chemiluminescence can be measured by means of modern equipment, especially when fluorescers are used to transform the electronic excitation energy of the triplet carbonyl compounds formed as primary reaction products. It is therefore possible to use it for analytical purposes [35], *e.g.* to measure the efficiency of inhibitors as well as initiators in autoxidation of polymer hydrocarbons [14], and in mechanistic studies of radical chain reactions.

R. F. Vasil'ev and coworkers [14] suggested that within the well-known general scheme of radical chain autoxidation reactions:

I. Start:

$$R'-R' \longrightarrow 2R'\cdot \quad \text{(initiator radical formation)}$$

II. Chain propagation:

$$R-CH_2-R + R'\cdot \longrightarrow R\overset{\cdot}{C}H-R + R'H$$

$$R-\overset{\cdot}{C}H-R + O_2 \longrightarrow \begin{array}{c} R-CH-R \\ O \\ O\cdot \end{array}$$

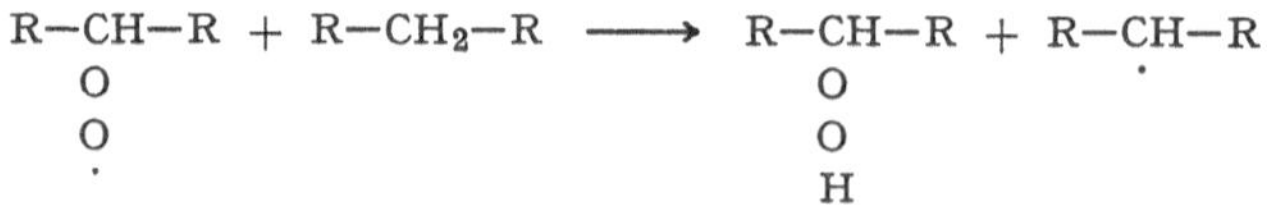

III. Chain termination:

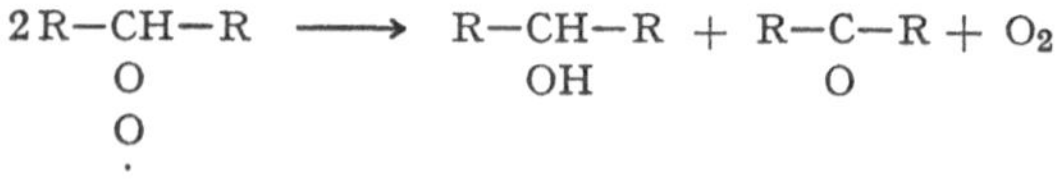

Only the chain-termination step provides sufficient energy for the production of electronically excited products. This step, consisting in the recombination of two peroxy radicals, yields energies of about 115—150 kcal/mole [36]. In the autoxidation of ethylbenzene to acetophenone, triplets were found to be the emitting species [37]. For luminescence excitation of carbonyl triplets, energies in the range of 75—80 kcal/mole are required [38]. Recent investigations have concentrated on the exploration of the recombination mechanism of the peroxy radicals and on finding out why this type of chemiluminescence is so weak. (See also [35a] for a recent review of autoxidation reactions.)

A. Hydrocarbons, Carbonyl Derivatives, Reactive Methylene Compounds

Peroxy radical recombination appears to be the most important source of the electronic excitation energy emitted during hydrocarbon autoxidation. In addition to the above-mentioned energetic considerations, this is clear from the following experimental facts: the termination rate for secondary peroxy radicals is 10^3 times faster than for tertiary peroxy radicals due to their having no α-hydrogen [14]; the termination rate constant decreases by 1.9 with α-deuteration [39–40].

The autoxidation of cyclohexane initiated by dicyclohexyl peroxydicarbonate (DCPD) takes this pathway [38]:

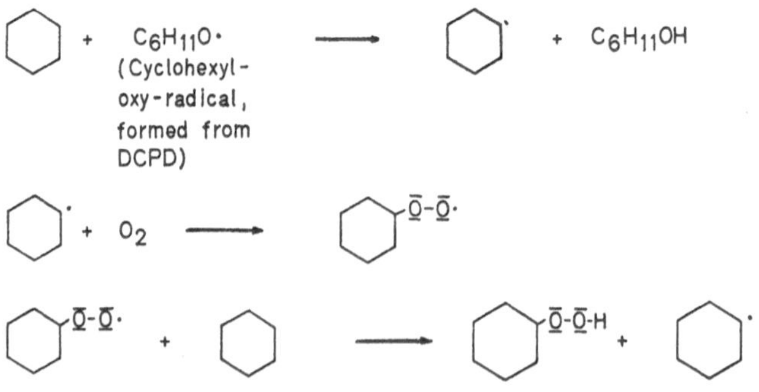
C6H11O·
(Cyclohexyl-oxy-radical, formed from DCPD)
C6H11OH
O2

Cyclohexyl radicals react with cyclohexyl hydroperoxide to yield Cyclohexane and the cyclohexyl peroxy radical:

The recombination of the cyclohexyl peroxy radicals produced in one of these two reaction pathways gives rise to cyclohexanol, cyclohexanone and oxygen ("disproportionation"):

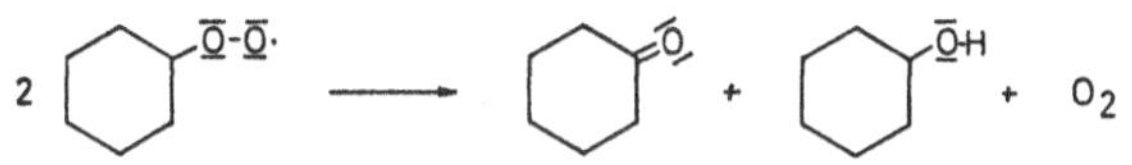

Quenching by adventitious oxygen occurs to a small extent only, as most of the oxygen is consumed in the chain-propagation reaction.

In a series of chemiluminescent autoxidation reactions the following quantum yields were measured [38]:

	Chemiluminescence quantum yields (Einsteins/mole)($\times 10^{-9}$)
n-heptane	0.8
n-octane	0.16
n-dodecane	0.15
cyclohexane	2.4
cyclododecane	1.1
ethylbenzene	0.9
2-butanone	7
2-heptanone	2.4
benzyl-phenylketone	8
cyclopentanone	40

The quantum yield is defined here as the ratio of photons emitted to the number of initiator molecules: one molecule of DCPD yields 2 radicals which in turn produce one molecule of ketone in the recombination reaction.

These results are discussed by R. E. Kellogg [38] in terms of the Russell diagram [39]:

For the production of an electronically excited carbonyl product molecule the "exothermicity" of the reaction must be concentrated in the carbonyl fragment according to the scheme:

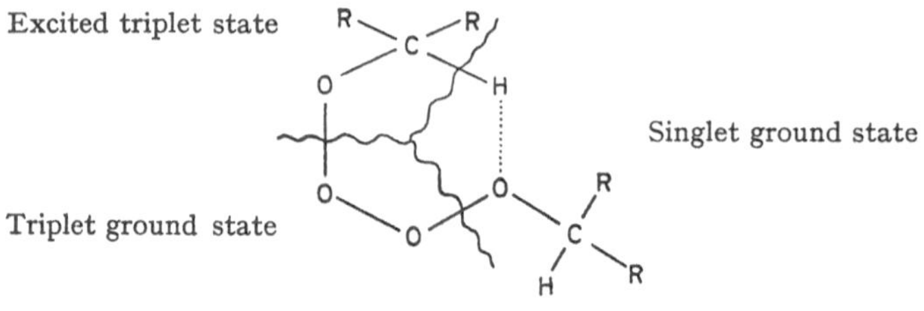

Excited triplet state

Singlet ground state

Triplet ground state

Concerted fragmentation of the transition state in the peroxy radical recombination yields carbonyl compound molecules in the excited triplet state, alcohol in its singlet ground state, and oxygen in its triplet ground state, in fulfilment of the spin selection rules.

The emission spectrum of the chemiluminescent autoxidation of 2-butanone was found to be identical with that of biacetyl phosphorescence [38], as reported earlier by Vasil'ev and coworkers [14]. A suggested solution of the problem of how phosphorescence can occur at all in an oxygen-saturated solution, in other words, why there is no considerable quenching by adventitious oxygen[a]), is as follows. There is no significant quenching by oxygen in the solution because the oxygen produced in the radical recombination reaction, which is retained in the solvent cage, quenches the excited triplet carbonyl molecules far more efficiently. The rate of triplet-triplet transfer quenching in the solvent cage by exchange interaction has been estimated to be about 10^{11} sec^{-1} [42]. Assuming a similar rate constant for oxygen quenching in the autoxidation reactions depicted, one can calculate that only one excited carbonyl molecule from 10^8 emits, the rest being quenched by oxygen in the solvent cage. As the highest quantum yields measured were 4×10^{-8}, the efficiency of the production of excited triplet carbonyl fragments in these radical recombination processes must be very high, reaching unity. That some quenching by adventitious oxygen is observed is due to a few excited carbonyl fragments escaping from the solvent cage before being quenched there.

---

[a]) In an experiment carried out in an argon atmosphere oxygen was present in the form of cyclohexyl hydroperoxide only: the chemiluminescence intensity increased by a factor of 2, which confirms the absence of dominant quenching by adventitious oxygen.

Beside the phosphorescence of the carbonyl compounds produced in autoxidation reactions, there is some additional luminescence by singlet oxygen [14,43]. It is sometimes difficult to differentiate between $^1\Sigma O_2$ emission and the longer-wavelength part of the ketone phosphorescence [38].

Recent detailed studies on autoxidation reactions have been published for tetralin [44-46] cumene and ethylbenzene [46,47], methyl oleate [48,49] and benzaldehyde [50].

The hypothesis of Kellogg [38] described above, that autoxidation reactions display low quantum yields in spite of high yields of excited products, due to oxygen quenching in the solvent cage, is criticized by J. Beutel [13] who very thoroughly investigated the chemiluminescent autoxidation of dimedone (1.1. dimethyl 3.5 cyclohexandione). Here the recombination of dimedone peroxy radicals should be the excitation step:

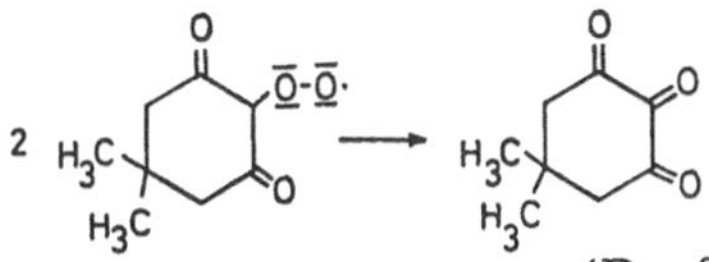

A detailed mechanism is proposed for this recombination process. On the basis of the experimental results obtained, Beutel (for details see [13]) comes to the conclusion that in the case of dimedone autoxidation the triplet triketone $D=O$ cannot be efficiently quenched by ground-state triplet oxygen formed in the decomposition of a "Russell tetroxide" which in this case should have the formula.

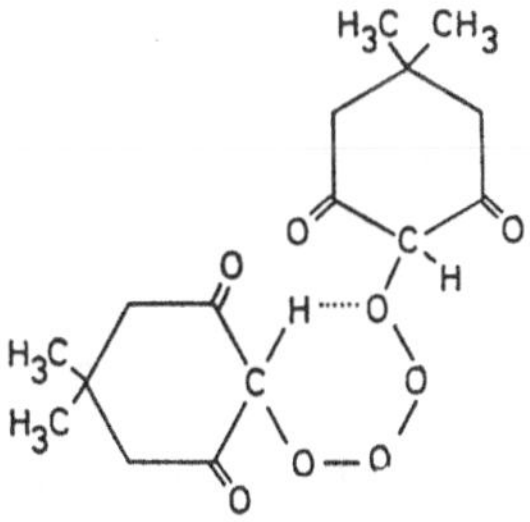

The low qantum efficiency (about 10^{-8}) therefore contradicts Kellogg's suggestions, due to the low excitation efficiency, which may be explained in one of two ways.

1. The triketone (D=O) is predominantly produced in a vibrationally excited state of the electronic ground state which may undergo an inefficient adiabatic transition to the triplet state. This transition competes with vibrational decay of the triketone molecule. The oxygen molecule, in contrast to Kellogg's suggestion formulated on p. 72, is formed in the excited singlet state. The few molecules of excited triplet triketone form a triplet charge-transfer complex [51] with the adjacent singlet oxygen. This complex is not quenched by oxygen and emits at 505 nm with an efficiency of ca. 4×10^{-3} or decomposes in solution to yield triplet triketone and singlet oxygen; the free triplet triketone molecules phosphoresce at ca. 615 nm with an efficiency of 8×10^{-4}.

2. There is occasional formation of the triketone $D = O$ in an excited singlet state (one of 10^8 molecules) while $D = O$ is usually formed in a vibrationally excited state. The excited singlet $D = O$ fluoresces (at 505 nm; efficiency: 4×10^{-3}) or undergoes intersystem crossing to the triplet state which emits at 615 nm.

The fact that the kinetic chain length of dimedone autoxidation is very low appears to indicate structural effects in autoxidation reactions. These may account for some of the discrepancies found in autoxidation chemiluminescence studies of different types of compounds.

The chemiluminescence occurring on thermolysis of dicyclohexyl peroxycarbonate in polystyrene, poly-methyl methacrylate, and poly-carbonate (poly-2.2-propane-bis-4-phenylcarbonate) was investigated by V. Ya. Shlyapintokh and coworkers [54]. In the polycarbonate and in polystyrene the emission has two maxima (450 and 530 nm) visible most distinctly about 60 sec after the beginning of thermolysis; the maximum at the longer wavelength decreases considerably faster than that at 450 nm. The latter is ascribed to excited triplet cyclohexanone, the re-combination reaction product of peroxy radicals derived from DCHP (see above). The excitation energy of the cyclohexanone molecules was transferred to anthracene or 9.10-dibromoanthracene, or to biacetyl, bibenzyl or naphthalene, so yielding the fluorescence or phosphorescence emission of the respective acceptors. The radii of the energy transfer from the primary excited product to the acceptors were calculated as 15.5 Å for anthracene and 30.5 Å for 9.10-dibromoanthracene in polycarbonate, and 35.5 Å for 9.10-dibromoanthracene in polystyrene, which is con-siderably more than the radii calculated on the basis of a Förster transfer mechanism [56].

The excited carbonyl compounds formed in autoxidation reactions are the primary source of chemiluminescence. However, it was reported recently that the recombination of formyl radicals can in turn yield an electronically excited product which was identified as glyoxal:

$$.CHO + .CHO \xrightarrow{\quad M \quad} O=C-C=O^* $$
$$\phantom{.CHO + .CHO \xrightarrow{\quad M \quad} O=C} H \quad H$$

M: inert molecules of the reaction medium

observed when formaldehyde was added to the products of microwave discharge in hydrogen diluted with inert gas (hydrogen concentration about 2%). In addition to spectroscopic evidence that glyoxal-excited singlet state molecules are indeed the emitting species, the chemiluminescence intensity was found to be dependent on the square of formaldehyde concentration. Strong quenching by nitrogen monoxide takes place, perhaps due to the formation of HC−NO [53].

$$\underset{O}{\overset{\|}{}}$$

B. Grignard Compounds

The weak chemiluminescence of Grignard compounds in air has been known since 1906. A radical chain mechanism similar to that of hydrocarbon autoxidation appears to provide the excitation energy of the emitting product. Until recently the relations between constitution and chemiluminescence in Grignard compounds were rather obscure; *p*-chloro-phenylmagnesium chloride was found to be the most efficient compound.

R. L. Bardsley and D. M. Hercules [59] reinvestigated the chemiluminescent oxidation of phenylmagnesium bromide. They measured the spectrum of the weak emission; the greatest part of it is situated between 300 and 420 nm, $\lambda_{max} = 357$ nm. The main product of the reaction is phenol (in benzene/triethylamine as solvent); numerous by-products are also formed: biphenyl, *p*-terphenyl, diphenyl ether, *p*-benzoquinone, and *p,p'*-dihydroxy-biphenyl. The most important of these products is *p*-terphenyl, 95% of the chemiluminescence emission being due to this compound which is formed in 0.1% yield only. The fluorescence spectrum of *p*-terphenyl matches well the emission spectrum of phenylmagnesium bromide autoxidation chemiluminescence. The terphenyl emission corresponds to an energy of ca. 92 kcal/mole. This means that the autoxidation of Grignard compounds represents one of the most exergonic chemiluminescent systems. It is suggested that the excitation energy is not simply transferred from a yet unknown primary product to *p*-terphenyl but that the latter is formed in an excited state via a peroxide intermediate.

C. Cysteine

A still more complicated reaction is the chemiluminescent oxidation of sodium hydrogen sulfide, cysteine, and gluthathione by oxygen in the presence of heavy metal catalysts, especially copper ions [60]. When copper is used in the form of the tetrammin complex $Cu(NH_3)_4^{2+}$, the chemiluminescence is due to excited-singlet oxygen; when the catalyst is copper flavin mononucleotide (Cu–FMN), additional emission occurs from excited flavin mononucleotide. From absorption spectroscopic measurements J. Stauff and F. Nimmerfall [60] concluded that the first reaction step consists in the addition of oxygen to the copper complex:

$$(CuX_n)^{2+} + O_2 \longrightarrow (CuX_nO_2)^{2+}$$

where $X = NH_3$ or FMN.

This is followed by insertion of cysteine into the complex where one of the former ligands is substituted:

$$(CuX_nO_2)^{2(+)} + \text{cysteine anion} \longrightarrow (CuX_{n-1} \text{cysteine}-O_2)^{(+)} + X$$

Oxygen radical anion $O_2^{(-)}$ is formed in an equilibrium reaction of the copper-cysteine-oxygen complex and a copper-cysteine complex:

$$(Cu X_{n-1} \text{cysteine} -O_2)^{(+)} \rightleftharpoons (CuX_{n-1} \text{cysteine})^{2+} + .O_2^{(-)}$$

Cystine formation is thought to be accompanied by formation of $HO_2^{(-)}$ anion:

$$(Cu X_{n-1} \text{cysteine}-O_2)^{(+)} + \text{cysteine} + X$$

$$\longrightarrow (Cu X_n-O_2H)^{(+)} + \text{cystine}$$

$$(Cu X_n-O_2H)^{(+)} \rightleftharpoons (Cu X_n)^{2(+)} + O_2H^{(-)}$$

Cysteine is evidently assumed to act as a monovalent ligand only.

A one-electron transfer is thus suggested from the cysteine sulfur to the oxygen molecule within the complex.

Oxygen radical anion forms excited-singlet oxygen in different pathways, e.g. by a reaction with copper-cysteine-oxygen complex to yield the excimer $(O_2)_2$. The computerized kinetic equations derived from this scheme allowed predictions in respect of the chemiluminescence intensity as a function of the oxygen and cysteine concentrations and as a function of time; these were satisfactorily confirmed by the experimental findings. The chemiluminescent autoxidation of cysteine to yield cystine thus appears to proceed via two one-electron abstraction steps.

Another non-enzymic chemiluminescence reaction in which copper ions play an important role is the system riboflavin/hydrogen peroxide / copper-II-sulfate/ascorbic acid investigated by R. H. Steele and coworkers [61]. This biochemical redox system simultaneously hydroxylates appropriate aromatic compounds. $.OH-$, $.O_2H-$ and $.O_2$ radicals appear to be involved in the chemiluminescent system. Substitution of ascorbic acid by β-mercaptoethanol results in a considerable increase of the chemiluminescence, which is similar to the results of Stauff and coworkers discussed above. The reaction of $.O_2H$ with riboflavin is assumed to yield a peroxide with unknown structure, the decomposition of which provides the excitation energy required. Light is emitted in the 400 nm range, corresponding to an energy of at least 72 kcal/mole; the decomposition process is thus very exergonic.

IV. Peroxide Decompositions (Except Dioxetanes)

In the preceding paragraph peroxides were described as key intermediates in autoxidation chemiluminescence. In most cases hydroperoxides were involved. The majority are well-defined compounds (*e.g.* cumene hydroperoxide), but autoxidation reactions are rather complex and peroxides are only one, though very important type of compound involved.

The rather stable diacyl peroxides such as dibenzoyl or phthaloyl peroxide have attracted special interest as some of their reactions, mostly not chain reactions, are chemiluminescent. Triplet-singlet energy transfer is very often involved, and emission generally occurs only when a fluorescer is present since the primary excited products cannot emit in the visible range of the spectrum.

A. Dibenzoyl Peroxide/Acridane

G. Lundeen and R. Livingston [62] observed weak chemiluminescence when dibenzoyl peroxide decomposed in chlorobenzene or *p*-dichlorobenzene.

When acridane *1* is oxidized by dibenzoyl peroxide in propanol/ water in acid or neutral medium, there occurs chemiluminescence whose emission spectrum matches the fluorescence spectrum of acridinium cation (protonated acridine) *2*. As radical scavengers have no influence on the chemiluminescence quantum yield, the reaction appears to be a 2-electron oxidation of acridane. The quantum yield is $2.5 - 3.0 \times 10^{-7}$ Einsteins/mole [67].

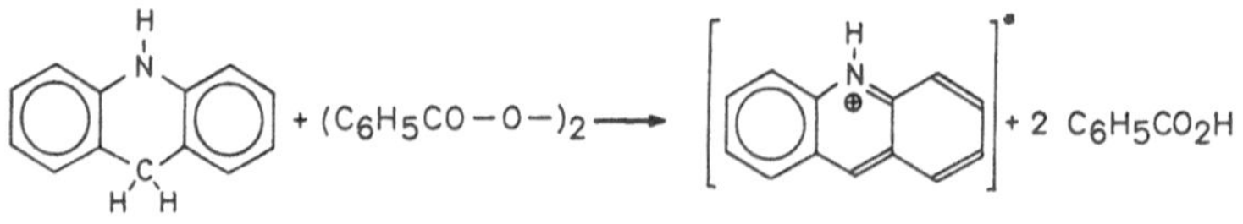

B. Phthaloyl Peroxide/Fluorescer

Phthaloyl peroxide 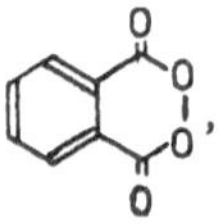 , on heating with 9.10-diphenylanthracene

(DPA) in dimethylphthalate at about 70°, yields a relatively strong blue ($\lambda_{max} = 435$ nm) chemiluminescence; the quantum yield is about 7% that of luminol [64]. The emission spectrum matches that of DPA fluorescence so that the available excitation energy is more than 70 kcal/mole. Energy transfer was observed on other fluorescers, *e.g.* rubrene and fluorescein. The mechansim of the phthaloyl peroxide/fluorescer chemiluminescence reaction very probably involves radicals. Luminol also chemiluminesces when heated with phthaloyl peroxide but only in the presence of base, which suggests another mechanism. The products of phthaloyl peroxide thermolysis are carbon dioxide, benzoic acid, phthalic anhydride, o-phenyl benzoic acid and some other compounds [65,66]. It is not yet known which of them is the key intermediate which transfers its excitation energy to the fluorescer.

C. Dibenzal Diperoxide/Fluorescer

When dibenzal diperoxide *3* was added, at 200°, to a solution of dibenzanthrone a red chemiluminescence was observed [67]; this is also seen in the reaction of violanthrone (dibenzanthrone) with alkaline hydrogen peroxide/chlorine.

It was suggested that dibenzal diperoxide first decomposes to yield singlet oxygen which reacts with dibenzanthrone to give an endo peroxide. Thermolysis of the latter should then produce triplet ground-state oxygen and dibenzanthrone in its excited triplet state [67]. However, *4*

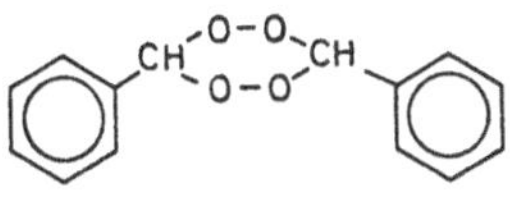

3

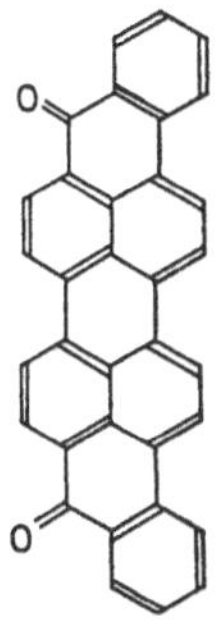

4

emits from its first excited singlet state [68]. The decomposition of dibenzal diperoxide provides sufficient energy either for the formation

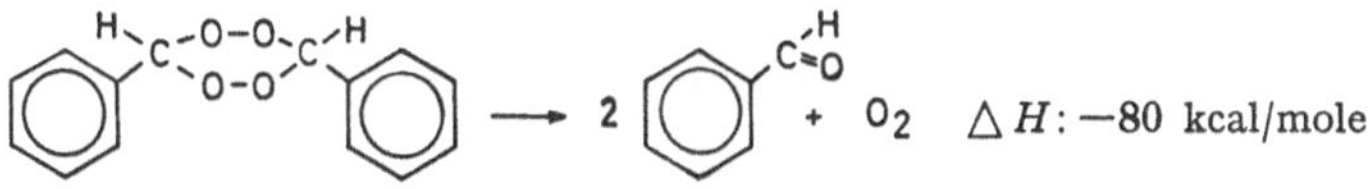

$$\longrightarrow\ 2\ \ + \ O_2 \quad \Delta H: -80 \text{ kcal/mole}$$

of benzaldehyde triplet (ca. 70 kcal/mole) or an excited singlet oxygen species ($1\,\Delta g = 22.5$ kcal/mole; $1\,\textstyle\sum g^{+} = 37.5$ kcal/mole) [69]. Hercules and coworkers [69] concluded from their investigation on the dibenzal diperoxide/dibenzanthrone chemiluminescence that the endo peroxide postulated by Kurtz is not a key intermediate in the excitation process but that triplet-singlet energy transfer is the most important process, as shown:

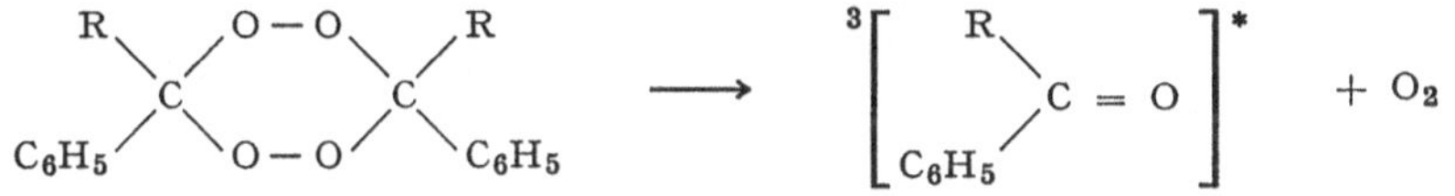

$$\longrightarrow\ {}^{3}\!\left[\ \genfrac{}{}{0pt}{}{R}{C_6H_5}\!\!\diagdown C = O\right]^{*} + \ O_2$$

5   *6*

$$6 + {}_{0}{}^{1}\text{Dibenzanthrone} \longrightarrow {}_{0}{}^{1}\!\left[\ \genfrac{}{}{0pt}{}{R}{C_6H_5}\!\!\diagdown C = O\right] + {}^{1}\!\left[\begin{array}{c}\text{Dibenz}\\\text{anthrone}\end{array}\right]^{*}$$

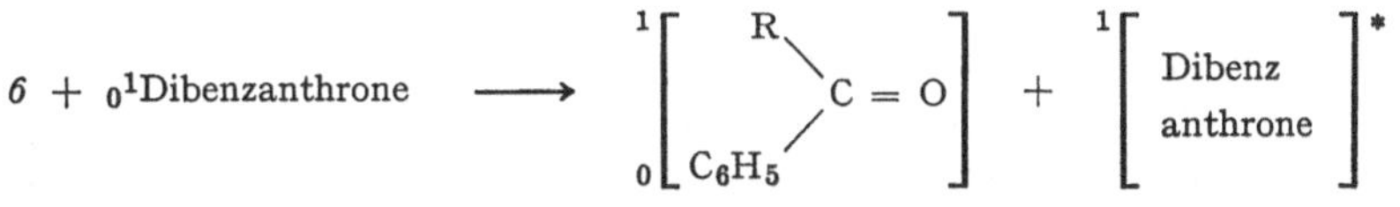

Similar results were obtained with the diperoxides *5* (R: phenyl) and *5a* (R: *p*-chlorophenyl) and dibenzanthrone or other fluorescers (perylene, rhodamine B, 9.10-diphenylanthracene, anthracene, fluorescein), with quantum yields of the respective chemiluminescence in the range $3.29 \times 10^{-8}\ldots\ldots5.26 \times 10^{-6}$.

79

The overlap integrals $\int (\tilde{\nu})$ which are essential for a Förster long-range dipole-dipole transfer mechanism [56] were obtained from the benzaldehyde phosphorescence spectrum and the absorption spectra of the fluorescers used. A linear relation was observed between the square root of the overlap integrals and the number of excited singlet fluorescer molecules produced per molecule of benzaldehyde triplet, which is a strong argument for the essential role of the proposed transfer mechanism.

Very weak chemiluminescence (quantum yields of $6.5 \ldots 9.1 \times 10^{-10}$) in the spectral ranges $400 \ldots 540$ nm (benzaldehyde phosphorescence) and 600 nm (emission from excited singlet oxygen collision pairs)) was also observed on thermolysis of 5 with no fluorescer present.

In the case of chemiluminescence occurring on treatment of dibenzanthrone with hypochlorite, as mentioned above, an endo peroxide might well be a key intermediate formed from 4 and singlet oxygen. The emitting species, however, is trichloro-dibenzanthrone, not 4 itself [70].

D. 1.4-Dimethoxy-9.10-Diphenyl-Anthracene-1.4-Endoperoxide

From all anthracene endo peroxides investigated so far ([71]; [1], p. 132) the compound 7 (1.4-dimethoxy-9.10-diphenylanthracene 1.4-endoperoxide) was found to exhibit the most efficient chemiluminescence [72] on

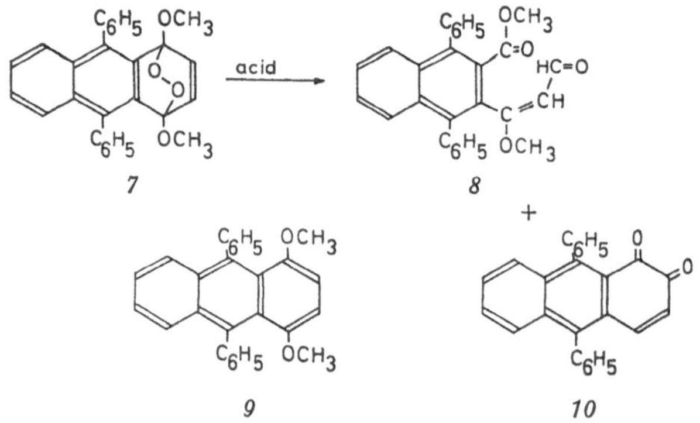

treatment with acids. It was shown [72] that most of the excitation energy is not produced by the decomposition of 7 into 1.4 dimethoxy-9.10-diphenylanthracene 9 and singlet oxygen (this is the mechanism usually formulated in endoperoxide chemiluminescence) with the anthracene 9 acting as fluorescer, but that the acid-catalyzed cleavage of the endo peroxide yielding 8 and 10 is the excitation-producing step. This cleavage was observed by J. Rigaudy and coworkers [73,74] and, independently, by J. E. Baldwin et al. [75]. The chemiluminescence quantum yield critically depends on the acidity of the medium: 7 is decomposed in the

presence of pyridine in a first-order reaction without significant chemiluminescence to 9 and oxygen in nearly quantitative yields. On heating
7 in toluene, xylene, or benzene 9 and oxygen are formed in minor
amounts, the aldehyde 8 being the main product: long-lasting chemiluminescence is observed which is quenched on addition of bases. The chemiluminescence spectrum of the acid-catalyzed decomposition of 7 matches
the fluorescence spectrum of 9. When the latter compound was added to
the solution of the peroxide 7, an increased quantum yield was obtained,
indicating that some intermediate had transferred the excitation energy
to 9. This key intermediate was thought to be the dioxetane derivative 11

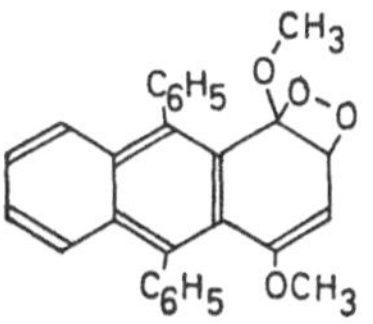

the concerted bond cleavage of which yields the aldehyde 8 in an excited
state; 8 in turn transfers its energy to 9. The quantum yields are 10^{-4}
Einsteins/mole peroxide 7. This relatively low quantum yield suggests
that some non-chemiluminescing side reactions occur during the peroxide decomposition. Although singlet oxygen is also produced, very
probably it does not appear to be cause of the chemiluminescence, as
singlet oxygen chemiluminescence quenchers like tetramethylethylene
do not influence the quantum yield. Moreover, 9 exhibits no immediate
chemiluminescence on treatment with gaseous singlet oxygen in hot
xylene solutions. Acids thus control the decomposition reaction of 7,
and the alternative pathways produce either the aldehyde 8, or the
anthracene 9 and singlet oxygen. G. W. Lundeen and A. H. Adelman [75a)]
performed a kinetic study concerning the chemiluminescence on decomposition of 7 in dioxane, in the presence of a carboxylic acid and a
fluorescent hydrocarbon. No emission was observed in absence of a
carboxylic acid; the latter could not be replaced by a mineral acid
(HCl or H_2SO_4). The kinetic data obtained are consistent with a mechanism in which an acid-induced non-reversible reaction is responsible for
chemiluminescence excitation:

$$AO_2 \xrightarrow{\;1\;} A + O_2$$

$$C + AO_2 \xrightarrow{\;2\;} X \qquad\qquad AO_2: \text{the peroxide } 7$$

$$X \xrightarrow{\;3\;} \alpha P^* + (1-\alpha)P \qquad\qquad A: \text{1.4-dimethoxy-9.10-diphenyl-anthracene } 9$$

$$P^* + A \xrightarrow{\;4\;} P + A^* \qquad\qquad P: \text{product}$$

$$A^* \xrightarrow{\ 5\ } A + \phi\, h\nu$$

α: fraction of P formed in the excited state

$$P^* \xrightarrow{\ 6\ } P$$

C: carboxylic acid

$$A^* + O_2 \xrightarrow{\ 7\ } A + O_2$$

ϕ: fluorescence efficiency of 9

$$P^* + O_2 \xrightarrow{\ 8\ } P + O_2$$

As reported by T. Wilson [71], the emitter is the anthracene derivative 9 which can be replaced by rubrene, but not by 9.10-diphenylanthracene.

The emission intensity is proportional to [A], $[AO_2]_0$, and [C]. The reaction product obtained in acidified dioxane was shown by high resolution mass spectroscopy to be 8 in confirmation of the results of Rigaudy [73] and Baldwin et al. [75].

E. Purpurogallin Oxidation

A still more complex reaction pattern underlies the oxidation of purpurogallin 12. The chemiluminescent oxidation of pyrogallol has been known for quite a long time.

$$R = H$$

More recent investigations [76] showed that the weak chemiluminescence observed on treatment of alkaline solutions of pyrogallol and formaldehyde with oxygen (Trautz-Schorigin reaction) was due to excited singlet oxygen formed on recombination of pyrogallol peroxy radicals. The products of this recombination reaction, beside singlet oxygen, are derivatives of purpurogallin and tropolone. Purpurogallin (2.3.4.6'-tetrahydroxy-5H-cycloheptabenzene-5-on) 12 is the main oxidation product of pyrogallol. When 12 is oxidized further by hydrogen peroxide at p_H 9-11, chemiluminescence in the spectral range 400...600 nm is observed, the emitters being carbonyl derivatives of α-tropolone. The quantum yields were measured as $10^{-7}...10^{-6}$. The oxidation reaction of purpurogallin produces purpurogallin quinone 13 as the first intermediate; this is further oxidized to yield tropolone α,β-anhydride 16 and oxalic acid (at room temperature and low alkali concentration) or α-carboxy-β-carboxymethyl-tropolone 15 (at higher temperature and higher alkali concentrations). Both pathways were found to contribute to the chemiluminescence; the energy contribution of the respective pathways is still unknown. Colchicine, which is structurally related to purpurogallin, also exhibits chemiluminescence on oxidation.

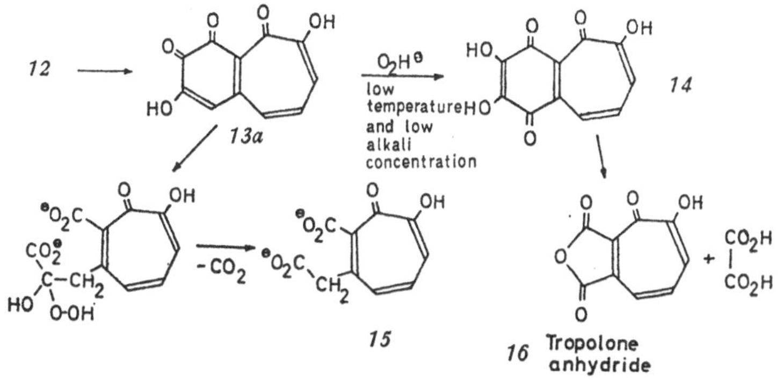

The oxidative ring cleavage undergone by purpurogallin quinone *14* is assumed to provide the excitation energy necessary to cause the tropolones to act as fluorescers. Perhaps dioxetane derivatives are key intermediates, as in the anthracene derivatives discussed in (IV.D.).

V. Dioxetanes

As mentioned in Section II. C., the concerted bond cleavage of 1.2-dioxetane derivatives has been proposed to be of general importance in respect of the excitation step of a large number of chemiluminescence reactions. The first experimental results concerning simple dioxetanes were obtained by M. M. Rauhut and coworkers in their work on activated oxalic ester chemiluminescence [24]. From experimental data on the reaction of *e.g.* bis (2.4-dinitrophenyl)oxalate with hydrogen peroxide in the presence of rubrene, they concluded that 1.2-dioxetanedione

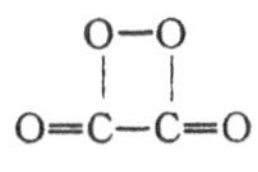

$$\begin{array}{c} O-O \\ | \quad | \\ O{=}C{-}C{=}O \end{array}$$

is formed and splits into two molecules of carbon dioxide from a donor-acceptor complex with the fluorescer. During recent years other simple dioxetanes have been synthesized, mostly by addition of singlet oxygen to activated olefinic double bonds [78–81].

A. Alkyl-1.2-Dioxetanes

Trimethyl-1.2-dioxetane *17*

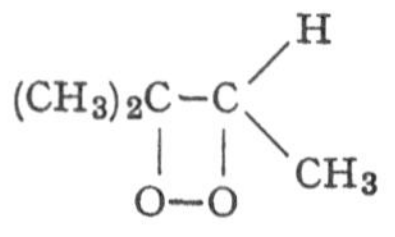

$$(CH_3)_2C-C\overset{H}{\underset{CH_3}{<}}$$
$$\underset{O-O}{|\quad|}$$

17

is thermally decomposed in the presence of certain lanthanide chelates in solution under emission of chemiluminescence light of nearly mono-chromatic character, *i.e.* in a very narrow spectral range: at least 80% of the light emitted appeared in a single band at 613 nm. The total width at half-height was about 5 nm. The quantum yield amounted to about 0.015 (based on dioxetane decomposed [82]). Lanthanide complexes are known to be very efficient acceptors of n, π^* lowest triplet energy [83]; the suggestion of P. D. Wildes and E. H. White [82] that triplet states are involved in dioxetane decomposition has very recently been confirmed experimentally by N. J. Turro and P. Letchken [34] who observed a triplet acetone yield of nearly 50% on thermolysis of tetramethyl di-oxetane. It was proved that these acetone triplets were not formed from a preexisting excited singlet acetone by trapping the decomposition product of tetramethyldioxetane with *trans*-1.2-dicyano-ethylene *18* because this very specifically yields the oxetane *19* with excited singlet acetone, whereas with triplet acetone isomerization to *cis*-1.2-dicyano-ethylene occurs:

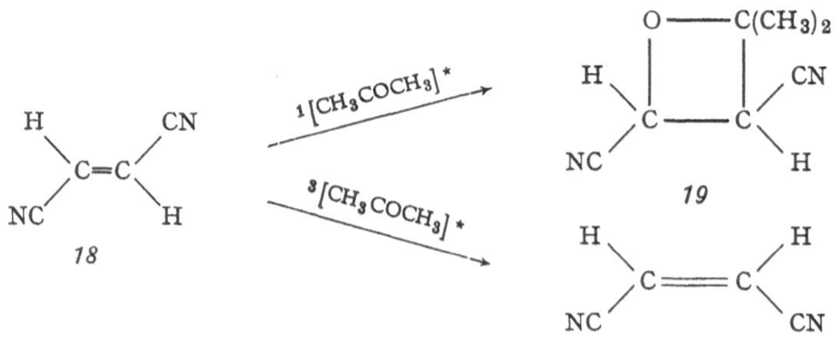

(For other "photochemical" reaction products from dioxetane decomposition, see Section X.).

B. Alkoxy-1.2-Dioxetanes

Alkoxy-dioxetanes undergo chemiluminescent thermolysis in the pres-

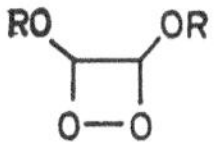

$$\begin{array}{c} RO \quad\quad OR \\ \boxed{} \\ O-O \end{array}$$

ence of appropriate fluorescers yielding excited esters as reaction products ($R: C_2H_5$, $-CH_2-$ (from singlet oxygen and 1.3-dioxolen), $-CH_2CH_2-$ (from singlet oxygen and p-dioxene) [84]). Fluorescers often used are anthracene and 9.10-dibromoanthracene. Very careful investigations by T. Wilson and P. Schaap [84] of the thermolysis of diethoxy-dioxetane showed that its cleavage into 2 moles of ethyl formate is practically quantitative; no "dark" reaction is observable to any significant extent:

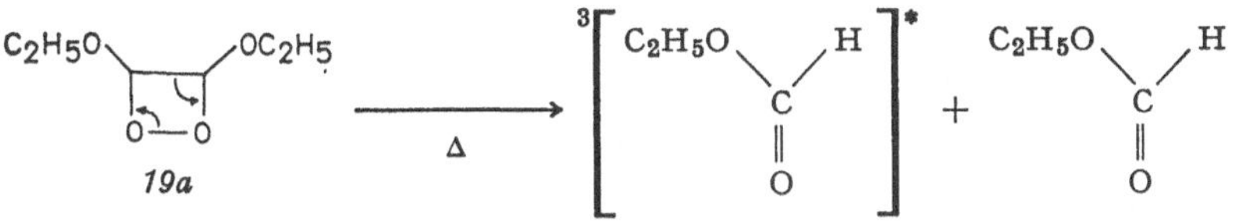

The activation energy of this decomposition was calculated to be about 24 kcal/mole, the free reaction enthalpy being $-80 \ldots -99$ kcal/mole. The energy transfer from the excited formate molecule to the fluorescer is a triplet-singlet transfer. The far higher quantum yield (about two orders of magnitude) obtained with 9.10-dibromoanthracene as fluorescer (5×10^{-4} Einsteins/mole dioxetane) as compared with 9.10-diphenylanthracene as fluorescer (7×10^{-6} Einsteins/mole dioxetane) thus has its rationale in the "heavy-atom" effect favoring intersystem crossing by spin-orbit coupling. The rate of decay of triplet ethyl formate is about 10^8 sec^{-1}.

A peculiar effect was observed in the decomposition of *19a* with anthracene as fluorescer when oxygen was carefully removed from the solutions: an increase of the chemiluminescence decay rate and of the dioxetane cleavage resulted. It was suggested that this was due to a catalytic effect of triplet anthracene (formed by energy transfer from triplet formate) on the decomposition of the dioxetane. When oxygen is present, triplet anthracene is quenched. Whether such a catalytic effect of triplet anthracene or similar compounds on dioxetane cleavage actually exists has not yet been fully established; positive effects were observed by M. M. Rauhut and coworkers [24] in oxalate chemiluminescence and by S. Mazur and C. S. Foote [80] in the chemiluminescent decomposition of tetramethoxy-dioxetane, where zinc tetraphenylporphyrin seems to exert a catalytic effect. However, the decomposition of trimethyl dioxetane exhibits no fluorescer catalysis [78].

C. Dioxetan-Dione (Carbon Dioxide Dimer)

Dioxetanedione *21*

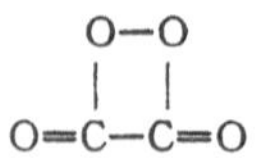

$$\begin{array}{c} O\!-\!O \\ |\quad| \\ O\!=\!C\!-\!C\!=\!O \end{array}$$

(carbon dioxide dimer), the cleavage of which is regarded as the energy source of oxalic ester chemiluminescence, was identified recently by mass spectroscopy (C. A. Heller et al.[85]), the gaseous products of the oxalic ester-hydrogen peroxide reaction being fed straight into the mass spetrometer.

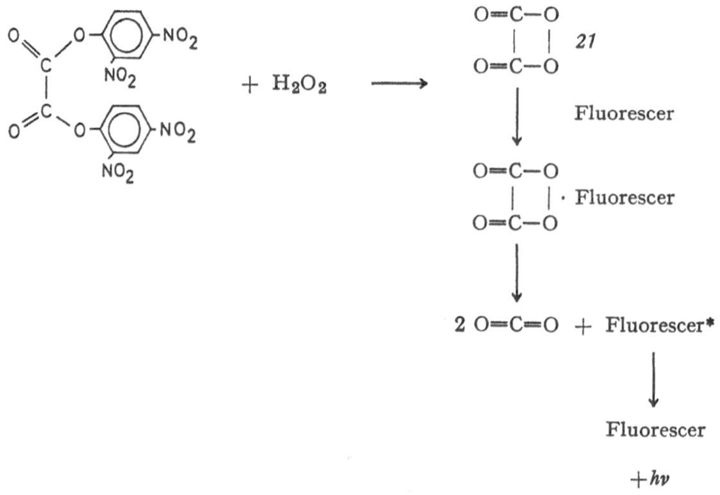

In a very recent investigation [86] J. J. De Corpo et al. confirmed the findings of these authors [85] and estimated the concentration of the dioxetandione molecule ions $C_2O_4^{(+)}$ formed. However, when this concentration was plotted against the ion residence time it was apparent that the concentration of $C_2O_4^{(+)}$ increased while the concentration of carbon dioxide molecule ions $CO_2^{(+)}$ decreased. (Carbon dioxide is, of course, the decomposition product of dioxetanedione in oxalate chemiluminescence and is therefore present in the gaseous products.) It is concluded, therefore, that the $C_2O_4^{(+)}$ detected in the mass spectrometer cannot have been formed in the chemiluminescent reaction. On the other hand $C_2O_4^{(+)}$ is not formed from carbon dioxide alone. Further investigations are evidently necessary (see also J. Stauff and U. Bergmann [196]).

Dioxetan-ones appear to be intermediates in the chemiluminescent reaction of singlet oxygen with ketenes, in the presence of fluorescers [81]:

$$
\begin{array}{c}
R^1 \\
\diagdown \\
\diagup \quad C{=}C{=}O \ + \ {}^1O_2 \\
R^2
\end{array}
\longrightarrow
\begin{array}{c}
\quad\; O{-}O \\
R^1 \quad |\quad| \\
\diagdown \\
\diagup \quad C{-}C{=}O \\
R^2
\end{array}
\longrightarrow
\begin{array}{c}
\text{Fluorescer-} \\
\text{dioxetane} \\
\text{complex}
\end{array}
$$

$$
+ \ \text{Fluorescer}
\longrightarrow
\begin{array}{c}
R^1 \\
\diagdown \\
\diagup \quad C{=}O \ + \ CO_2 \ + \ \text{Fluorescer*} \\
R_2
\end{array}
$$

The ketene (ketene, $R^1 = R^2 = H$, diphenylketene, $R^1 = R^2 = C_6H_5$) was mixed with triphenyl phosphite ozonite (as singlet oxygen source [87]) at $-70°$ in methylene chloride and allowed to warm to room temperature in the presence of the fluorescer (9.10-diphenylanthracene (DPA), 9.10-bis(phenylethynyl)anthracene (BPEA)). The emission spectrum is identical with the fluorescence spectrum of the fluorescer used. Quantum yields with diphenylketene and DPA used as fluorescer are about 1.9×10^{-4} Einsteins/mole; the yield decreases slightly with increasing concentration of the ketene. Some energy transfer from singlet oxygen to the fluorescer is involved, as the latter gives a faint emission when treated with singlet oxygen in the absence of a ketene (quantum yield ≤ 0.001).

When a ketene acetal is used instead of a ketene, e.g. diphenyl-ketene glycol acetal 22

$$
\begin{array}{c}
C_6H_5 \qquad\qquad O{-}CH_2 \\
\diagdown \qquad\qquad\quad\diagup \qquad| \\
\quad C{=}C \qquad\qquad\quad| \\
\diagup \qquad\qquad\quad\diagdown \qquad| \\
C_6H_5 \qquad\qquad O{-}CH_2 \\
\\
22
\end{array}
\qquad\qquad
\begin{array}{c}
\qquad\quad O{-}O \\
C_6H_5 \quad|\quad| \quad O{-}CH_2 \\
\diagdown \qquad\quad\diagup\quad\diagup \qquad| \\
\quad C{-}C \qquad\qquad| \\
\diagup \qquad\qquad\diagdown\quad| \\
C_6H_5 \qquad\qquad O{-}CH_2 \\
\\
23
\end{array}
$$

the chemiluminescence is very weak. In this case the dioxetane intermediate should have the formula 23 which would obviously give rise to a significantly lower free energy on decomposition. The structural influence of the dioxetane on their decomposition energies was calculated (see [81]) on the basis of EHMO treatment [88]:

Decomposition energies of some dioxetanes and their dimers [in kcal/mole] (after[81])

$\begin{array}{c} O{-}O \\ \| \quad \| \\ H_2C{-}CH_2 \end{array} \longrightarrow \ 2\ CH_2O$		-197

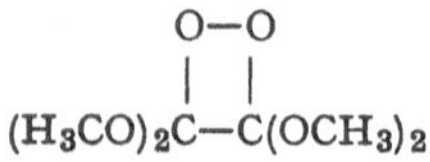 $\underset{\text{(H}_3\text{CO)}_2\text{C}-\text{C(OCH}_3)_2}{\overset{\text{O}-\text{O}}{\vert\quad\vert}}$	$\longrightarrow$ $2\,\text{O}=\text{C(OCH}_3)_2$	-206
$\underset{\text{H}_2\text{C}-\text{C}=\text{O}}{\overset{\text{O}-\text{O}}{\vert\quad\vert}}$	$\longrightarrow$ $\text{CH}_2\text{O} + \text{CO}_2$	-184
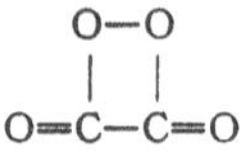 $\underset{\text{O}=\text{C}-\text{C}=\text{O}}{\overset{\text{O}-\text{O}}{\vert\quad\vert}}$	$\longrightarrow$ $2\,\text{CO}_2$	-132

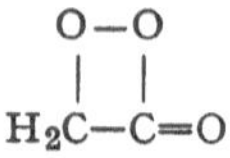

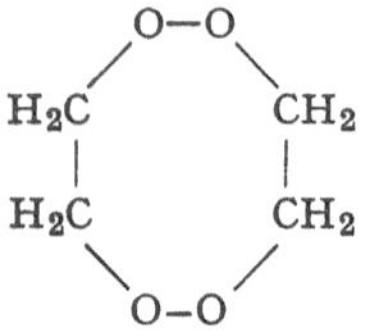

	$\longrightarrow$ $4\,\text{CH}_2\text{O}$	-380

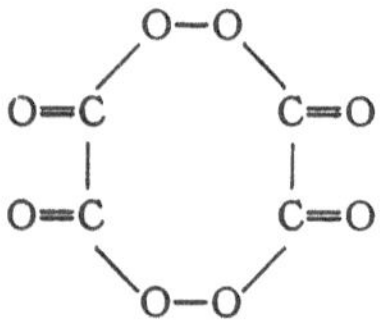	$\longrightarrow$ $4\,\text{CO}_2$	-267

It was pointed out that these values are only approximate but they nevertheless demonstrate the high exergonicity of dioxetane decomposition. The sp^2 carbons in the dioxetanones and -diones appear to stabilize the 4-membered ring peroxide.

Another method of preparing a dioxetanone is ring closure in a α-hydroperoxy-carboxylic acid, as was demonstrated very recently [89]:

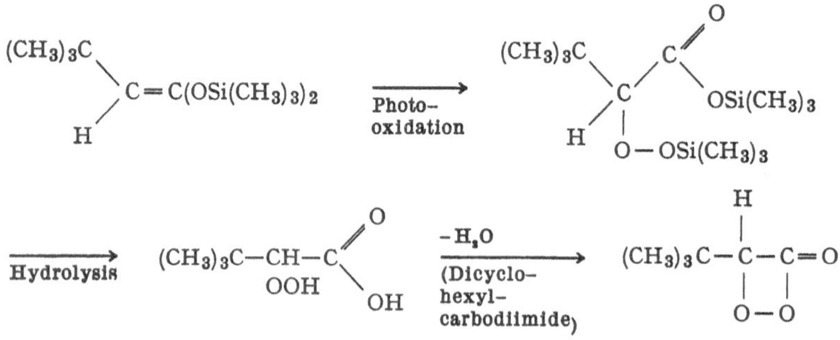

This is the first example of the isolation of a compound of the type postulated as intermediate in the ketene-singlet oxygen reaction. Instead of classifying it as a dioxetanone, we can also call it a peroxy-α-lactone. The compound decomposes to pivalaldehyde and carbon dioxide with the emission of light:

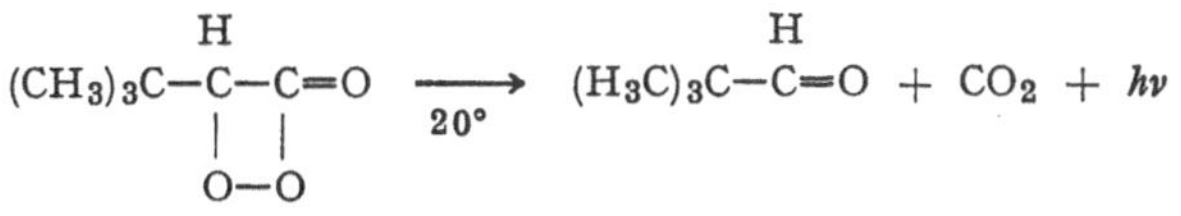

$$(CH_3)_3C-\underset{\underset{O-O}{|}}{\overset{\overset{H}{|}}{C}}-C=O \xrightarrow{20°} (H_3C)_3C-\overset{\overset{H}{|}}{C}=O + CO_2 + h\nu$$

VI. Hydrazide Chemiluminescence

Hydrazide chemiluminescence has been investigated very intensively during recent years (for reviews, see [1], p. 63, [2], [90]). Main topics in this field are: synthesis of highly chemiluminescent cyclic diacyl hydrazides derived from aromatic hydrocarbons, relations between chemiluminescence quantum yield and fluorescence efficiency of the dicarboxylates produced in the reaction, studies concerning the mechanism of luminol type chemiluminescence, and energy-transfer problems.

A. Constitution and Chemiluminescence

As outlined above, the quantum yield of a chemiluminescent reaction depends on

a) the number of molecules taking the "correct" chemical pathway,

b) the fraction of the latter crossing over to the excited state,

c) the fluorescence efficiency of the emitting species.

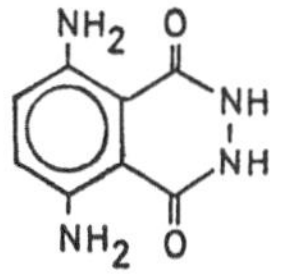

Fluorescence is by far the most important emission in hydrazide chemiluminescence; the possible involvement of triplet-singlet transfer in some cases is discussed below. The "correct" chemical pathway can be understood to mean in part that a hydrazide must be stable enough under the oxidative reaction conditions usually applied to ensure that only a small proportion of the hydrazide molecules are consumed in non-chemiluminescent oxidation reactions. 3.6-diamino-phthalic hydrazide 24, for example, should yield a highly fluorescent dicarboxylate dianion and therefore exhibit strong chemiluminescence. However, being a derivative of *p*-phenylenediamine, it is very rapidly oxidized to yield a quinone derivative which is not capable of chemiluminescence.

E. H. White and coworkers [12] stated that the fluorescence efficiency of the dicarboxylate formed in the general reaction is not the factor

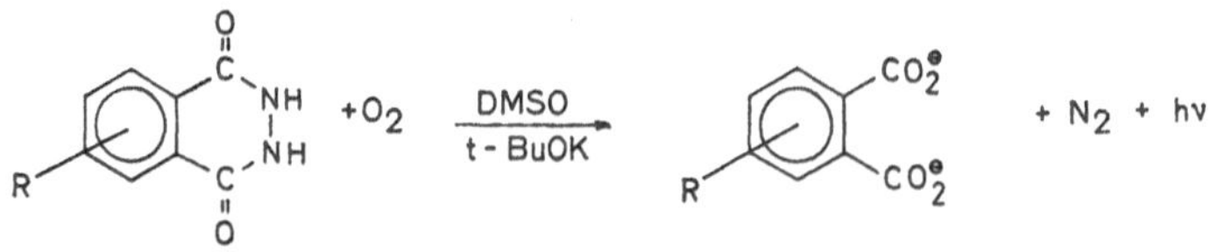

which predominantly determines the quantum yield. In the series of hydrazides listed in Table 1 they observed differences in chemiluminescence quantum yield of up to 600-fold, whereas the fluorescence yields of the corresponding dicarboxylates differ by a factor of 10 only.

It was stated that ϕ_{es} increases with decreasing excited state energy (frequency of fluorescence); this agrees with observations of McCapra [93] with indolenyl peroxides.

Concerning the estimation of ϕ_{es}, *i.e.* the yield of excited product molecules, it was calculated that in the equation

$$\phi_{CL} = \phi_r \times \phi_{es} \times \phi_{fl}$$

ϕ_r is approximately unity since the yield of the reaction products (substituted phthalic acids) in preparative isolation is more than 85%.

Another rather striking example demonstrates that the fluorescence efficiency of the respective dicarboxylates is not the most important factor in determining the chemiluminescence efficiencies of the hydrazides: 9.10-diphenylanthracene-2.3-dicarbonic acid *25* has a fluorescence efficiency of about 0.9 (as has the parent compound 9.10-diphenylanthracene) [94]. The corresponding hydrazide *26*, however, gives a quantum yield of 48% that of luminol only (in DMSO/t-BuOK/O₂) [95] although 3-aminophtalic acid has a fluorescence efficiency of about 0.3 only.

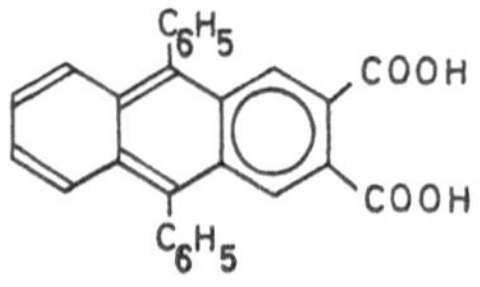

25 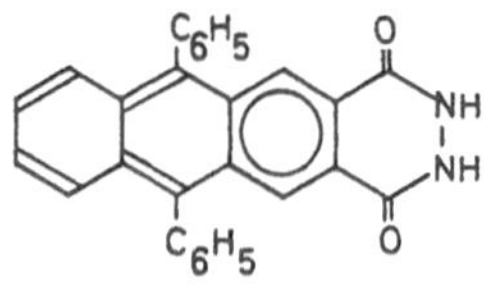   26

Table 1. Chemiluminescence and fluorescence efficiencies of some substituted phthalic hydrazides (after Brundrett, Roswell, and White [12])

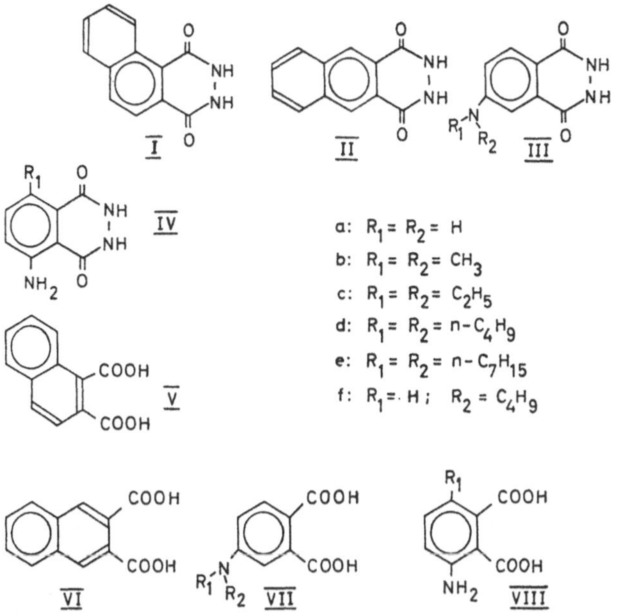

a: $R_1 = R_2 = H$
b: $R_1 = R_2 = CH_3$
c: $R_1 = R_2 = C_2H_5$
d: $R_1 = R_2 = n\text{-}C_4H_9$
e: $R_1 = R_2 = n\text{-}C_7H_{15}$
f: $R_1 = H$; $R_2 = C_4H_9$

Compound	ϕ Chemi-luminescence[1]	ϕ Fluorescence of the corresponding acids[2] (in 0.1 M K_2CO_3)		ϕ_{es} (efficiency of excited product formation)
I	0.000047	0.029	(V)	0.0017
II	0.00052	0.12	(VI)	0.0043
III a	0.0012	0.13	(VII a)	0.0092
III b	0.0075	0.25	(VII b)	0.030
III c	0.0094	0.31	(VII c)	0.030
III d	0.0140	0.30	(VII d)	0.046
III e	0.0038	0.26	(VII e)	0.015
III f	0.0058	0.28	(VII f)	0.021
IV a	0.015	0.30	(VIII a)	0.050
IV b	0.028	0.28	(VIII b)	0.10

[1]) Obtained in aqueous 0.1 M K_2CO_3 solution ($p_H = 11.4$) with hydrogen peroxide and hemin; the values are relative to luminol (0.0125 according to Lee and Seliger [91]). The precision of the measurements is about $\pm 10\%$.

[2]) Values relative to quinine bisulfate (0.55 after [92]). It should be mentioned that even small changes in reaction conditions (e.g. change of p_H) give rise to changes in ϕ_{CL}.

This is not due to the relatively extended aromatic system in *25*, for C. C. Wei and E. H. White [96)] recently succeeded in synthesizing the benzoperylene compound *27* which is the most efficient hydrazide yet known with a chemiluminescence quantum yield of 7.3% (in DMSO/t-BuOK/O$_2$). The corresponding dicarboxylate has a fluorescence efficiency of 14% and emits at 420 and 450 nm which matches the chemiluminescence emission of *27* [96)].

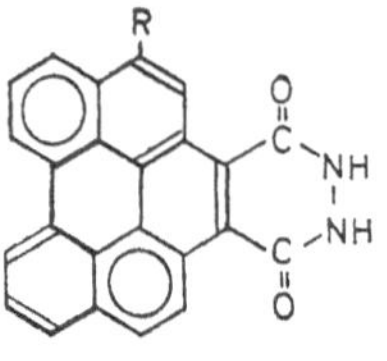

27   27: R=H
        27a: R=N(CH3)2

The dimethylamino derivative *27a* was found to give no higher quantum yield in chemiluminescence than *27*. This may be due to a steric hindrance of the dimethylamino group by the neighboring peri-hydrogen atom. However, 8-dimethylamino-naphthalene-2.3-dicarboxylic acid hydrazide [97)] *28* where a similar steric effect should be expected, is far more efficient than the parent compound (naphthalene 2.3-dicarboxylic hydrazide), the ratio of their quantum yields being about 40:1.

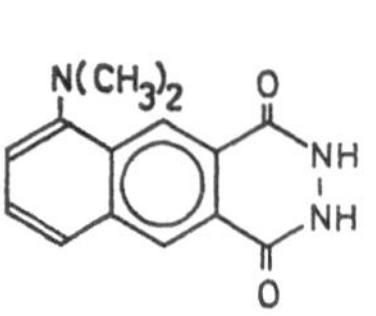

28 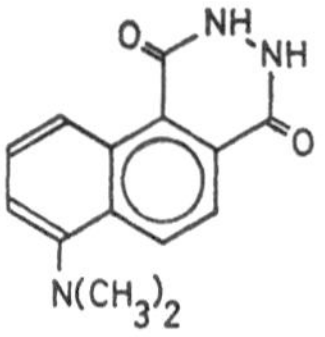   29 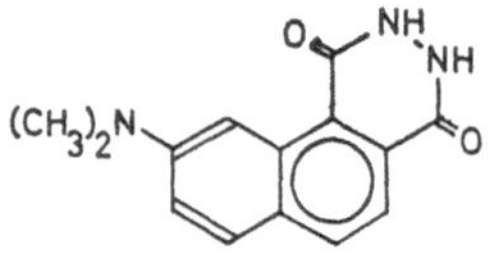   30

Finally, the hydrazide *29* [98)] is strongly fluorescent in neutral solution (*e.g.* in dioxane), the fluorescence intensity amounting to about 200% of that of 7-dimethylamino-naphthalene-1.2 dicarboxylic hydrazide *30*, which is one of the best chemiluminescent hydrazides [97)]. The 5-isomer, however, is very poor in chemiluminescence in an aqueous system (hemin-catalyzed oxidation with aqueous alkaline hydrogen peroxide), the light yield being only 1% of that of the 7-isomer; in DMSO/t-BuOK/O$_2$ its quantum yield is slightly better but very distinctly below that of *30* [98,99)]. It should be mentioned that in aqueous alkaline solution the fluorescence of the hydrazide *29* "breaks down" to a very small fraction of that in neutral solution [99].

Some cyclic diacyl hydrazides synthesized recently are listed in Table 2.

Table 2. Luminol-type hydrazide chemiluminescence

No.	Compound	Ref.	Chemi-luminescence λ_{max}[nm]	Relative Quantum Yield (Luminol = 100) DMSO	H$_2$O
31	R = (CH$_3$)$_2$N—	100)	510—540[1]	2	0,75
32	R = (C$_2$H$_5$)$_2$N —	100)	460—490[1]	4	0,3
33	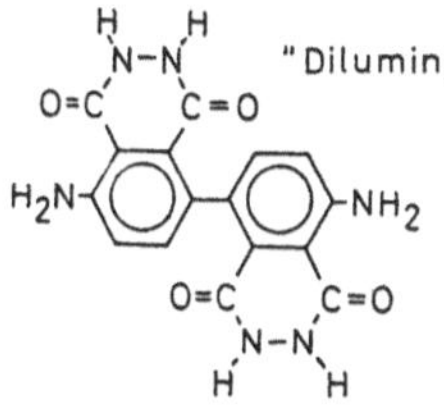 "Diluminyl"	100)	510		6
		100)	450[1]	30	
34	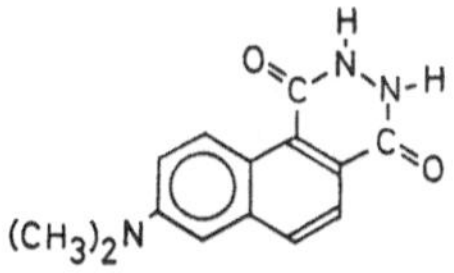	98)	485	14	150
29	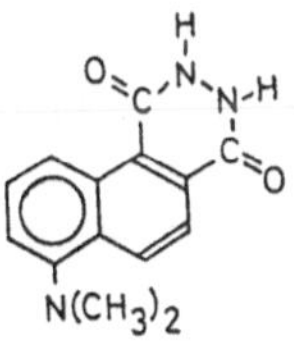	98) 99)	550	2,1	3
35	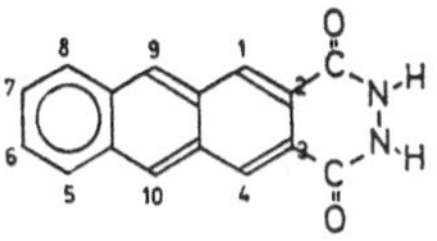	101)	430	3,3	25
36	6 - Methoxy	95)	457	47	75
37	6-(CH$_3$)$_2$N	102)	470	—	50

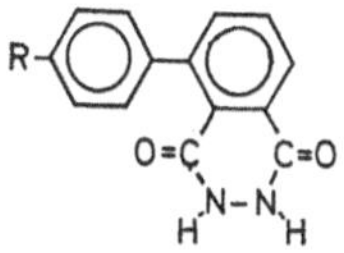

Table 2. (continued)

No.	Compound	Ref.	Chemi-luminescence	Relative Quantum Yield (Luminol = 100)	
			λ_{max}[nm]	DMSO	H$_2$O
38	9 – Br	95)	—	5,3	—
39	9,10 –Diphenyl	95)	450	48	38
40	R = H	103)	368	—	—
41	R = NH$_2$	103)	490	—	2,1
42	R = OCH$_3$	103)	480	—	26
43	R = OH	103)	525	—	44
44	R = (CH$_3$)$_2$N	103)	520	—	115
45	R = (C$_2$H$_5$)$_2$N	103)	510	—	129
46	R = NH$_2$	103)	450^1)	—	0,03
47	R = N(CH$_3$)$_2$	103)	450^1)	—	0,10
48	R = N(C$_2$H$_5$)$_2$	103)	500^1)	—	0,15

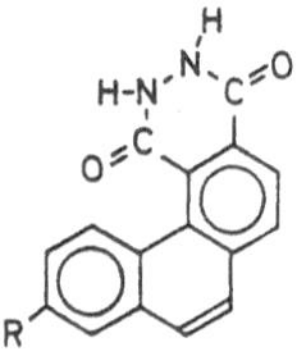

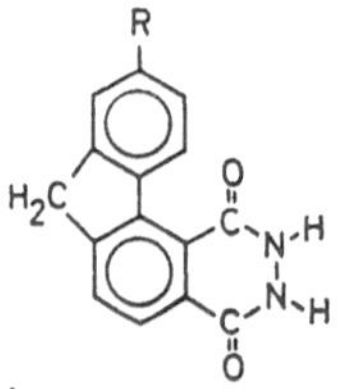

^{1}) In DMSO.

As can be seen from Table 2, certain phenanthrene derivatives produce more chemiluminescence than luminol. The diphenyl derivative *33* "diluminyl" evidently has its benzene rings in a non-coplanar position; this is apparent from the fact that this compound does not behave as a benzidine derivative under the oxidative conditions of chemiluminescence reactions. The chemiluminescence quantum yield is only about one third that of luminol [100].

E. H. White and coworkers [104] found two monoacyl hydrazides (linear hydrazides) of remarkably high chemiluminescence efficiency, namely *49* (R=NHNH$_2$) and *50* (R=NHNH$_2$).

The monoacyl hydrazides so far investigated gave very low quantum yields in most cases (for references see [104, 1] p. 67).

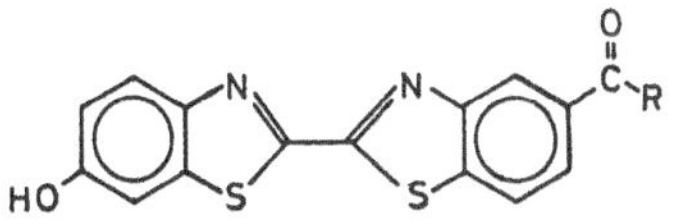

49 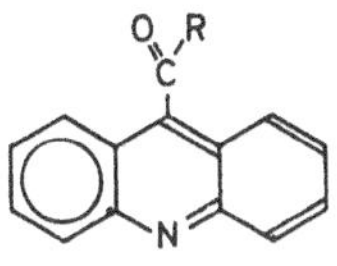   50

In DMSO/tert.BuOK/O$_2$ quantum yields found were 3×10^{-4} for 50 and 5×10^{-4} for 49. These values are not regarded as maximum values, as a strong influence of the reaction conditions was noted [104]. Whereas, in luminol-type compounds, any substitution in the cyclic hydrazide ring system renders the compound non-chemiluminescent [105], this is not the case with these open-chain hydrazides: the methylated compounds 49, 50 (R=N(CH$_3$)NH$_2$) and 50 (R=N(CH$_3$)NH(CH$_3$)) were also found to be chemiluminescent, though less so than the unsubstituted hydrazide. E. H. White and coworkers [104] therefore suggest a mechanism via acyl anions for the non-cyclic hydrazides (see Section VI. C.).

B. The Emitting Species

It has been established [106] that 3-aminophthalate is the emitting species in luminol chemiluminescence and that in all known cases of other cyclic hydrazides the corresponding dicarboxylate is the emitter (e.g. [97,107]).

The maximum of luminol chemiluminescence emission is at 425 nm in aqueous and at 480 nm in DMSO-containing solvent. It was suggested that different anions of 3-aminophthalate were responsible for this phenomenon, namely 51 (in water) and 52 (in aprotic solvents).

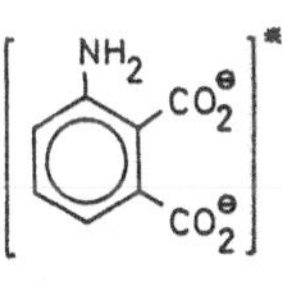

51 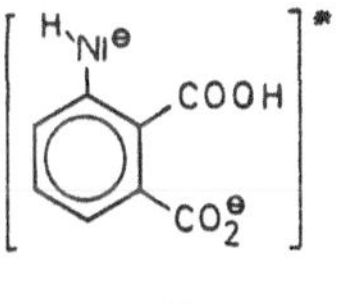   52

This would mean that in DMSO and other aprotic solvents a proton is transferred from the amino group to the ortho-carboxylate group.

D. S. Bersis and J. Nikokavouras [108] came to the conclusion, on the basis of their chromatographic investigations of luminol preparations, that these two emission maxima of luminol chemiluminescence were actually due to impurities in luminol, not to different states of 3-aminophthalate. Recent investigations of 3-aminophthalate in solution [109] and of luminol in respect of photochemical changes [110] have shown that 3-amino-phthalic acid rather quickly undergoes chemical reactions leading to secondary products, *e.g.* formation of *53* [109].

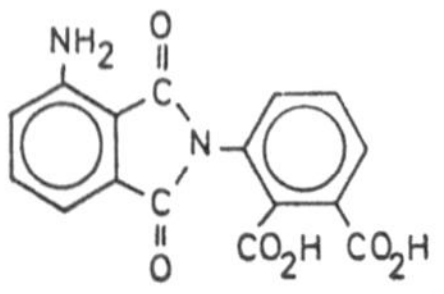

or *54* from luminol according to [110]:

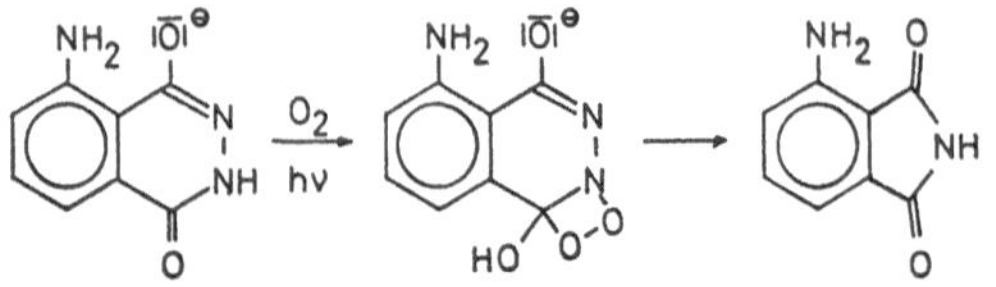

However, the formation of these products does not appear to play a critical role in the decision as to whether the 425 nm and 480 nm maxima are due to different states of the same molecule or to different compounds. It was reported that special care was taken to ensure the purity of luminol and of 3-aminophthalate [109]. In commercially available 3-amino-phthalic acid a yellowish impurity exhibiting brilliant green fluorescence was detected [109]; this substance also formed in neutral solutions of pure 3-amino phthalic acid and crystallized from these solutions in yellow crystals. The structure of this substance was determined to be *53*; its absorption spectrum has a maximum at 388 nm; the fluorescence maximum is at 475 nm, with a fluorescence quantum yield of about 0.75 in DMF [109].

The fluorescence of 3-amino-phthalic acid has been investigated very thoroughly by Lee and Seliger [111] and by Gorsuch and Hercules [109] in protic and in aprotic solutions. 3-Aminophthalate dianion exhibits fluorescence maxima at 490 nm and 475 nm which are of similar intensity in DMSO solutions containing 41—50 mole % water, whereas the 490 nm maximum predominates in nearly pure DMSO (water content ~0.01 mole %) and the 425 nm maximum in pure water. The quantum yields are 12 and 26%, respectively [109]. The well-known p_H dependence of luminol chemiluminescence is in part due to the strong influence of p_H on 3-aminophthalate fluorescence [111]. In the range p_H 7—11 in aqueous solution the maximum fluorescence quantum yield of 3-aminophthalate is about 30%, decreasing very remarkably at p_H values above and below this.

The absorption spectra of 3-amino-phthalic acid (APH_2) measured at different pH values indicate that different ion species are the respective fluorescence emitters which fluoresce with very different quantum yields. The following dissociation equilibria were stated:

$APH_3^{(+)}$	$\rightleftharpoons$	APH_2	$\rightleftharpoons$	$APH^{(-)}$	$\rightleftharpoons$	$AP^{(--)}$
cation		"neutral" acid		mono-anion		dianion

	pK 1.5	pK 3—4		pK 5.0	
Fluorescence quantum yields		1—4%			30%

The pK values of 1.5 and 5.0 could be determined well by isosbestic points of the absorption curves whereas this was difficult in the range between 1.5 and 5.0. pK values in this range belong to the ground state of APH_2. In the excited state the pK values change because the basicity of the amino group decreases while that of the carboxyl groups increases [112,113]. The strong decrease in 3-aminophthalate fluorescence at pH values higher than 11.0 is probably due to a further deprotonation with foundation of a tris-anion which is very poorly fluorescent.

4-Dialkylamino phthalic acids resulting from the chemiluminescence reaction of 4-dialkylamino-phthalic hydrazides cannot easily form a tris anion because deprotonation of the amino group is impossible. They should therefore not exhibit such a strong decrease in fluorescence efficiency at higher pH values. This can actually be concluded from the pH dependence of the chemiluminescence of the 4-dialkylamino-phthalic hydrazides and related compounds [97].

The red shift of 3-amino-phthalate fluorescence when the solvent is changed from water to DMSO or another aprotic solvent is due to different hydrogen bonding; above all, however, it is due to proton transfer from the amino group of the excited $AP^{(--)}$ to the neighboring carboxylate yielding the species $(-)AP(-)$, so that the emitters are *51 a* in water and *52 a* in DMSO [111]:

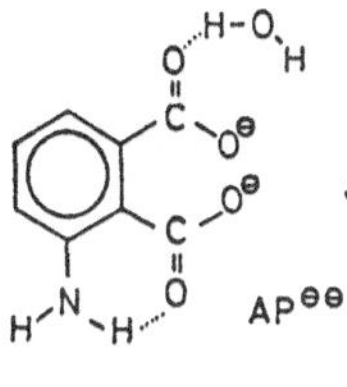

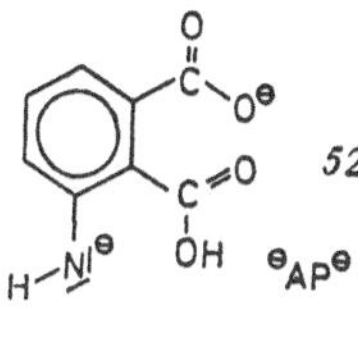

(in water)   (in DMSO)

It is suggested that the proton transfer which occurs with DMSO is impossible in water [111].

Gorsuch and Hercules [109] stated that certain discrepancies between the fluorescence spectrum of 3-amino-phthalate dianion and the chemiluminescence spectrum of luminol are partly due to reabsorption of the shorter-wavelength chemiluminescence light by the luminol monoanion. These authors confirmed the results of E. H. White and M. M. Bursey [114] concerning the very essential solvent effect on luminol chemiluminescence: the relative intensity of the latter in anhydrous DMSO/t-BuOK/oxygen was found to be about 30,000 times that in DMSO/28 mole % water/potassium hydroxide/oxygen.

The emission maxima of luminol chemiluminescence and of 3-amino-phthalate dianion in some solvents are listed in Table 3:

Table 3. Chemiluminescence maxima of luminol and fluorescence maxima of 3-amino-phthalate dianion (after J. Lee and H. H. Seliger [115])

Solvent	Chemiluminescence of luminol		Fluorescence of 3-amino-phthalate	
	λ_{max} (nm)	ν'	λ_{max} (nm)	ν'
Water	431	23250	431	23250
DMSO	502	19900	495	20200
DMF	499	20050	497	20100
Acetonitrile (AN)	500	20000	500	20000
THF	496	20150	500	20000
DMF:AN (7:3 vol%, −50 °C)	472	21200	478	20900

C. Mechanistic Studies

E. H. White and coworkers [2] have proposed a mechanism of luminol (Lum) chemiluminescence in aprotic media in which the luminol dianion ($Lum^{2(-)}$) is a key intermediate:

$$Lum^{(-)} + Base \underset{K_{-1}}{\overset{k_1}{\rightleftharpoons}} Lum^{2(-)}$$

$$Lum^{2(-)} + O_2 \xrightarrow{k_2} LO_2^{2(-)}$$

$$LO_2^{2(-)} \longrightarrow (AP^{2(-)})^x + N_2$$

$$(AP^{2(-)})^* \longrightarrow AP^{2(-)} + h\nu$$

($Lum^{(-)}$: luminol monoanion; $LO_2^{2(-)}$: peroxy intermediate; $AP^{2(-)}$: 3-amino-phthalate dianion)

The luminol dianion $Lum^{2(-)}$ does not exist in appreciable quantities in aqueous solvents; hydrogen peroxide and a catalyst such as hemin are required. Thus another mechanism seems to be at work here. Perhaps a hydrogen atom is abstracted from the luminol monoanion $Lum^{(-)}$ to yield a luminol radical anion *55* which then reacts with oxygen or a radical ion derived from hydrogen peroxide according to [3,4,109]

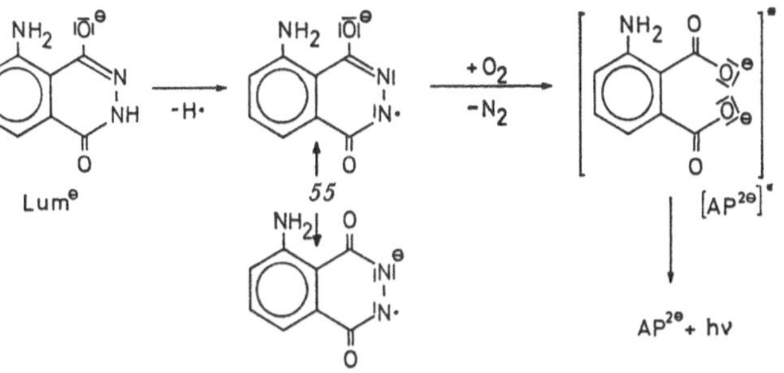

The ultraviolet spectrum of the monosodium salt of luminol shows the absorptions of both the mono- and the dianion of luminol; on addition of potassium tert. butylate the equilibrium is shifted to the dianion[109]. On the other hand, even small quantities of water shift the equilibrium back to the monoanion. The luminol dianion $Lum^{2(-)}$ was found to have a higher fluorescence efficiency than the monoanion. Absorption and fluorescence data for luminol, $Lum^{(-)}$ and $Lum^{2(-)}$ are listed in Table 4.

Table 4. Absorption and fluorescence of luminol and its ions in DMSO (from Gorsuch and Hercules [109])

	Absorption λ_{max}[nm], (ε)				Fluorescence λ_{max}[nm]
Luminol (neutral)	297	(7650);	360	(7720)	410
Luminol-Monoanion[1])	333	(6370);	370	(6320)	500
Luminol-Dianion[2])	310	(3250);	393	(6500)	520

[1]) Sodium salt with a trace of potassium-t.-butylate.
[2]) tert. butylate used as base.

Stopped-flow experiments of luminol chemiluminescence in the system luminol/pure DMSO/tert.butylate/oxygen [109] with independent variations of the concentrations of reactants confirmed the results obtained previously by E. H. White and coworkers [117] as to pseudo-first-order dependence of the chemiluminescence intensity upon each of the reactants. Moreover, the shapes of the decay curves obtained permitted some qualitative interpretations, the most important of which appears to be that the decomposition of the peroxide intermediate in a first-order reaction $LO_2{}^{2(-)}$ is probably the rate-controlling step. The rate constant for the reaction

$$LO_2^{2(-)} \longrightarrow (AP^{2(-)})^* + N_2$$

has been proposed to be $1.2 \pm 0.3 \times 10^{-1}$ sec^{-1} [109].

With mixed DMSO-water solvents the stopped-flow experiments resulted in a strong decrease of maximum chemiluminescence intensities but a considerable increase in decay times as a function of the water content of the mixture. Under certain reaction conditions the time required to reach half the intensity was one second in pure DMSO but 4.2 min in DMSO containing 9% water, with the light emission ending after 20 min. This strong influence of water on the decay time may well be the reason for the different results obtained in kinetic experiments by White et al. [117] and by Gorsuch and Hercules [109]. The former found a pseudo-first-order rate constant for luminol chemiluminescence (at low luminol concentrations 4.0×10^{-4} M — and excess base — 6.0×10^{-2} M and oxygen — 1.6×10^{-3} M) of $k' = 2.5 \times 10^{-3}$ sec^{-1}; this is considerably smaller than that observed by Gorsuch and Hercules [109] who found a k' of 1.4×10^{-1} sec^{-1}. This may be explained by the fact that the solvent used by White et al. was a mixture of 90% DMSO −10% water, so that decay occurred far more slowly than in the water-free DMSO used by Hercules and Gorsuch.

The luminol chemiluminescence reactions can be divided into two classes depending on the oxidation conditions [115]:

a) In aprotic solvents (DMSO, DMF, tetrahydrofurane) high quantum yields are obtained. The excitation efficiency ϕ_{es} which is essential for the fluorescence quantum yield of $AP^{2(-)}$ amounts to 0.09. This efficiency is not influenced by changes in solution temperature nor by changes in solvent polarity, nor by quenchers. In aqueous solutions, at optimal p_H 11—13 hemin-catalyzed hydrogen peroxide oxidation produces high chemiluminescence quantum yields with a $\phi_{es} = 0.04$ which is independent of temperature, viscosity, and quenchers. No appreciable influence of the luminol concentration was observed.

b) Other oxidation conditions give far lower quantum yields, probably due to competitive non-chemiluminescent reactions. A distinct influence of temperature and concentration on chemiluminescence quantum yields was observed.

Table 5. Luminol chemiluminescence quantum yields in different solvents with some oxidative systems (after J. Lee and H. H. Seliger [115])

Luminol concentration (M)	Solvent	T [°C]	p_H	Reaction conditions	Chemiluminescence quantum yields Q_c
—	Water	3	11,6	H_2O_2/Hemin	0.0123
—	Water	20	11,6	H_2O_2/Hemin	0.0124
—	Water	40	11,6	H_2O_2/Hemin	0.0135
—	Water	54	11,6	H_2O_2/Hemin	0.0116
—	DMSO	25	—	O_2/t-BuOK	0.0124
—	DMSO	45	—	O_2/t-BuOK	0.0123
10^{-5}	Water	20	12,2	$K_3Fe(CN)_6$	0.00010 ± 0.00001
10^{-5}	Water	20	11,6	NaOCl	0.004 ± 0.001
10^{-3}	Water	20	11,6	$K_2S_2O_8$	0.007 ± 0.001
10^{-5}	Water	20	11,6	Photosensitized with methylene	0.003
10^{-6}	Water	50	11,6	blue/O_2	0.01

The photosensitized results are from I. B. C. Matheson and J. Lee [118]. It is seen that the quantum yields in photosensitized oxidation depend on the concentrations of luminol and base, and on temperature. At higher temperature (50°) and low luminol concentrations, the quantum yields reached those of hemin-catalyzed hydrogen peroxide oxidation of luminol in aqueous-alkaline solution. Primary products of the photosensitized oxidation are singlet oxygen ($^1\Delta gO_2$) or a photoperoxide derived from methylene blue, but neither of these is directly responsible for the luminol chemiluminescence.

Finally, we mention another experimental result which points to a radical mechanism in luminol chemiluminescence.

Aqueous alkaline luminol solutions can be excited to chemiluminescence by pulse radiolysis, the only additional requirement being oxygen [119]. The suggested mechanism is that hydroxyl radicals attacking luminol monoanions, followed by reaction of the luminol radical anion thus formed with oxygen:

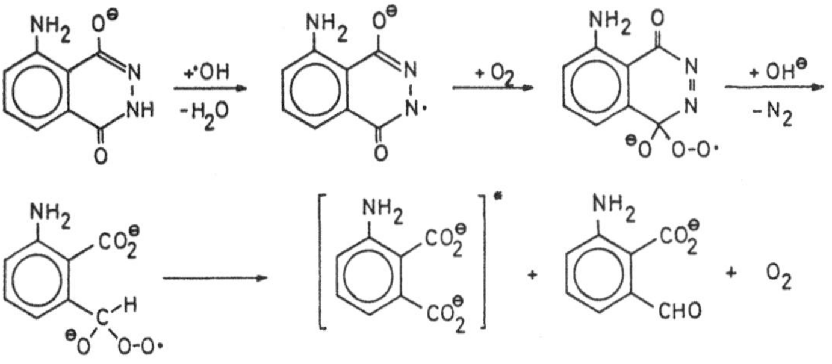

P. B. Shevlin and H. A. Neufeld [192] studied the mechanism of luminol chemiluminescence obtained by oxygen and ferricyanide in aqueous alkali, in absence of hydrogen peroxide. (For an earlier investigation of the luminol ferricyanide reaction see Stross and Branch [193]). In the presence of ferricyanide the emission maximum is slightly shifted to a longer wavelength, λ_{max} being 450 nm. The reaction was found to be first-order in respect of luminol and hydroxide ion concentration; as regards ferricyanide and oxygen concentration, however, the intensity depends approximately on the square of the concentrations, levelling off at higher concentrations. The mechanism proposed is very similar to that outlined on p. 101: the ferricyanide acts as one-electron oxidizing agent, forming the luminol radical anion ("semidione"), which yields a peroxide anion with oxygen via a peroxide radical which accepts an electron from a ferrocyanide ion. The peroxide anion is thought to be decomposed in a concerted mechanism to yield aminophthalate and nitrogen. Two-electron oxidation of luminol dianion to yield the diazaquinone or a N-hydroxy luminol 55a is proposed as a competing nonchemiluminescent reaction.

55a

The quantum yields were 3.8×10^{-7} (for solutions 4.62×10^{-5} M in luminol, 5.14×10^{-4} M in $K_3Fe(CN)_6$, and 8×10^{-2} in hydroxide; oxygen-saturated) and 3.25×10^{-6} (for similar solutions 8×10^{-3} in hydrogen peroxide, additionally). The authors point out that in chemiluminescent one-electron oxidations of luminol what determines light production is the outcome of competion between the reaction of the radical anion with oxygen to yield the peroxide and non-chemiluminescent further oxidation of the radical anion. The key parameter here is the stability of the radical anion, which should be increased by electron-donating substituents. This is certainly true when further oxidation of the radical anion leads to destruction of the whole luminol molecule, *e.g.* by oxidation of the primary amino group, etc. If, however, luminol undergoes only two-electron oxidation to yield the diazaquinone, further chemiluminescence will be produced (see Section VI. F.). Thus the conclusion that the fraction of luminol involved in diazaquinone formation does not contribute to chemiluminescence is not correct (see also M. M. Rauhut et al. [129]).

Chemiluminescence can also be produced in acoustically cavitated alkaline luminol solutions [194]. The emission spectrum is nearly the same as that of chemically induced chemiluminescence, λ_{max} being 433 nm. The lifetime of the emitting species was measured as 140 microseconds, in contrast to previous reports [195] of a far longer lifetime (50 msec). OH radicals appear to be the essential agent, forming hydrogen peroxide as oxidant as well as luminol radicals. It seems rather improbable that luminol dianion plays a role under the reaction conditions suggested by the authors [194], for in aqueous solutions the mechanism pathway should involve (cf. Gorsuch and Hercules [109]) direct recombination of a luminol radical anion and a hydrogen peroxide radical anion. The lifetime of the emitting species far exceeds the time taken for a cavitation bubble to implode, so it is suggested that the light emission in acoustically induced luminol chemiluminescence originates from the liquid surrounding the bubbles.

There is, however, no non-kinetic evidence put forward in favour of this mechanism. It should be kept in mind, moreover, that oxidation of luminol by $^{18}O_2$, in DMSO/t-BuOK, was found to yield 3-aminophthalate containing ^{18}O in *each* of the carboxyl groups [117]. This fact, together with some energetic considerations [3], suggests that a cyclic peroxide 56 may be the intermediate $LO_2^{2(-)}$ discussed above, which has not yet been isolated or synthesized. This cyclic peroxide could be formed, for example, by oxygen attacking a diazaquinone (see Section VI. F.) or a luminol radical anion. Concerning monoacyl hydrazides

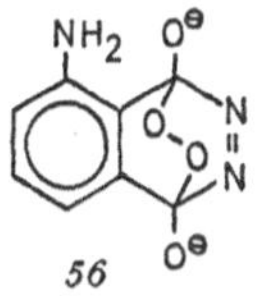

56

(linear hydrazides), there is evidence that acyl anions are key intermediates in chemiluminescence [120]

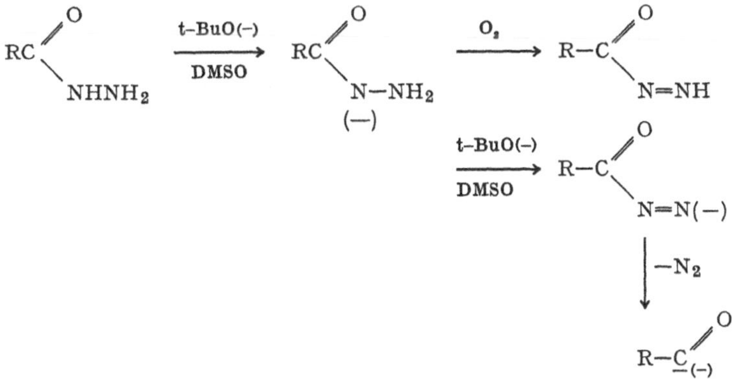

Attack of molecular oxygen on this carbanion should yield acyl radicals which could react with loss of carbon monoxide, yielding the respective radicals, or with the formation of acyl peroxide radicals:

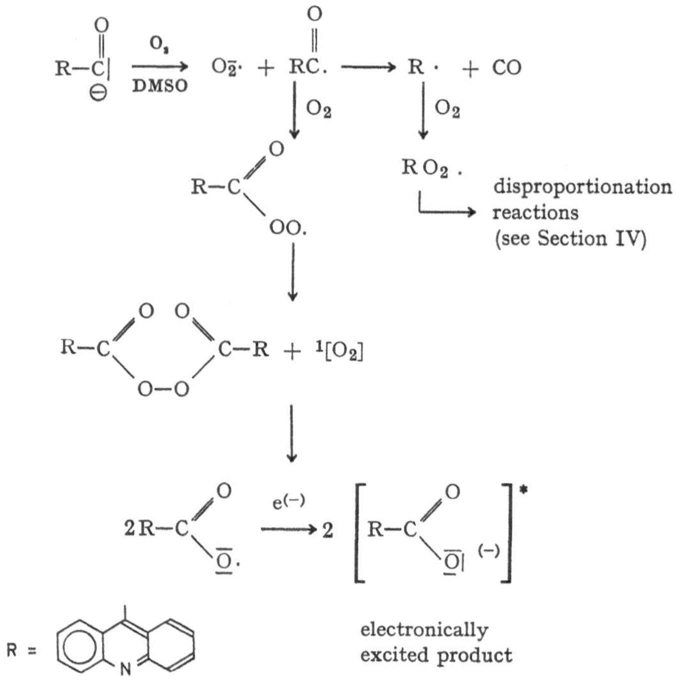

The electron transfer involved in the formation of excited carboxylate anion, as formulated above, should occur in such a way that the lowest antibonding orbital of the carboxylate group is occupied first (see Section II. B.).

Diacyl azo compounds such as azo-bis-9-acridinoyl *57* were found to be brightly chemiluminescing in DMSO, with t.-butoxide and oxygen.

The alkoxide anion is postulated to react with the azo compound, giving ester and acyl anion:

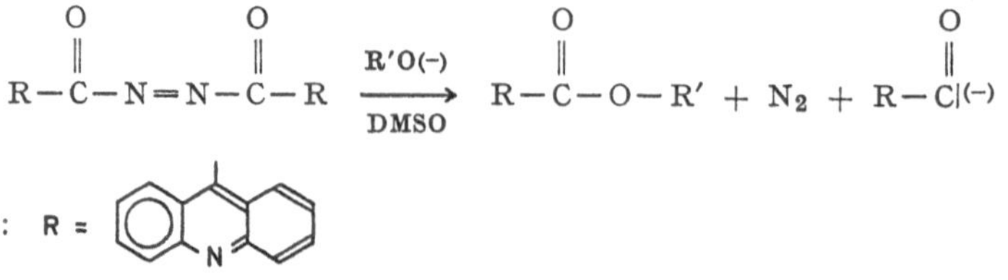

Whereas substitution of the hydrazide ring in luminol leads to complete loss of chemiluminescence capacity [121], N-methylated linear hydrazides such as *57a* and *57b* were found to exhibit chemiluminescence on oxidation with oxygen in DMSO [120].

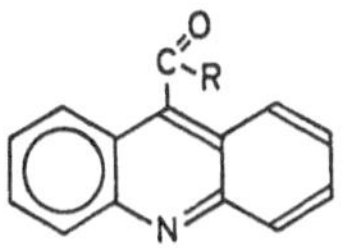

$$57a: \quad R = -\overset{\overset{\displaystyle CH_3}{|}}{N} - NH_2$$

$$57b: \quad R = -\overset{\overset{\displaystyle CH_3}{|}}{N} \overset{\overset{\displaystyle CH_3}{|}}{} - NH$$

The emission spectra match the fluorescence of the corresponding acid. Methane was detected as a major product in the chemiluminescent oxidation of *57a* and it was suggested that it resulted from the decomposition of methyl-diimine formed after dehydrogenation of the hydrazide *57a*:

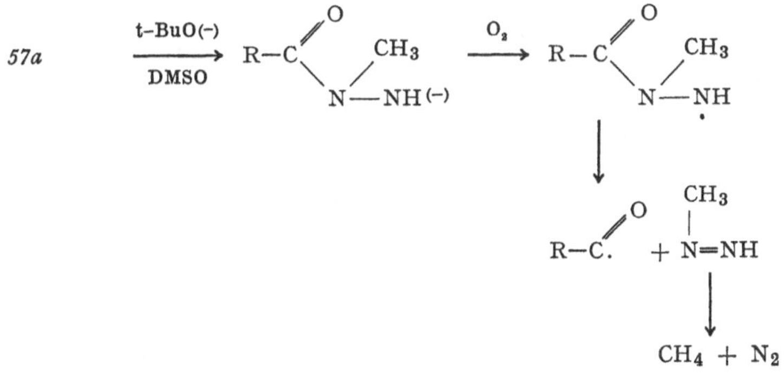

$$57a \xrightarrow[\text{DMSO}]{\text{t--BuO}^{(-)}} R-C \overset{\displaystyle O}{\underset{N-NH^{(-)}}{\diagdown CH_3}} \xrightarrow{O_2} R-C \overset{\displaystyle O}{\underset{N-NH}{\diagdown CH_3}}$$

$$\downarrow$$

$$R-C \overset{\displaystyle O}{\diagdown} \quad + \quad N=NH \overset{CH_3}{|}$$

$$\downarrow$$

$$CH_4 + N_2$$

D. Enzyme-Catalyzed Luminol Chemiluminescence

In the reaction of luminol, hydrogen peroxide, and horse radish peroxidase [122] the chemiluminescence intensity is proportional to the square of luminol radical concentration. The lifetime of these luminol radicals was found by ESR techniques to be about 10 sec. Titration studies revealed that luminol acts as two-electron donor during the reduction of a hydrogen peroxide-horseradish peroxidase complex. The enzyme is not involved in the reaction step leading directly to light emission. This step is formulated as

$$2 \text{ luminol radical} + H_2O_2 \longrightarrow 3\text{-amino-phthalate} + h\nu$$

which means that the total reaction rate should be proportional to the root of luminol concentration. This was observed at low luminol concentrations.

E. Intramolecular Energy Transfer in Luminol-Type Hydrazide Chemiluminescence

Phthalate dianion does not fluoresce [124] and naphthalene-2.3-dicar-boxylate has its fluorescence maximum at 365 nm with a small part of the fluorescence emission extending into the visible (see also p. 91). In compounds like *58* and *59* the cyclic hydrazides of these dicarboxylic acids are linked, via a methylene group, with highly fluorescent struc-tures such as 9.10-diphenylanthracene (DPA) or N-methyl-acridone. Direct interaction of the π-electron systems of the latter groups and those of the hydrazide remnant is not possible. Both of the compounds are chemiluminescent on oxidation in aqueous or DMSO solvent. The quantum yield is 2.6×10^{-3} for the DPA derivative *58* [123], whereas that of naphthalene-2.3-dicarboxylic hydrazide itself is only about 5×10^{-4} [125]. A mixture of DPA and of naphthalene-2.3-dicarboxylic hydrazide gave no more light, visually, than the hydrazide alone, and the chemiluminescence spectrum of the mixture was identical with that of naphthalene-2.3-dicarboxylic hydrazide. However, the DPA derivative *58* exhibited a chemiluminescence spectrum consisting of the DPA *and* naphthalene-2.3-dicarboxylate fluorescence in water (on oxidation by hydrogen peroxide/hemin) and DPA fluorescence alone in DMSO/tert.BuOK/O$_2$.

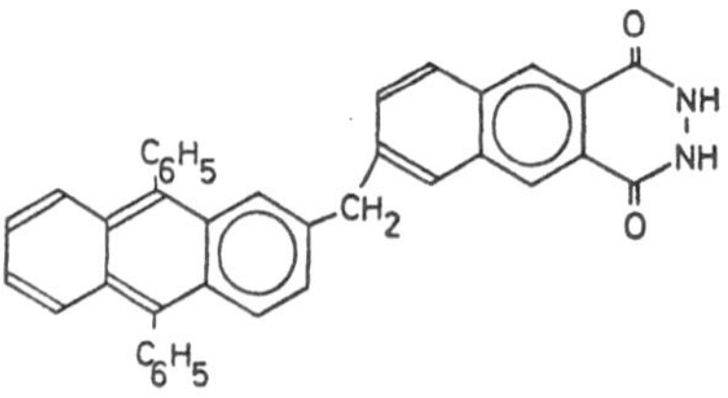

58 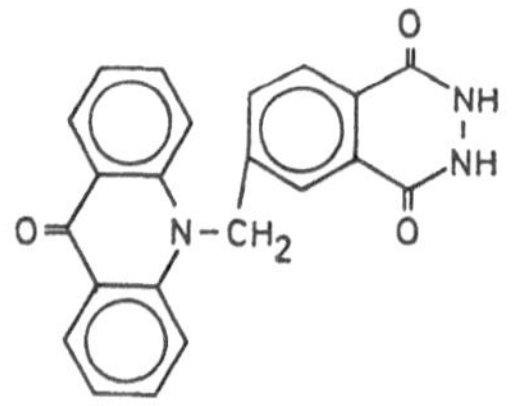   *59*

It was concluded, therefore, that intramolecular energy transfer is involved in the chemiluminescence of *58*. The rate of energy transfer was calculated to be 2.3×10^7 sec^{-1} [126].

Similarly the N-methylacridone derivative *59* yielded chemilumi-nescence which was about a third that of *58*. In this case the "donor part" of the molecule, *i.e.* the phthalic hydrazide, cannot produce a fluorescent product at all, but on its oxidation excited phthalate dianions very probably are produced [124]. From a series of other compounds of this type (*60, 61, 62*) the conclusion was drawn that the relative chemiluminescence quantum yields were not governed by the fluorescence efficiencies of the "fluorescer" part of the molecules. This fact, together with considerations of overlap integrals of phthalate and carbazole moieties, and furthermore the complete non-fluorescence of phthalate dianion are the basis of the suggestion of E. H. White and coworkers [2,127] that this intramolecular energy transfer is not a singlet-singlet, but a triplet-singlet energy transfer of a mixed dipole-dipole and exchange character.

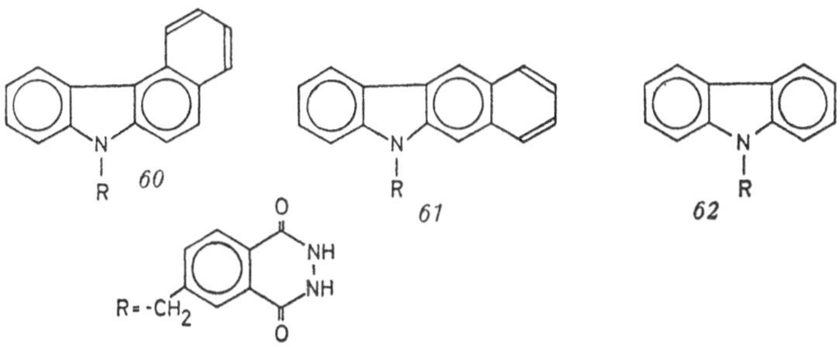

*Inter*molecular energy transfer is apparently involved in the "anomalous" chemiluminescence of phthalic hydrazide in aprotic solvent (DMSO/tert.BuOK/O_2) [124]: the energy of excited phthalated ianion is transferred to phthal-hydrazide monoanion which then emits at 525 nm with relatively low quantum yield. This phenomenon has not been observed in aqueous systems [2].

In the following scheme the difference between intra- and intermolecular electronic excitation energy transfer is summarized (as formulated in [2]):

Intermolecular

$$B + C \longrightarrow D^x$$
$$D^x + A \longrightarrow D + A^x \longrightarrow A + h\nu$$

B, C: reactants; D^x electronically excited product; A: acceptor molecule with high fluorescence efficiency

Intramolecular

$$A-C \xrightarrow[\text{oxidation}]{} A-D^x \longrightarrow A^x-D \longrightarrow A-D + h\nu$$

A: fluorescent, light-emitting part of a molecule; C: hydrazide energy-generating part of the molecule, transformed into the primary excited product.

F. Diazaquinone Chemiluminescence

As early as in his first paper on luminol chemiluminescence H. O. Albrecht [128] postulated that diazaquinone 63 ("dehydroluminol") is involved in the chemiluminescence mechanism. Kinetic experiments [129, 130] suggested that in fact a two-electron oxidation product (corresponding to the diazaquinone) of luminol or a similar hydrazide was a key intermediate. The importance of luminol dianion for luminol chemiluminescence in aprotic solvents (see Section VI. C.) is not inconsistent with the possibility of a diazaquinone as further intermediate (see [2]).

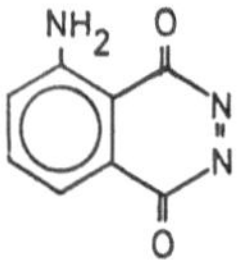

63 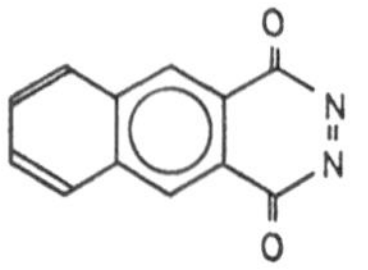   64 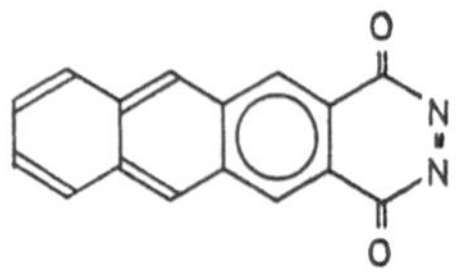   65

At present, one can state that diazaquinones of appropriate structure are in fact chemiluminescent, that the emission spectra and the reaction products are the same as those of the corresponding hydrazides, and that their light yields are of similar magnitude. Whereas the luminol diazaquinone 63 has not yet been obtained in a chemically pure state[b]), probably due to the sensitivity of the amino group towards the oxidative reagents required for the dehydrogenation of luminol, the diazaquinones 64 [107] and 65 [101] are reasonably stable and so may be investigated qualitatively and quantitatively in chemiluminescence experiments. It was found that these compounds chemiluminescence in aqueous and in non-aqueous (e.g. dimethyl phthalate) systems yielding the corresponding dicarboxylates; the chemiluminescence emission spectrum matches the fluorescence of the latter. In aprotic as well as in aqueous solvents hydrogen peroxide or at least oxygen is required [101,107] but hemin has no influence, in contrast to hydrazide chemiluminescence in aqueous solvent [101].

The quantum yield of 65 was found to be about 4.3 times that of the corresponding hydrazide in its hemin-catalyzed oxidation [131].

The requirement of hydrogen peroxide or oxygen in diazaquinone chemiluminescence appears to be another argument against the hypothesis of Albrecht [128] or of Kautsky and coworkers [132] that the diazaquinone is simply hydrolyzed to yield aminophthalate and diimine which in turn reduces other diazaquinone molecules to luminol in an excited state so that it then emits.

b) A Diels-Alder adduct of 63 and cyclopentadiene was detected by chromatography of luminol solutions oxydized by $K_3(Fe(CN))_6$ in the presence of cyclopentadiene [133] (see also [101]).

The question is whether diazaquinones are in fact "in the main stream" [2] of cyclic diacyl hydrazide chemiluminescence or whether they are only one of the possible intermediates or, finally, whether diazaquinone chemiluminescence is a special type of chemiluminescence reaction although very closely related to the chemiluminescence of the corresponding hydrazides. E. H. White and D. F. Roswell [2] point out that the respective diazaquinone may actually be one of the possible intermediates formed either from the hydrazide dianion (essential in aprotic solvents) or from a hydrazide radical anion (which is the usual species in aqueous media). This is outlined in the following scheme:

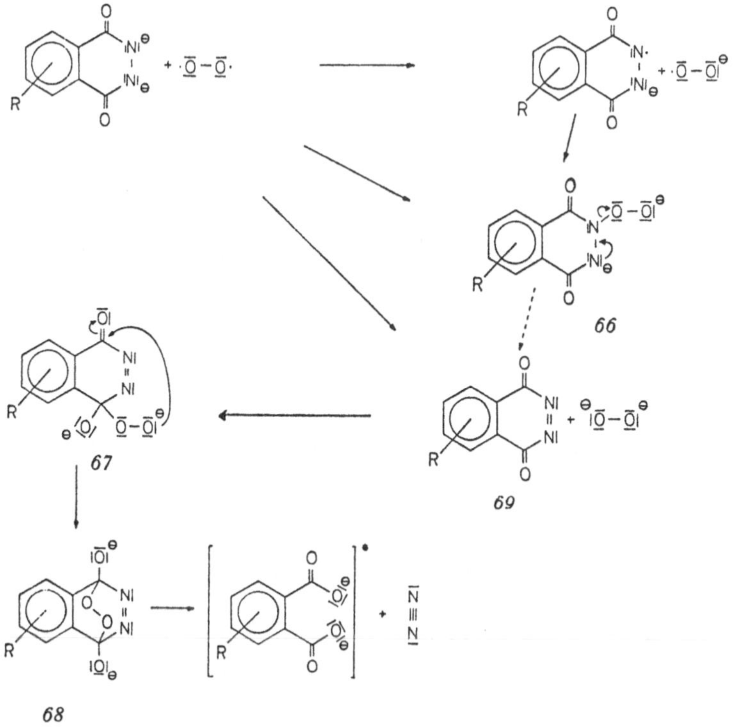

66

67

69

68

That a bridged peroxide 68 (for a more detailed discussion, see [1], p. 84) arises from the intermediate open-chain peroxide formed either by recombination of hydrazide radical ion with $.O_2^{(-)}$ radical ion, such as 66, or by nucleophilic attack of $O_2^{2(-)}$ ion on a carbonyl group of a diazaquinone, such as 67, appears to be very plausible. One of the most important experimental reasons for this assumption has been put forward by E. H. White and M. M. Bursey [106] who, on oxidation of luminol with

$^{18}O_2$ in DMSO/t-butylate, isolated 3-amino-phthalate containing ^{18}O atoms in each of the carboxyl groups. This should be possible only via an intermediate which has both its carbonyl carbon atoms bonded with the oxygen-oxygen grouping. This requirement also seems to rule out the possibility of a dioxetane intermediate.

Attempts to synthesize the compound *70* via a nucleophilic substitution reaction of an activated azodicarboxylate and hydrogen peroxide according to

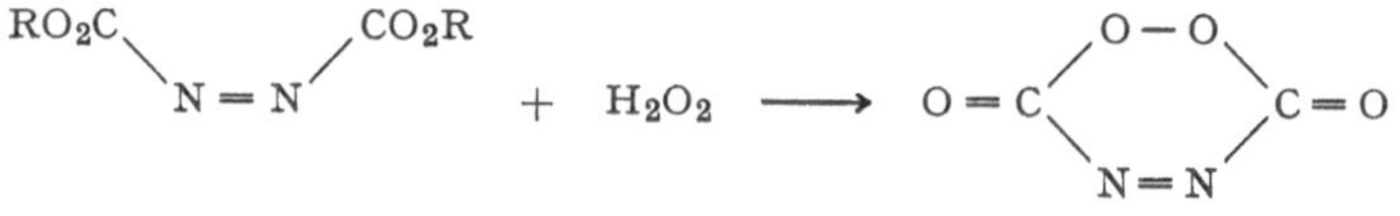

70

have so far been unsuccessful, as *trans* azodicarboxylates containing an appropriate ester group, *e.g.* R=2.4.6-trichlorophenyl, could not be isomerized to the *cis* form [134]. However, the reaction of *trans*-bis (2.4.6-trichlorophenyl)-azodicarboxylate with hydrogen peroxide and a base is chemiluminescent in aprotic solvents, rubrene or other fluorescers being required:

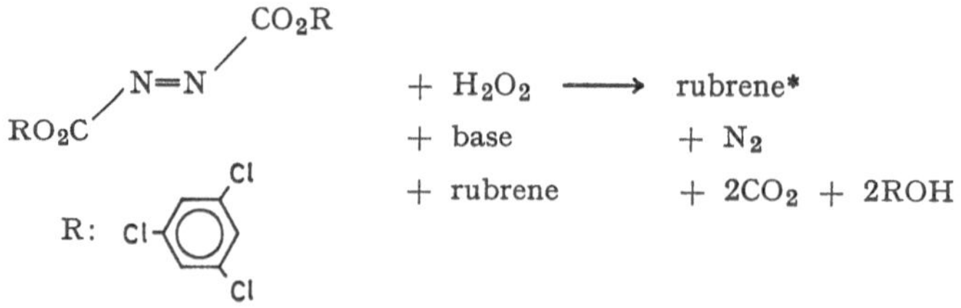

This reaction produces fluorescer molecules in an excited state as the emission corresponds to the fluorescence of the former.

The light output is visually distinctly lower than that of a diazaquinone, or a hydrazide of comparable structure. A bridged peroxide is not possible as intermediate for steric reasons.

VII. Lucigenin Chemiluminescence: Role of Radical Ions

Lucigenin *71* chemiluminescence is more complicated than that of luminol and related compounds due to the presence in the reaction mixture of several species capable of emission [135]. N-Methylacridone *72*, however, has been established as the primary excited product (for references see [1] p. 90) from which energy transfer occurs to the other species [3].

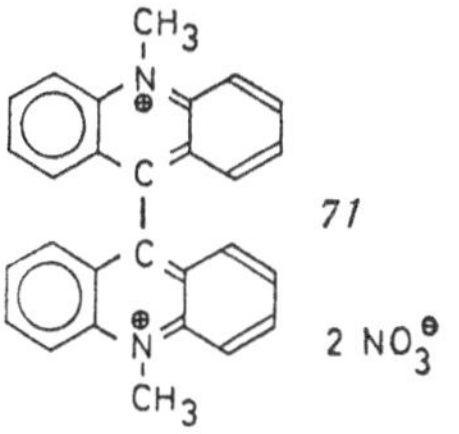

Because of the close relations between lucigenin, N-alkylacridones, and other 9-substituted acridines, it appears appropriate to consider the chemiluminescence reactions of all these compounds simultaneously.

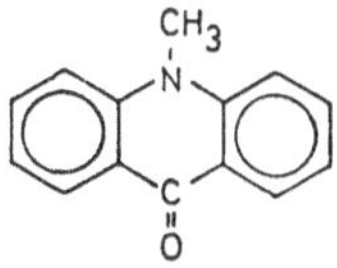

In view of the increasing interest in the radical ion chemiluminescence 71 has also been investigated to see whether radical ions play an essential role here, especially by E. G. Janzen et al. [136,137,138]. The reaction pathway to N-methylacridone 72 was thought to proceed perhaps via a homolytic cleavage of the lucigenin carbinol base 74, yielding N-methylacridone radical anion 75.

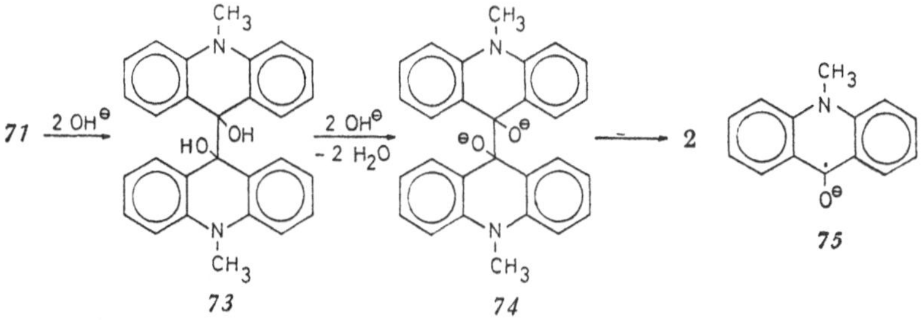

ESR spectroscopy, however, revealed that 75, which one may call a ketyl radical, was not present in alkaline solutions of 71 in absence of oxidizing agents. The signal actually observed was that of 9-hydroxy-N,N'-dimethyl-9,9'-biacridan radical 76. It was suggested that this radical was formed by one-electron transfer from N-methyl-acridone radical anion 75 to lucigenin monocarbinol 77, or by addition of hydroxide ion to lucigenin radical cation 78 (formed from lucigenin by one-electron transfer by the ketyl 75).

Small amounts of light were observed, obviously due to this one-electron transfer, even in absence of oxygen, hydrogen peroxide or other oxidants from basic aqueous DMSO solutions of lucigenin; the intensity of the light increased with base concentration, but also when no base was present. The bright chemiluminescence which usually occurs on treatment of lucigenin with base and an oxidant is the result of several possible mechanisms, the key intermediate being a four-membered cyclic peroxide 79 which is decomposed in a concerted bond-cleavage

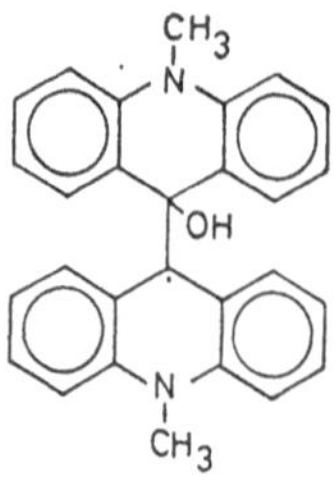

76

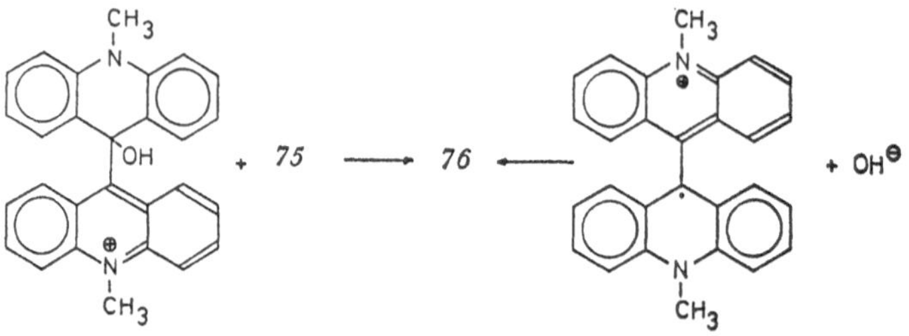

77   + 75 ⟶ 76 ⟵   + OH⊖

78

mechanism, as proposed by McCapra and Richardson [139] and Rauhut and coworkers [140].

The lucigenin radical *78* is not involved, according to Janzen and coworkers [136], in the direct formation of this cyclic peroxide; one would, however, expect a reaction of *78* with oxygen radical anion to be a possible way of forming the cyclic peroxide, although lucigenin radicals were not detected in the presence of hydrogen peroxide.

The dioxetane derivative *79* may be formed as intermediate in the brilliant chemiluminescence reaction between 10,10'-dimethyl-9,9'-bi-acridylidene and excited-singlet oxygen [125]. Chemiluminescence also occurs when potassium cyanide is added to lucigenin solutions in the

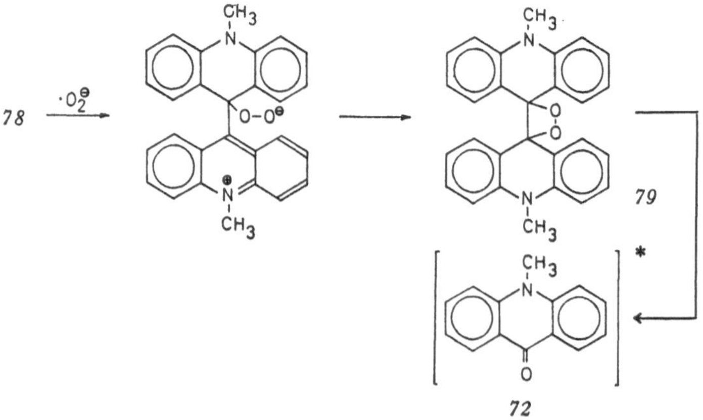

78 $\xrightarrow{\cdot O_2^{\ominus}}$

79

72

presence of oxygen: N-methylacridone *72* and potassium cyanante are the products [137)]

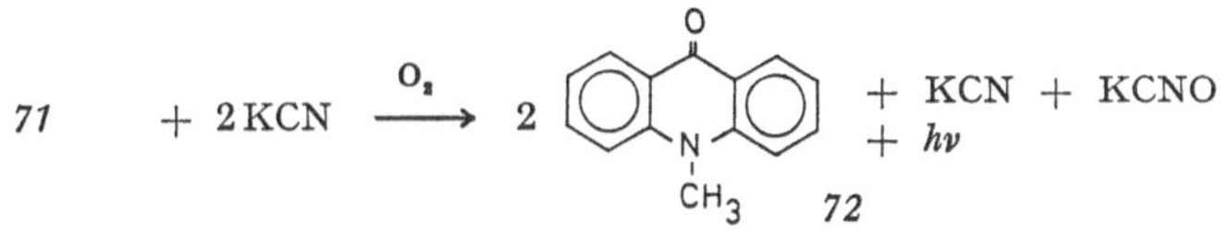

$$71 \quad + 2\,KCN \xrightarrow{\text{O}_2} 2 \quad\fbox{} \quad + KCN + KCNO$$
$$+ h\nu$$
$$72$$

In oxygen-free solutions, N-methyl-9-cyano-acridanyl radical *81* was detected; this stems from a homolytic cleavage of the biacridanyl-9,9'-dicyanide *80*.

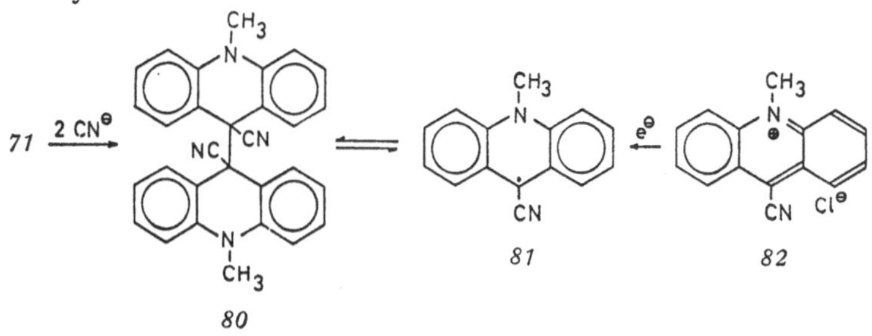

The acridanyl radical is also obtained from N-methylacridinium chloride *83* and potassium cyanide in air-saturated DMF, DMSO, or DMSO/water mixtures. It is a remarkably stable radical [137)]; when *83* is treated with excess cyanide and oxygen N-methylacridone and cyanate are produced with light emission:

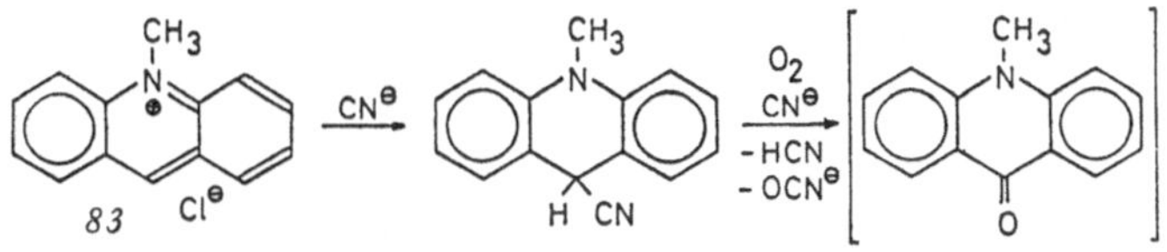

In this case the radical *85* is probably formed from the N-methyl-9-cyanoacridanyl anion *84* by electron abstraction by oxygen:

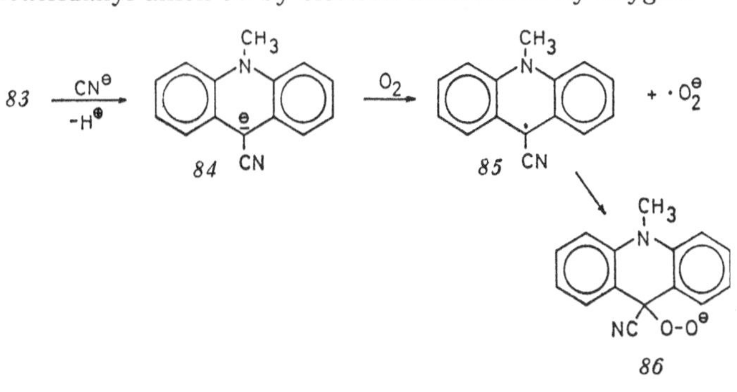

113

The radical *85* yields the peroxide anion *86* the decomposition of which gives rise to excited N-methylacridone and cyanate. In addition to previous suggestions put forward in respect of this exergonic decomposition [3,139] (see also [1] p. 97), a recombination of the radical *85* and the peroxy radical *87* (from *86* and O_2) according to

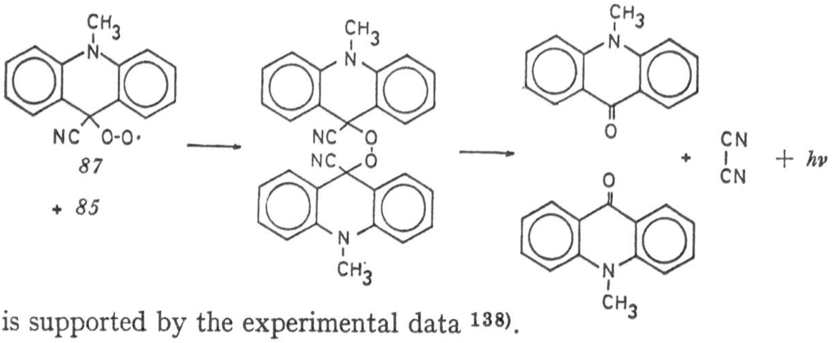

is supported by the experimental data [138].

Other chemiluminescent reactions yielding excited N-methyl acridone

The reaction of 9-carboxy-N-methylacridinium chloride *88* with potassium persulfate in aqueous base is chemiluminescent, the quantum yields approaching 0.02 at high base concentrations. N-methylacridone is the emitter [141].

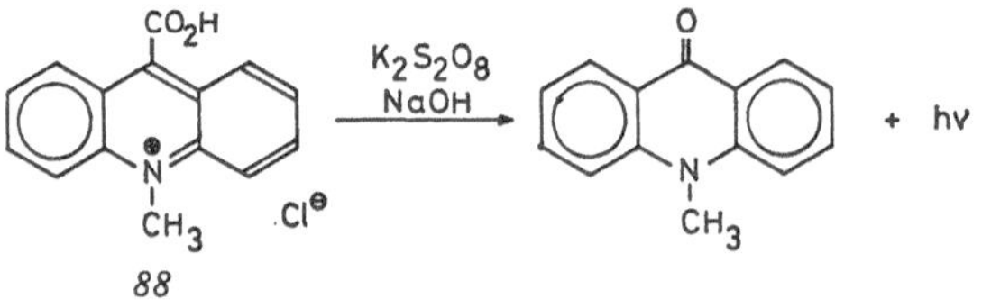

The reaction is first-order in *88* and in persulfate. It proceeds very probably via a monopersulfate ester *90* and a dioxetanone derivative *91*:

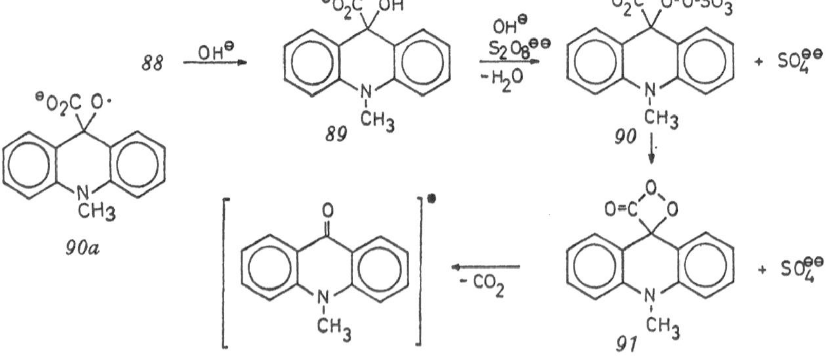

A homolytic cleavage of the peroxy bond in *90* to yield the radical anion *90 a* with subsequent elimination of CO_2 and formation of the N-methylacridone radical anion (see p. 111) should be a possible pathway, too, but as radical scavengers *enhance* the *88* chemiluminescence [141] it appears rather improbable.

Similar mechanisms are suggested for the chemiluminescence of 9-formylacridine *92*, 9-formyl-10-methylacridinium methosulfate *95* 9-benzoylacridine *93* and 9(4-nitrobenzoyl)-acridine *94* [142] which occurs on oxidation of these compounds with base and oxygen in DMSO.

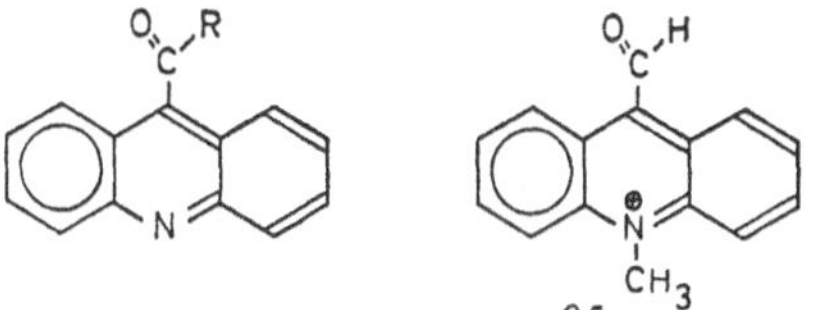

92	R=H
93	$R=C_6H_5$
94	$R=C_6H_4 (NO_2-p)$

The quantum yields are about 10^{-4} with the aldehydes and 10^{-3} with the ketones *93* and *94*. The respective acridones were found to be the emitters when potassium hydroxide was used as base in the oxidation of *92* and *95* and t-butoxide, in the oxidation of the ketones *93* and *94*.

It is assumed that the formyl derivatives form the 9-carbanions, *e.g. 96*, which then react with oxygen in the way outlined in VI.C. for linear hydrazides.

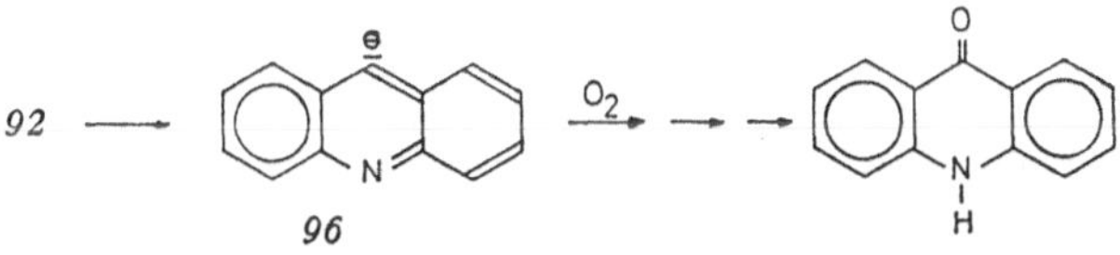

When the aldehydes *92* and *95* are treated with oxygen and t-butoxide the emission spectrum of the chemiluminescence matches the fluorescence of a mixture of the acridone and the 9-carboxylate: the latter being formed via an acyl anion.

Quenching in lucigenin fluorescence (and similar effects should operate in lucigenin and acridine chemiluminescence in general) has recently been investigated by Hercules [143].

VIII. Radical-Ion Chemiluminescence

The importance of radical ions and electron-transfer reactions has been pointed out in the preceding sections (see also [1] p. 128). Thus, in linear hydrazide chemiluminescence (p. 103) or acridine aldehyde or ketone chemiluminescence, the excitation steps consist in an electron transfer from a donor of appropriate reduction potential to an acceptor in such a way that the electron first occupies the lowest antibonding orbital, as in the reaction of 9-anthranoyl peroxide 96 with naphthalene radical anion 97 [142]:

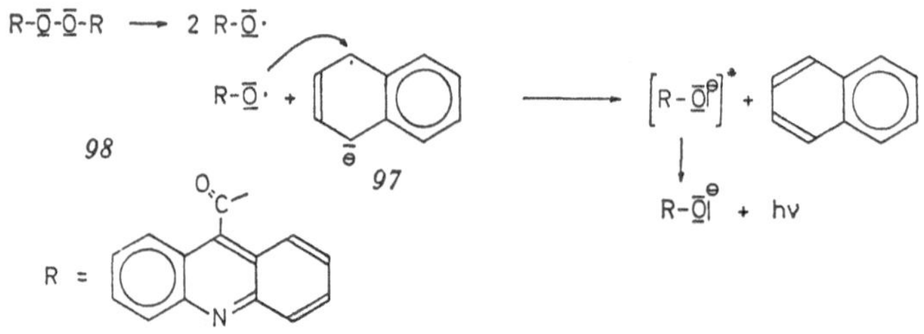

The simplest systems where electron-transfer chemiluminescence occurs on interaction of radical ions are radical-anion and radical-cation recombination reactions in which the radical ions are produced from the same aromatic hydrocarbon (see [1], p. 128) by electrolysis: this type of chemiluminescence is also called electro-chemiluminescence. The systems consisting of e.g. a radical anion of an aromatic hydrocarbon and some other electron acceptor such as Wurster's red are more complicated. Recent investigations have concentrated mainly on the energetic requirements for light production and on the primary excited species.

A. Electrogenerated Chemiluminescence

In electrogenerated chemiluminescence, light emission occurs not on the electrode surface but in the solution. Oxygen has to be excluded [5]. In the usual form a one-electrode technique is applied; the potential of the electrode is changed periodically. At cathodic potential the radical anion is produced, at anodic potential the radical cation. These two radical ions react in the diffusion layer near the electrode surface; the electron transfer from the radical anion to the radical cation causes the light emission [5,20].

The quantum yield in the reaction between rubrene radical ions was found to be in the range 0.006....0.015, depending on whether the radical anion or the radical cation was produced first [144]. The latter is, in general, less stable.

A recent investigation of the rubrene radical-ion electrochemi-luminescence carried out by means of the rotating ring-disc electrode (rrde) method [145,146] gives a quantum yield of about 7×10^{-4}. This technique uses a rotating two-electrodes system allowing a very rapid encounter of the simultaneously generated radical anion and cation. (For a detailed description of the apparatus, see [145]). This technique is superior to the one-electrode technique, as non-chemiluminescent side reactions of the very reactive radical ions, especially the radical ca-tions [147], are considerably reduced. This makes quantitative measure-ments of the electrochemiluminescence efficiency far more reliable. "Pre-annihilation" chemiluminescence [21] could be excluded when the rrde technique was used with very carefully purified solvents and re-agents [146].

The species produced in radical anion-radical cation recombination reactions are described in Section II. B. (p. 66). Concerning the con-stitutional and energetic premises of electrochemiluminescence, the following qualitative and quantitative relationship has been stated: electrochemiluminescence is relatively strong when the radical ions producing it are relatively stable. The aromatic hydrocarbon or other organic compound from which the radical ions are derived must, of course, be fluorescent [21]. The stability of the radical ions is more im-portant than the fluorescence efficiency of the parent compound: this is apparent from the poor correlation between fluorescence efficiency and electrochemiluminescence quantum yield with the various compounds. Electron-donating substituents in aromatic hydrocarbons increase the stability of the radical cation, which is the compound critical for electro-chemiluminescence because it is less stable than the corresponding radical anion [21]. Electron-donating substituents can, however, reduce the stability of the radical anion [21].

Concerning the energetic conditions of electrochemiluminescence, it is assumed that either the free energy of radical ion recombination and/or that of triplet-triplet annihilation are equal or higher than the energy of the first excited singlet state of the emitter [5]. Some additional thermal energy is sometimes required, when the recombination energy of the radical ion is not quite sufficient.

According to Hercules [5] a measure of the relationship between direct excitation of the first excited singlet state by radical-ion re-combination and triplet-triplet annihilation is the entropy factor $T \Delta S$, estimated to be on average 0.2 eV. The enthalpy of the radical cation-radical anion recombination can be measured as the difference between the redox potentials $E_{1/2}$ Ar—Ar$^{(\dot{+})}$ (oxidation) and $E_{1/2}$ Ar—Ar$^{(\dot{-})}$ (reduction). This difference has to be corrected by the entropy term. If this corrected radical-ion recombination enthalpy is equal to or larger than the energy of the first excited singlet state of the respective emitting molecule, direct singlet-state excitation is possible: the system is energy-sufficient. If, however, the corrected radical-ion recombination energy is lower than the energy of the first excited singlet state, the system is energy-deficient [5,148].

An example of an "energy-sufficient" system is the chemiluminescent recombination of 9.10-diphenylanthracene (DPA) radical ions produced by electrolysis of DPA in dimethylformamide [148]. $E_{1/2}$ DPA$-$DPA$^{(+)}$ $= + 1.35$V and $E_{1/2}$ DPA$-$DPA$^{(-)} = -1.89$ V (vs. sce) which gives a corrected enthalpy of ca. 3.04 V. The energy of the first excited singlet state of DPA is 3.0 V.

The relations between the redox potentials and the energy of the first excited singlet states of aromatic compounds are a convenient basis for predicting their electrochemiluminescence. It has to be repeated that their energy alone is not sufficient to produce electrochemiluminescence; the activation energy is of decisive importance, too. This depends on the configuration changes of the reacting molecules, as outlined in Section II. B. (see also the work of J. Hoytink [149], one of the first investigators of electrochemiluminescence).

In energy-deficient systems triplet-triplet annihilation is involved. This was proved by several experimental facts.

1. If chemiluminescence is to occur, a critical electron-transfer enthalpy value must be exceeded. This value is similar to the energy of the lowest excited triplet state, as was found in the case of fluoranthene [150], for example.

2. Chemically inert triplet quenchers e.g. trans-stilbene, anthracene, or pyrene, suppress the characteristic chemiluminescence of radical-ion recombination. When these quenchers are capable of fluorescence, as are anthracene and pyrene, the energy of the radical-ion recombination reaction is used for the excitation of the quencher fluorescence [150]. Trans-stilbene is a chemically inert [162] triplet quencher which is especially efficient where the energy of the first excited triplet state of a primary product is about 0.2 eV above that of trans-stilbene [163]. This condition is realized, for example, in the energy-deficient chemiluminescent system 10-methyl-phenothiazian radical cation and fluoranthene radical anion [164].

3. The intensity of electrogenerated chemiluminescence is influenced by an external magnetic field [148,150,152].

Energy-deficient radical-ion reactions generally exhibit low quantum yields due to the rather "circumstantial" triplet-triplet annihilation mechanism [150].

The electrogenerated radical anions of aromatic hydrocarbons, *e.g.* DPA, rubrene, fluorene, can also act as "reductants" towards electrochemically obtained radical cations which are derivatives of other aromatic compounds such as N,N-dimethyl-*p*-phenylenediamine (Wurster's red) [150] (see Section VIII. B.). When a mixture of DPA and a halide such as *99* (DPACl$_2$) or *100* is electrolysed, a bright chemiluminescence is observed; the quantum yields are about two orders of magnitude higher than that of the DPA radical anion-radical cation reaction [153].

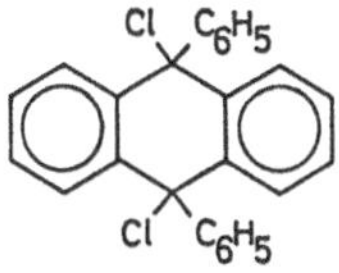

99 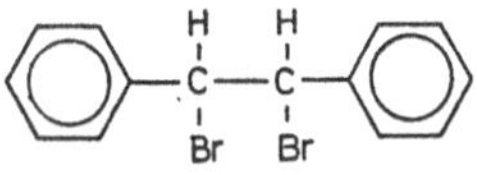　　　　　　　　　　　　　　100

Chemiluminescence also occurs during electrolysis of mixtures of DPACl$_2$ *99* and rubrene or perylene: In the case of rubrene the chemiluminescence matches the fluorescence of the latter at the reduction potential of rubrene radical anion formation (-1.4 V); at -1.9 V, the reduction potential of DPA radical anion, a mixed emission is observed consisting of rubrene and DPA fluorescence. Similar results were obtained with the dibromide *100* and DPA and/or rubrene. An energy-transfer mechanism from excited DPA to rubrene could not be detected under the reaction conditions (see also [154]). There seems to be no explanation yet as to why, in mixtures of halides like DPACl$_2$ and aromatic hydrocarbons, electrogenerated chemiluminescence always stems from that hydrocarbon which is most easily reduced. A great number of aryl and alkyl halides is reported to exhibit this type of rather efficient chemiluminescence [155].

An energy-sufficient mixed chemiluminescent radical-ion reaction is that of thianthrene (TH) radical cation *101* and 2.5-diphenyl-1.3.4-oxadiazole (DPO) radical anion *102* [156]:

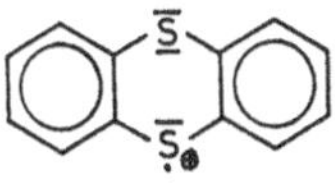

101 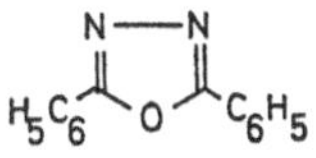　　　　　　　　　　102

The chemiluminescence spectrum matches the fluorescence of both thianthrene and 2.5-diphenyl-1.3.4-oxadiazole (430 and 340 nm, respectively).

An external magnetic field was observed to have practically no effect on the intensity of the thianthrene (430 nm) emission, indicating that no triplet states are involved. The shorter-wavelength emission of the oxadiazole *102*, however, is probably due to a triplet-triplet annihilation reaction of diphenyloxadiazole triplets. These are produced in the radical-ion reaction between *101* and *102*, yielding thianthrene excited-singlet molecules and diphenyl-oxadiazole excited-triplet molecules:

$$TH^{+}_{\cdot} + DPO^{-}_{\cdot} \longrightarrow {}^{1}(TH)^{*} + {}^{3}(DPO)^{*}$$

B. Reactions of Radical Anions of Aromatic Hydrocarbons with Wurster's Red Type Compounds or Aromatic Amines

Light emission occurs during the reaction of numerous radical anions of aromatic hydrocarbons with radical cations such as Wurster's red *103*, Wurster's blue *104* or radical cations derived from triarylamines of the type *105*, *106*.

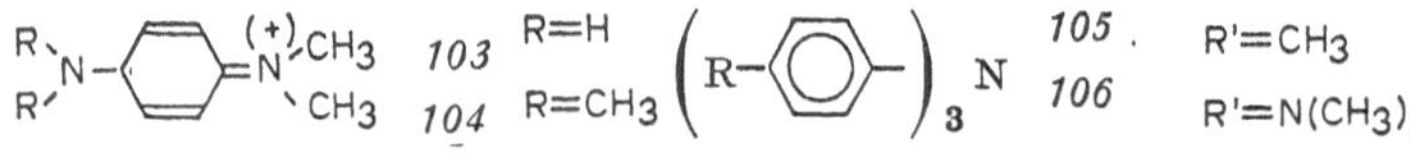

A. Weller and K. Zachariasse [157–160] thoroughly investigated this radical-ion reaction, starting from the observation that the fluorescence of aromatic hydrocarbons is quenched very efficiently by electron donors such as N,N diethylaniline which results in a new, red-shifted emission in nonpolar solvents: This emission was ascribed to an excited charge-transfer complex ${}^{1}(Ar^{(-)} D^{(+)})$, designated heteroexcimer, with a dipole moment of 10 D. In polar solvents, however, quenching of aromatic hydrocarbon fluorescence by diethylaniline is not accompanied by hetero-excimer emission; in this case the "free" radical anions $Ar^{(-)}$ and cations $D^{(+)}$ were formed.

The reverse should be chemiluminescence via recombination reactions of radical ions derived from aromatic hydrocarbons and amine-type compounds, as was shown by experimental evidence. The energetic requirements are analogous to those reported in VIII. A. For example, the reaction *104* with chrysene anion gives only 2.66 eV, whereas the chrysene fluorescence requires the energy of chrysene first excited singlet state of 3.43 eV: it is thus an energy-deficient system. The energy of chrysene triplets, however, is only 2.44 eV, so that triplet-triplet annihilation is regarded as the excitation mechanism for the observed chemiluminescence. In such a case the quantum yields are relatively low ($10^{-5} \ldots 10^{-4}$ Einsteins/mol [160]). Higher quantum yields were observed in the reactions of aromatic radical anions with triarylamines; they exceed those of the reactions with Wurster's red by an order of magnitude. The heteroexcimer, formed directly from the radical ions, is the main emitter at low temperatures [160].

C. Ruthenium Complexes

F. E. Lyttle and D. M. Hercules [161] investigated the energetic relations in the chemiluminescent reaction of ruthenium chelate complexes (Ru-III-2.2'-bipyridyl-, — 5-methyl-1.10-phenanthrolin — and other complexes) with hydroxyl ion or hydrazide. The general reaction is

$$\text{Metal (ligand)}_x^{(n+1)\,+} \quad + \; e^{(-)} \quad \text{(from the reductant)}$$
$$\longrightarrow \quad \text{metal (ligand)}_x^{n\,+} \quad + \; h\nu$$

The redox energy is not sufficient, in general, to produce the first excited singlet state of the Ru-II-complex formed, since the redox energy is 0.0—0.9 V for hydroxyl ion and 1.4—1.5 V for hydrazine acting as reductant, whereas the energy of the first excited singlet state of the Ru-II complex is about 2.23 eV. Therefore, either triplet-triplet annihilation occurs, or certain highly energetic oxidation products of hydrazine are involved in the reaction producing Ru-II-complex molecules in their first excited singlet state. Hydrazyl radicals (N_2H_3.) and diimine ($HN{=}NH$) seem to be possible agents, by analogy with the reaction of hydrazine with Fe-III compounds [164,165]. At low hydrazine concentrations the following mechanism accounts for the kinetics observed [161,166]:

$$\text{Ru(III)} + N_2H_4 \longrightarrow (\text{Ru(II)})^* + N_2H_3.$$
$$2\,N_2H_3. \longrightarrow N_2 + 2\,NH_3$$
$$\text{Ru(III)} + N_2H_3. \longrightarrow HN{=}NH + \text{Ru(II)}$$
$$\text{Ru(III)} + HN{=}NH \longrightarrow (\text{Ru(II)})^*$$
$$(\text{Ru(II)})^* \longrightarrow (\text{Ru(II)}) + h\nu$$

(Ru(II))*: excited ruthenium chelate, ligands omitted as well as in Ru(III).

The reaction of Ru(III) chelate with diimine is about 99 times more efficient than that of Ru(III) with hydrazine. Computer-simulated chemiluminescence time curves based on the kinetic data of the above reaction scheme exactly matched light-intensity time curves recorded in a stopped-flow spectrophotometer [166]. At high hydrazine concentrations, however, only a qualitatively correct result was obtained, which suggests that an unknown reaction is occurring here.

Ruthenium chelate chemiluminescence can also be produced electrochemically [167]: the oxidoreduction of $Ru(2,2'bipyridine)_3Cl_2$, in acetonitrile with tetra-n-butyl ammonium fluoborate as supporting electrolyte yields an orange chemiluminescence in the spectral range 450....800 nm of an intensity visually similar to that of rubrene radical-ion chemiluminescence. The reaction marginally belongs to the group of energy-sufficient systems, as the free enthalpy of 2.6 eV is just sufficient to populate the first excited triplet state, but barely high enough to reach excited singlet states. In the reducing step a Ru(I), and in the oxidating step a Ru(III) chelate ion is produced; in terms of radical anion-radical cation electron transfer, this yields one molecule of excited Ru(II) and one Ru(II) in the ground state.

IX. Firefly Chemiluminescence and Bioluminescence

The bioluminescence of the American firefly (Photinus pyralis) is certainly the best-known bioluminescent reaction, thanks to the work of McElroy and coworkers and E. H. White and his group (for references see [1], p. 138, [6,168,169]). The substrate of this enzyme-catalyzed chemiluminescent oxidation is the benzothiazole derivative 107 (Photinus luciferin) which yields the ketone 109 in a decarboxylation reaction. The concept of a concerted cleavage of a dioxetane derivative has been applied to this reaction [170] (see Section II. C.). Recent experiments with $^{18}O_2$ have challenged this concept, as no ^{18}O-containing carbon dioxide was detected from the oxidation of 107 [171].

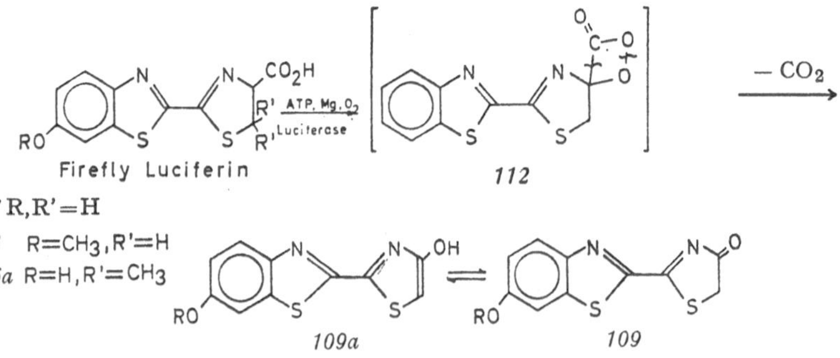

Firefly Luciferin   112

107 R,R'=H
108 R=CH₃,R'=H
116a R=H,R'=CH₃

109a   109

Derivatives of synthetic firefly luciferin [172] and analogs thereof [173] were found to exhibit strong chemiluminescence (i.e. in contrast to bioluminescence, no enzyme was present as catalyst) in anhydrous DMSO/base oxidation:

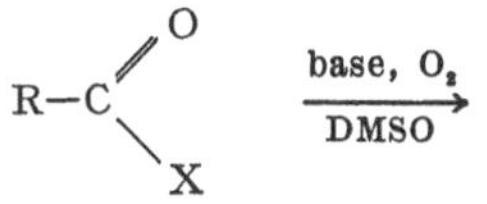

$$R-C{\overset{\displaystyle O}{\underset{\displaystyle X}{\Big\langle}}} \quad \xrightarrow[\text{DMSO}]{\text{base, O}_2}$$

110

red light (at low base concentrations)
yellow-green light (at higher base concentrations)

R:

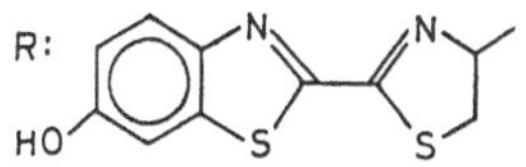

X: OC_6H_5 (phenyl ester)

AMP (mixed anhydride with adenosine mono-phosphate)

The emission is base-dependent in that with low base concentrations ($\leq 0,05$ M) red light (15850/cm; 626 nm) and with higher base concentrations (ca. 0.5 M) yellow-green light (18000/cm; 562 nm) is produced. The thiazolin-4-one *109* has not yet been isolated from the spent reaction mixtures of firefly bioluminescence [174] nor has it in luciferin *107* chemiluminescence [6]. This is certainly due to self-condensation reactions of compounds of the type *109*. However, an analog of firefly luciferin, the dimethyl derivative *116a* which has the active methylene group blocked and which undergoes a chemiluminescent oxidation with the same red emission as *107* (at low base concentrations), has been found to yield the corresponding dimethylthiazolinone as emitter [6]. It is clear that the normal firefly luciferin emission requires ionization of the phenolic hydroxyl group in *109* because methylation of this group to yield O-methyl-luciferin *108* leads to a product whose esters and adenosine-monophosphate mixed anhydride chemiluminesce very weakly.

The dioxetane intermediate *112* should be formed via an oxygen attack on the carbanion formed by deprotonation in α-position of the luciferin carboxyl group. That this proton can be easily removed was proved by deuterium exchange [175] and by very easy racemization of luciferin at C-4 (= the carbon atom bearing carboxyl group in *113*), e.g. during hydrolysis of luciferin adenylate at p_H values below neutrality. The carbanion is assumed to react with oxygen to give the peroxide anion *114*:

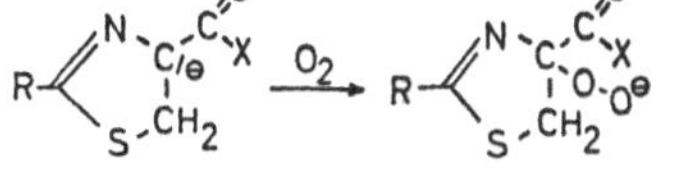

113   *114*

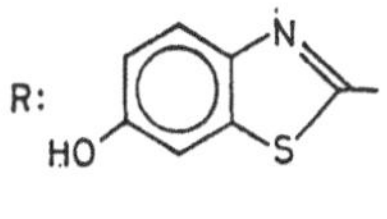

R:

The light yields decrease in the order $X = AMP \sim OC_6H_5 > OCH_3$
$> OH$ [6] corresponding to decreasing carbonyl activity of the $-C\overset{\displaystyle O}{\underset{\displaystyle X}{<}}$
group in 110 in respect of the intramolecular cyclization reaction to give the dioxetane 112.

The yellow-green chemiluminescence of firefly luciferin is evidently dependent on the enol form of the thiazolinone 109a, for 5.5-dimethyl-luciferin 116a which does not yield an enolizable ketone does not exhibit a yellow-greenish emission on addition of excess base; only red emission is observed.

It is therefore suggested that the emitting species are the monoanion 115 in red, and the dianion 117 in yellow-green chemiluminescence.

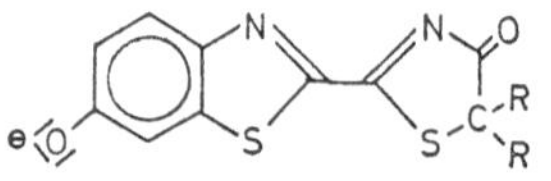

115: R = H
116: R = CH₃

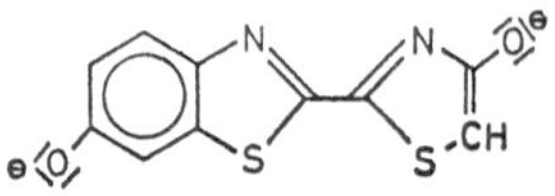

The thiazolinone 117 was synthesized by Goto and coworkers [176]; its fluorescence is yellow-green in basic media.

The dianion 117 bonded to enzyme (firefly luciferase) appears to be the emitter in blue-green firefly bioluminescence as the emission spectrum exactly matches the fluorescence of 117, and in an analogous way in the case of red bioluminescence the emitter is 115.

The chemiluminescence of the firefly luciferin analogue 116a on treatment with oxygen and base in DMSO was independently reported by F. McCapra, Y. C. Chang, and V. P. Francois [177]. These authors also described a negative example of the importance of a good leaving group in the carboxy derivative 110: the N,N'-dicyclohexyl ureide 118 is oxidized to the thiazolinone without chemiluminescence, as the peroxide anion evidently cannot easily replace the urea grouping to give the dioxetane intermediate.

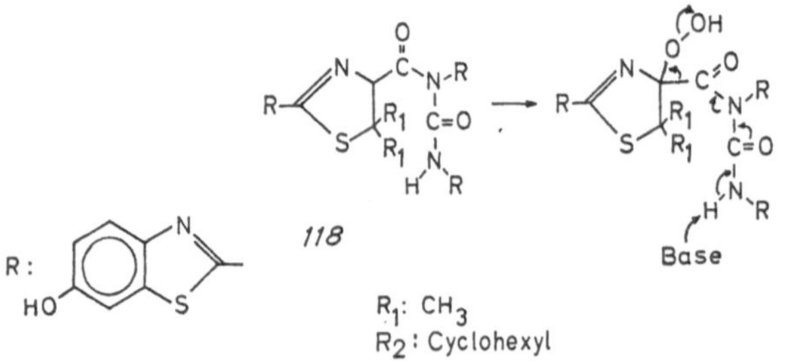

118

R:

HO

R₁: CH₃
R₂: Cyclohexyl

As mentioned in the introductory section, other bioluminescent systems will not be discussed in detail in this review. However, the dioxetane approach has been validated in the bioluminescence mechanism of Cypridina hilgendorfii, Latia, and bacteria [90,178,179,180]. As in Cypridina luciferin *119*, a Schiff's base grouping is apparently involved in the bioluminescent oxidation; the same may be true of Latia neritoides *120*.

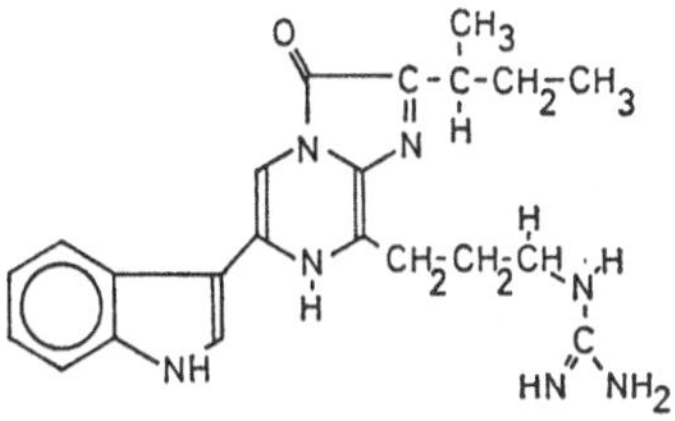

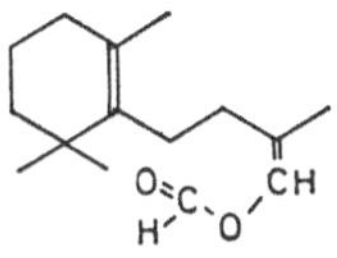

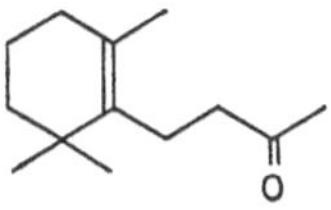

Certain Schiff bases, *i.e.* *122*, were synthesized as model compounds for Latia luciferin. This compound exhibits strong blue chemiluminescence (λ_{max} 385 nm) on oxidation with oxygen in DMSO/potassium t.-butylate, the main products being acetone and 2-formamido pyridine *124*. The mechanism suggested by McCapra and Wrigglesworth includes the concerted bond cleavage of a dioxetane derivative *123*.

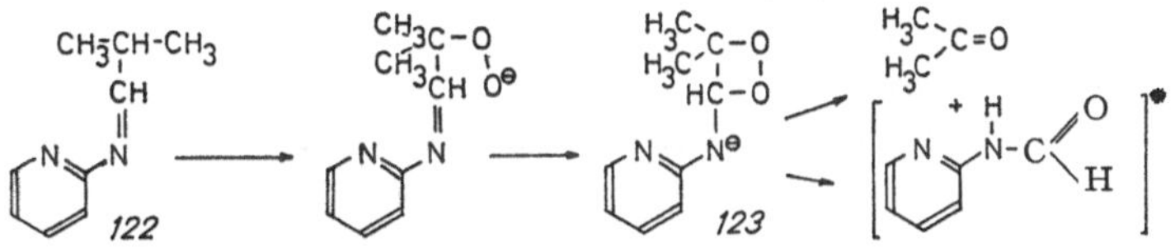

In Latia bioluminescence the ketone *121* was detected as product. Similarly, the chemiluminescence of imidazol-pyridinones *125* represents a model for Cypridina bioluminescence [181].

$R_1 = C_6 H_5$
$R_2, R_3 = CH_3$

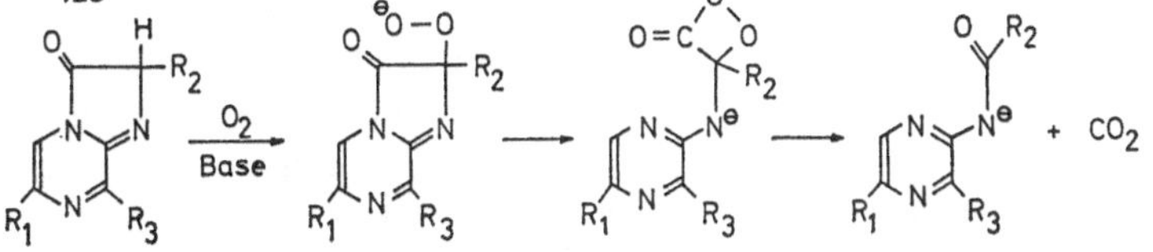

The Schiff base hydroperoxide *126* (which is not related to any bioluminescent system known so far) exhibits weak bluish chemiluminescence on heating in toluene or benzene, benzophenone and benzoyl-*p*-toluidine being the products [104]. As no photoproduct (*e.g.* benzpinacol)

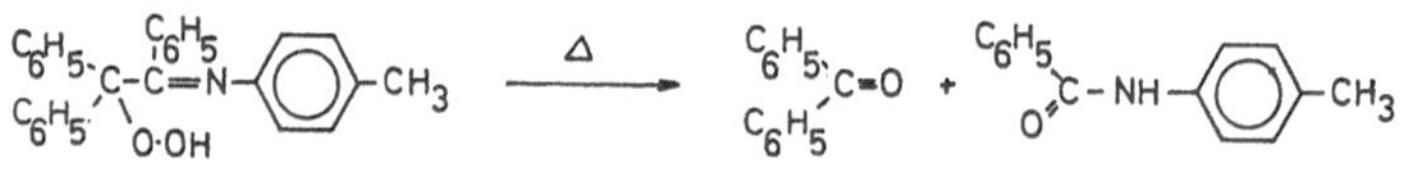

was detected, a heterolytic cleavage via a dioxetane has been regarded as unlikely [104], although one should keep in mind that a base is required for the dioxetane pathway (see above).

Indole chemiluminescence ([1] p. 112, [180,181a]) can also be regarded as a special type of Schiff's base chemiluminescence, for the indole compounds very probably react via a hydroperoxide derived from the respective indolenine form.

X. Photochemistry without Light

Energy is transferred from molecules electronically excited in a chemical reaction to other molecules which emit the accepted excitation energy in the form of light; alternatively the accepting molecules can undergo photochemical transformations. First examples of this "photochemistry without light" were described by E. H. White and coworkers [182]. Thus the trans-stilbene hydrazide *127*, on oxidation, yielded small amounts of the cis- *128* beside the trans-stilbene dicarboxylate in a luminol-type reaction.

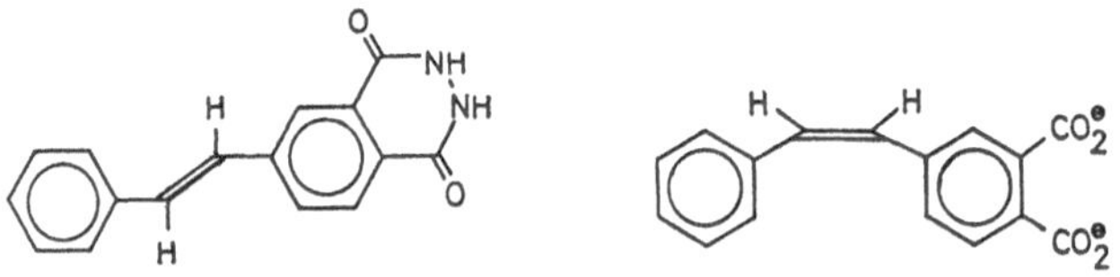

Up to about 10 percent of *cis*-stilbene was obtained when trimethyl-dioxetane *129* was decomposed in the presence of *trans*-stilbene [182]: the electronic excitation energy of the excited carbonyl compounds formed in the cleavage of *129* (see Section V.) was transferred to *trans*-stilbene, so effecting the photochemical *trans-cis* isomerization. When bis (2.4-dinitrophenyl) oxalate reacted with hydrogen peroxide (see Section V.C. in the presence of *o*-tolyl-propane-1.2-dione *130*, 2-methyl-2-

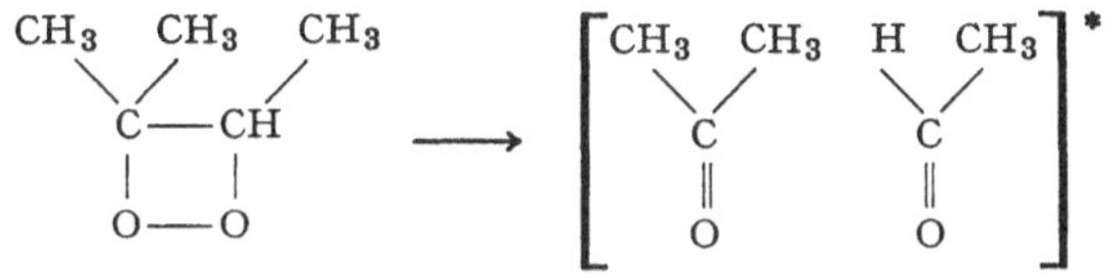

129

hydroxy-indanone *131* was formed in 3 percent yield. Here dioxetane-dione is probably the "photochemical" reagent [183]:

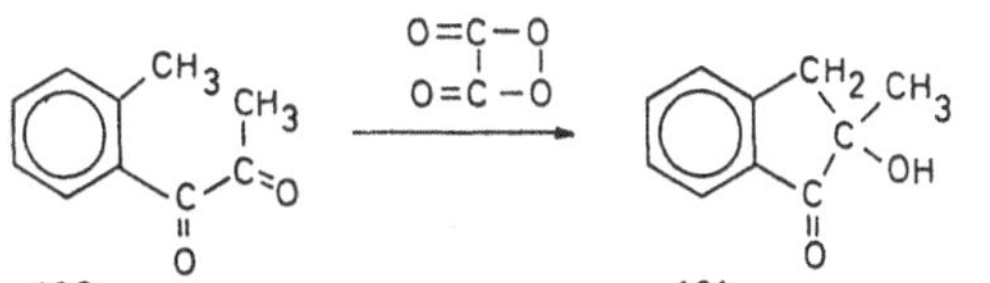

130   *131*

The chemical sensitization effect was 0.006 (calculated from the quantum yield of the photochemical transformation of *130* to *131*, the yield of *131* obtained with the oxalate/hydrogen peroxide reaction, and the moles of oxalate employed). Higher chemical sensitization efficiencies (about 0.04) were observed when the oxalate/hydrogen-peroxide system was used in the addition of ethyl vinyl ether onto phenanthrene quinone

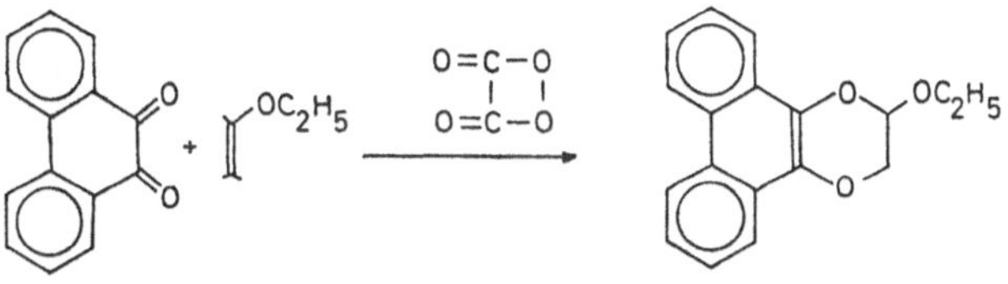

and in the isomerization of trans-4-methoxy-4'nitro-stilbene. In the latter case, part of the excitation energy is emitted as light in the form of the fluorescence of the trans-isomer. No trans-cis isomerization of unsubstituted stilbene could be detected. This was suggested to be due to energetic and perhaps steric reasons [183].

XI. Analytical Applications; Use in Instrumentation

The chemiluminescence of luminol in aqueous alkaline hydrogen-peroxide oxidation is catalyzed by many metal ions (see K. Weber and coworkers [184], A. K. Babko and coworkers [185]; [1] p. 159). This catalytic effect is not only observed down to very small quantities of the metal ions but is also sharply dependent on the valence state of the metal. Both these properties have been used in qualitative and quantitative analytical applications of the luminol reaction, for in the presence of excess reactants the light intensity is proportional to the metal-ion concentration: this is the basis of chemiluminescence methods in trace-metal analysis.

Seitz, Suydam, and Hercules [186] recently developed on the basis of luminol chemiluminescence a method for chromium-III ion determination which has a detection limit of about 0.025 ppb. The method is specific for "free" chromium-III ions as chromium-VI compounds have no catalytic effect and other metal ions can be converted to a non-catalytic form by complexing with EDTA, since the chromium-III complex of EDTA, which is in any case not catalytically active, is formed kinetically slowly [186]. To detect extremely small light emissions, and hence very small metal concentrations, a flow system was used which allows the reactants to be mixed directly in front of a multiplier. (For a detailed description of the apparatus, see [186]).

A spectro-radiometer-luminometer for chemiluminescence and fluorescence quantum yield studies has been described by B. G. Roberts and H. C. Hirt [187]. To obtain emission spectra from very weak chemiluminescence reactions, a large-aperture spectrograph combined with a sensitive image-intensifier tube has been used [68]; this was developed from a device previously constructed by Bass and Kessler [188]. With it, it was possible to record the very weak emission of singlet oxygen dimer $(^1\Sigma gO_2)_2^*$ [68].

R. Bezman and L. R. Faulkner [189] developed methods for defining a concise set of parameters which quantitatively describe the efficiencies of chemiluminescent electron-transfer reactions (see Section VIII. A.) by means of analysis of chemiluminescence decay curves.

Luminol amidine *132*, synthesized from luminol and the Vilsmeier reagent from DMF and thionyl chloride, has been proposed as a suitable luminol derivative for analytical purposes because, unlike luminol, it can be easily purified by recrystallization from water. *132* exhibits a chemiluminescence quantum yield of about 20% of luminol in ferricyanide-catalyzed oxidation by aqueous alkaline hydrogen peroxide; λ_{max} of the emission is 452 nm [196].

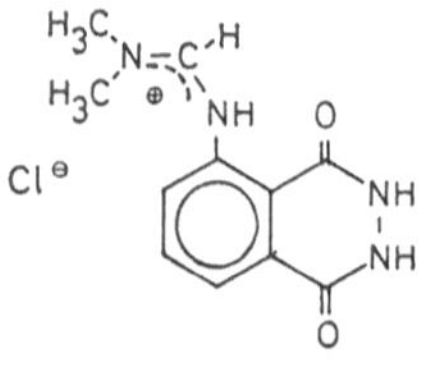

132

Chemiluminescence can be a nuisance in liquid scintillation counting. This was shown by D. A. Kalbhen [191] with reference to solutions containing the widely used hyamine 10-X (*p*-(di-isobutylcresoxyethoxyethyl)-dimethyl-benzylammonium chloride) as solubilizing agent for biological materials. In alkaline media hyamine undergoes chemiluminescent oxidation by oxygen or peroxides (very difficult to exclude totally from scintillation mixtures) and so interferes with the light emission by scintillation. This chemiluminescence can be suppressed by working in neutral or slightly acid hyamine solutions.

XII. Note Added in Proof

After the manuscript of this article was completed (summer 1972) a considerable number of papers in the rapidly developing field has been published a large part of which was presented at the International Conference on Chemiluminescence held at the University of Georgia, Athens, October 10—13, (see M. J. Cormier, D. M. Hercules, and J. Lee [197]).

J. Stauff and U. Bergmann [196] described a blue and a yellow greenish chemiluminescence occurring on oxidation of oxalic acid with permanganate; this chemiluminescence is ascribed to HCO_3. and $CO_3^{(-)}$. radicals formed from singlet oxygen and hydrogen carbonate or carbonate anion, respectively. The carbonate or hydrogen carbonate radicals are assumed to form di- or monoperoxy oxalate the decomposition of which yields light. These compounds are also proposed as precursors of 1.2-doxetane dione (see also M. M. Rauhut [2]; M. M. Rauhut, B. G. Roberts, and A. M. Semsel [198]; J. Stauff, U. Sander, and W. Jaeschke [199]).

In the field of the dioxetanes not only a series of new compounds was synthesized and investigations of their thermal stability were performed — a particularly stable compound being the bis adamantyl derivative

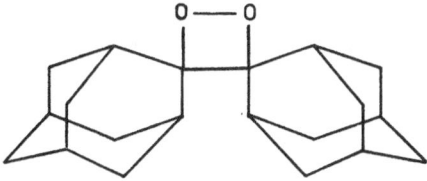

having a melting point of 163 °C; visible light is emitted at 200 °C — but also the question has been discussed whether a concerted cleavage of the dioxetane ring (see p. 68) or a stepwise cleavage via a diradical is affording the excited carbonyl compounds being the primary chemi-luminescing species [201].

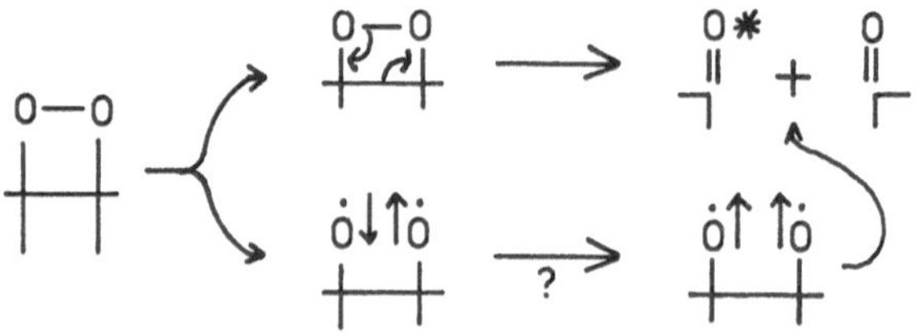

A special problem is the high yield of triplet carbonyl compounds being formed — neither the concerted nor the diradical mechanism are fully explaining this fact. Further data on the identities and yields of excited products from different dioxetanes are needed.

As mentioned earlier (see p. 122) the previously postulated dioxetane intermediate in firefly bioluminescence has been challenged as no ^{18}O is in-corporated in the carbon dioxide released during oxidation of firefly luciferin with $^{18}O_2$. In view of the crucial significance of the ^{18}O. experiments De Luca and Dempsey [202] rigorously examined the relia-bility of their tracer method. They conclude from their experiments that all available evidence is in favour of a linear, not a cyclic peroxide inter-mediate — in contrast to Cypridina bioluminescence where at least part of the reaction proceeds via a cyclic peroxide (dioxetane) as con-cluded from the incorporation of ^{18}O into the carbon dioxide evolv-ed [202,203]. However, the dioxetane intermediate is not absolutely excluded as there is the possibility of a non-chemiluminescent hydro-lytic cleavage of the four-membered ring [204].

The synthesis of α-peroxy-lactones has been intensively studied and remarkably improved by W. Adam and coworkers [205]. These compounds are very appropriate models for the bioluminescence of firefly, cypridina and other organisms.

Tetrachloro ethylene carbonate *134*, on treatment with hydrogen peroxide and base, was found to exhibit very strong chemiluminescence with quantum yields of 6.5% under optimum conditions. (D. R. Maulding

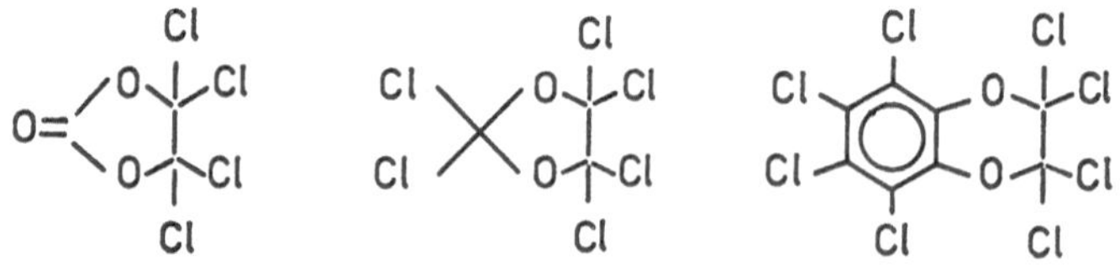

134	*135*	*136*

and B. G. Roberts [217]).

The cyclic peroxide *137* is proposed as intermediate. Similar high efficiencies are observed with the compounds *135* and *136*.

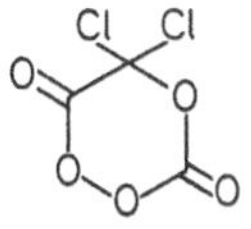

137

The observation of Bersis and Nikokavouras [108] that if a rapid stream of carbon dioxide is passed through a basic solution of luminol containing sodium chloride, hydrogen peroxide, and manganous chloride the intensity of light passes through four maxima has been confirmed by E. H. White and R. B. Brundrett [206]. These four maxima are, however, not considered to be due to special species of luminol but to four different catalytically active forms of the manganese differing in their coordination spheres.

P. Wildes and E. H. White [207] investigated the differences between excited states produced chemically (by oxidation of luminol in basic aqueous dimethyl sulfoxide) and photochemically (by absorption of light) of aminophthalate dianion. The cause for these differences is assumed to be specific interactions of the aminophthalate ions with alkali metal ions: the fluorescence spectrum of aminophthalate ion is determined largely by the degree of association in the ground state prior to excitation; in chemiluminescence the association between sodium ions and amino phthalate ions is determined largely at the transition state of the reaction. By this ion pair formation the photoisomerization of the $AP^{2(-)}$ ion to yield $^{(-)}AP^{(-)}$ (which spectroscopically results in the red shift of the aminophthalate ion from 400—425 nm to ca. 510 nm) is inhibited by the ion pair formation because of the resulting lower net charge on the o-carboxylate group:

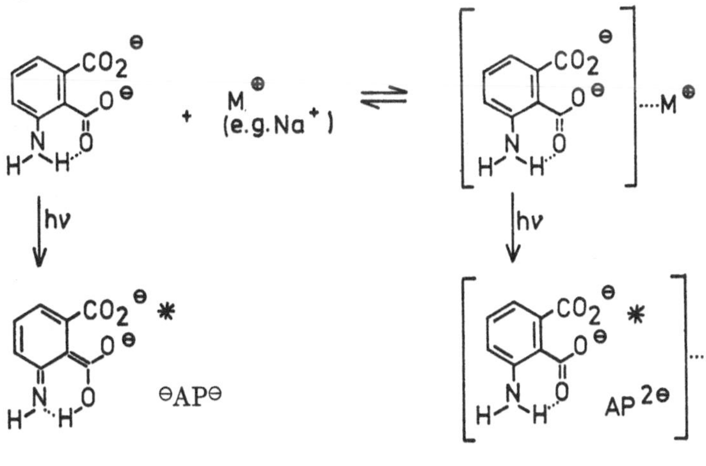

therefore in DMSO-water mixtures (*e.g.* 30 mol % water-70 mol % DMSO) where aminophthalate dianion normally exhibits both fluorescence maxima (425 and 510 nm) increasing concentrations of sodium ions produce a decrease of the longer wavelength emission in favour of the shorter wavelength emission. This effect is also dependent on the metal ions present in the reaction mixture sodium ions being about twice as effective as potassium ions, for example. In luminol chemiluminescence with DMSO-water mixtures as solvent the spectroscopic shifts reported above occur similarly but require far higher concentrations of alkali ion.

Concerning the mechanism of the chemiluminescent reaction between luminol and molecular oxygen in DMSO M. T. Beck and F. Joo [209] performed kinetic experiments leading them to the conclusion that the formation of an oxygen containing intermediate (see p. 109) is a reversible step, in contrast to the opinion of E. H. White and M. M. Rauhut, D. M. Hercules and their coworkers, and others [2,3,109].

The first chemiluminescent paracyclophanes have been described recently [208]: the compounds *138* and *139* both contain a phthalhydrazide group as that part of the molecule producing the excitation energy which is transferred to the substituted benzene resp. anthracene moiety. *139* chemiluminesces with about double the amount of 2.3-anthracene dicarboxylic hydrazide on oxidation by oxygen/potassium tert. butoxide in DMSO.

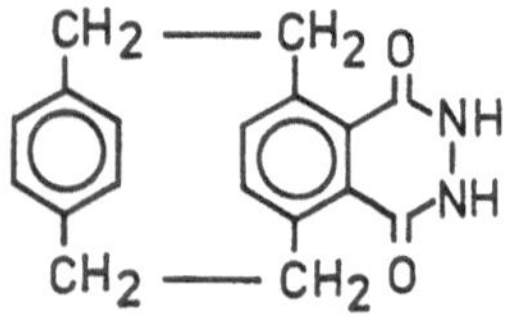

138	139

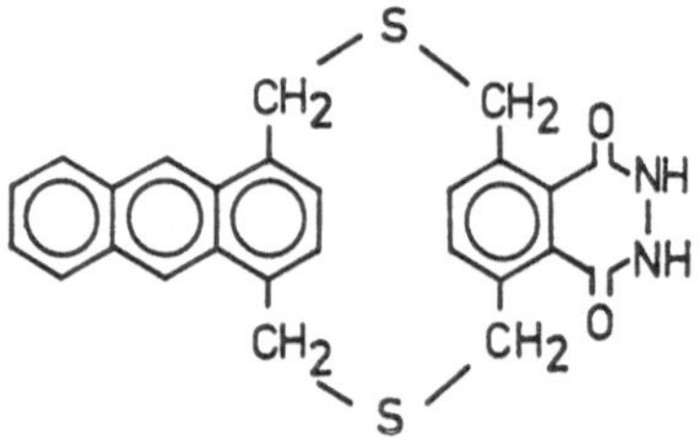

A general theory of the aromatic hydrocarbon radical cation and anion annihilation reactions has been forwarded by G. J. Hoytink [210] which in particular deals with a resonance or a non-resonance electron transfer mechanism leading to excited singlet or triplet states. The radical ion chemiluminescence reactions of naphthalene, anthracene, and tetracene are used as examples.

R. Bezman and L. R. Faulkner discussed theoretical and practical aspects for measurements of the efficiencies of chemiluminescent electron transfer reactions [189]. They also performed absolute measurements of the chemiluminescence of the fluoranthene-10-methyl-phenothiazine system [211].

The most simple case of a electron-transfer chemiluminescence has been realized recently by the reaction of hydrated electrons with tri-(bipyridyl)ruthenium(III) (J. E. Martin, E. J. Hart, A. W. Adamson, H. Gafney, and J. Halpern [212]):

$$
\text{Ru(bipy)}_3{}^{3+} + e_{aq} \Big\langle \begin{array}{l} (\text{Ru(bipy)}_3{}^{2+})^x \longrightarrow \text{Ru(bipy)}_3{}^{2+} + h\nu \\[6pt] (\text{transition state}) \longrightarrow \text{Ru(bipy)}_3{}^{2+} \end{array}
$$

The hydrated electrons were produced in a matrix consisting of triply distilled water, 0.2 M in tert. butyl alcohol (for rapid scavenging of OH-radicals), and acetate buffer, by 4—40 nsec single pulses of 15 MeV electrons.

The mechanism of the chemiluminescent reactions between alkyl halides and electrogenerated aromatic hydrocarbon radicals (cf. p. 119) has been elucidated in more detail [213]. The proposed general mechanism is consistent with the observed experimental results:

$$
\begin{array}{lll}
(1) & R + e^{(-)} & \rightleftharpoons R^{\overline{\cdot}} \\[4pt]
(2) & R^{\overline{\cdot}} + AX_2 & \rightleftharpoons AX_2{}^{\overline{\cdot}} + R \\[4pt]
(3) & AX_2{}^{\overline{\cdot}} & \longrightarrow AX\cdot + X^{(-)} \\[4pt]
(4) & R^{\overline{\cdot}} + AX\cdot & \rightleftharpoons {}^1R^x + AX^{(-)} \\[4pt]
(5) & R + AX\cdot & \rightleftharpoons R^+_{\cdot} + AX^{(-)} \\[4pt]
(6) & AX^{(-)} & \longrightarrow A + X^{(-)} \\[4pt]
(7) & R^{\overline{\cdot}} + R^+_{\cdot} & \longrightarrow {}^1R^x + R \\[4pt]
(8) & {}^1R^x & \longrightarrow R + h\nu
\end{array}
$$

(R = aromatic hydrocarbon, AX$_2$ = alkyl halide).

At least experiments with fluoranthene and other aromatic hydro-carbons having lower triplet energies than fluoranthene resulted only in fluoranthene singlet emission- this excited singlet appears to be generated directly and not by triplet-triplet annihilation.

Hetero-excimer chemiluminescence yields were measured by A. Weller and K. Zachariasse [214]; the system dimethylanthracene anion radical/tri-p-tolylaminium perchlorate in tetrahydrofurane exhibits particularly strong chemiluminescence with quantum yields of about 7.5×10^{-2} [215]. A. J. Bard and coworkers [216] very thoroughly investigated the influence of several parameters, e.g. supporting electrolyte concentration, on the efficiency of electrogenerated chemiluminescence.

K.-D. Gundermann

XIII. References

1) Gundermann, K.-D.: Chemilumineszenz organischer Verbindungen. Berlin–Heidelberg–New York: Springer 1968.
2) White, E. H., Roswell, D. F.: Accounts Chem. Res. *3*, 54 (1970).
3) Rauhut, M. M.: Accounts Chem. Res. *2*, 80 (1969).
4) Hercules, D. M.: Accounts Chem. Res. *2*, 301 (1969).
5) Hercules, D. M.: In: A. Weissberger and B. Ross (ed.), Physical methods in chemistry, Vol. 1, B, p. 257. New York: J. Wiley 1971.
6) White, E. H., Rapaport, E., Seliger, H. H., Hopkins, T. A.: Bioorganic Chem. *1*, 92 (1971).
7) Evans, M. G., Eyring, E., Kincaid, J. F.: J. Chem. Phys. *6*, 349 (1938).
8) Johnson, F. H., Eyring, H., Polissar, M. J.: The kinetic Basis of biology. New York: J. Wiley 1954.
9) Mayer, J.: In: F. H. Johnson, The luminescence of biological systems. Washington: American Association for the Advancement of Science 1955.
10) Marcus, R. A.: J. Chem. Phys. *43*, 2654 (1965).
11) McCapra, F.: Chem. Commun. *1968*, 155.
12) Brundrett, R. B., Roswell, D. F., White, E. H.: J. Am. Chem. Soc., in press.
13) Beutel, J.: J. Am. Chem. Soc. *93*, 2615 (1971).
14) Vasil'ev, R. F.: Progr. Reaction Kinetics *4*, 305 (1967).
15) Visco, R. E., Chandross, E. A.: J. Am. Chem. Soc. *86*, 5350 (1964).
16) Hercules, D. M., Lansbury, R. C., Roe, D. K.: J. Am. Chem. Soc. *88*, 4578 (1966).
17) Weller, A., Zachariasse, K.: J. Chem. Phys. *46*, 4984 (1967).
18) Parker, C. A., Short, G. D.: Trans. Faraday Soc. *63*, 2618 (1967).
19) Chang, J., Werner, T. C., Hercules, D. M.: Abstracts 155th National Meeting of the American Chemical Society, San Francisco 1968, No. R 42.
20) Bard, A. J., Santhanam, K. S. V., Cruser, S. A., Faulkner, L. R.: In: G. Guilbault, Fluorescence, S. 627 (Kap. 14). New York: M. Dekker 1967.
21) Maricle, D. L., Maurer, A.: J. Am. Chem. Soc. *89*, 188 (1967).
22) Hoffmann, R., Woodward, R. B.: Accounts Chem. Res. *1*, 17 (1968); see also: Nguyen Trong Anh, Die Woodward-Hoffmann-Regeln und ihre Anwendung. Weinheim: Verlag Chemie 1972.
23) White, E. H., Harding, M. J. C.: Harding, Photochem. Photobiol. *4*, 1129 (1965).
24) Rauhut, M. M., Bollyky, L. J., Roberts, B. G., Loy, M., Whitman, R. H., Iannotta, A. V., Semsel, A. M., Clarke, R. A.: J. Am. Chem. Soc. *89*, 6515 (1967).
25) McCapra, F., Richardson, D. G., Chang, Y. C.: Photochem. Photobiol. *4*, 1111 (1965).
26) Rauhut, M. M., McCapra, F., Sheehan, D., Clarke, R. A.: Quart. Reviews *20*, 485 (1966).
27) Roberts, B. G., Semsel, A. M.: J. Org. Chem. *30*, 3587 (1965).
28) McCapra, F., Chang, Y. C.: Chem. Commun. *1966*, 522.
29) Rauhut, M. M.: American Chemical Society, 153rd Meeting Abstracts 1967, 0—169; recent review: see 3).
30) Trozzolo, A. M., Murray, R. W., Wasserman, E.: J. Am. Chem. Soc. *84*, 4991 (1962).
31) Fletcher, A. N., Heller, C. A.: J. Phys. Chem. *71*, 1507 (1967).
32) Urry, W. H., Sheeto, J.: Photochem. Photobiol. *4*, 1067 (1965).
33) Totter, J. R., Philbrook, G. E.: Photochem. Photobiol. *5*, 177 (1965).

34) Turro, N. J., Letchken, P.: J. Am. Chem. Soc. *94*, 2886 (1972).
35) Shlyapintokh, V. Ya., Karpukhin, O. N., Postnikov, L. M., Tsepalov, V. F., Vichutinskii, A. A., Zakharov, I. V.: Chemiluminescence Techniques in chemical reactions. New York: Consultants Bureau 1968.
35a) Ingold, K. U.: Accounts Chem. Res. *2*, 1 (1969).
36) Hine, J.: Physical organic chemistry, p. 34. New York: McGraw Hill, 1956.
37) Vasil'ev, R. F., Rusina, I.: Dokl. Akad. Nauk SSSR *153*, 1101 (1963).
38) Kellogg, R. E.: J. Am. Chem. Soc. *91*, 5433 (1969).
39) Russell, G. A.: Am. Chem. Soc. *79*, 3871 (1957).
40) Howard, J. A., Ingold, K. V.: Can. J. Chem. *43*, 2737 (1965).
41) Howard, J. A., Ingold, K. V.: J. Am. Chem. Soc. *90*, 1058 (1968).
42) Wagner, P. J., Kochevar, I.: J. Am. Chem. Soc. *90*, 2232 (1968).
43) Howard, J. A., Ingold, K. V.: J. Am. Chem. Soc. *90*, 1058 (1968).
44) Howard, J. A., Schwalin, W. J., Ingold, K. V.: Advan. Chem., Series No. *75*, 6 (1968).
45) Howard, J. A., Ingold, K. V., Symonds, M.: Can. Chem. J. *46*, 1017 (1968).
46) Middleton, B. S., Ingold, K. V.: Can. Chem. J. *45*, 191 (1967).
47) Tsepalov, V. F., Shlyapintokh, V. Ya.: Kinetika i Kataliz *3*, 870 (1962).
48) Bateman, L., Gee, G., Norris, A. L., Watson, W. F.: Discussions Faraday Soc. *10*, 250 (1951).
49) Boland, J. L., Gee, G.: Trans. Faraday Soc. *42*, 236 (1946).
50) Ingles, T. A., Melville, H. W.: Proc. Roy. Soc. (London), Ser. A, *218*, 163 (1953).
51) Tsobumura, H., Mulliken, R. S.: J. Am. Chem. Soc. *82*, 5966 (1960).
52) Hartley, D. B.: Chem. Commun. *1967*, 1281.
53) Napier, I. M., Norrish, R. G. W.: Proc. Roy. Soc. (London) *299*, 337 (1967).
54) Phillips, D., Anissimov, V., Kharpukhin, O., Shlyapintokh, V. Ya.: Photochem. Photobiol. *9*, 183 (1969).
55) Anissimov, V. M., Phillips, D., Kharpukhin, O. N., Shlyapintokh, V. Ya.: Izv. Akad. Nauk SSSR, Ser. Khim. *7*, 1529 (1970).
56) Förster, T.: Discussions Faraday Soc. *27*, 7 (1959).
57) Wedekind, E.: Z. Wiss. Phot. *5*, 29 (1907).
58) Evans, W. V., Diepenhorst, E. M.: J. Am. Chem. Soc. *48*, 715 (1926).
59) Bardsley, R. L., Hercules, D. M.: J. Am. Chem. Soc. *90*, 4545 (1968).
60) Stauff, J., Nimmerfall, F.: Z. Naturforsch. *24b*, 852, 1009 (1969); see also: [1] p. 9ff.
61) Vorhaben, J. E., Steele, R. H.: Biochemistry *6*, 1404 (1967), and previous papers referred there.
62) Lundeen, G., Livingston, R.: Photochem. Photobiol. *4*, 1085, (1965); see also [75a].
63) Steenken, St.: Photochem. Photobiol. *11*, 279 (1970).
64) Gundermann, K.-D., Steinfatt, M., Fiege, H.: Angew. Chem. *83*, 43 (1971).
65) Russel, K. E.: J. Am. Chem. Soc. *77*, 4814 (1955).
66) Greene, F. D.: J. Am. Chem. Soc. *78*, 2246 (1956).
67) Kurtz, R. B.: Ann. N. Y. Acad. Sci. *16*, 399 (1954).
68) Ness, S., Hercules, D. M.: Anal. Chem. *41*, 1467 (1969).
69) Abbot, S. R., Ness, S., Hercules, D. M.: J. Am. Chem. Soc. *92*, 1128 (1970).
70) Hercules, D. M., Chang, J.: unpublished results.
71) Wasserman, H. H., Scheffer, J. R.: J. Am. Chem. Soc. *89*, 3073 (1967).
72) Wilson, T.: Photochem. Photobiol. *10*, 441 (1969).
73) Rigaudy, J.: Pure Appl. Chem. *16*, 169 (1968).

74) Rigaudy, J., Deletang, C., Spartel, D., Cuong, N. K.: Compt. Rend. *267*, 1714 (1968).

75) Baldwin, J. E., Basson, H. H., Krauss, H., Jr.: Chem. Commun. *1968*, 984.

75a) Lundeen, G. W., Adelman, A. H.: J. Am. Chem. Soc. *92*, 3914 (1970).

76) Bowen, E. J., Lloyd, R. A.: Proc. Chem. Soc. (London) *1963*, 305.

77) Slawinski, J.: Photochem. Photobiol. *13*, 489 (1971).

78) Kopecky, K. R., Mumford, C.: Abstracts 51th. Annual Conference of the Chemical Institute of Canada, Vancouver 1968, p. 41; Can. J. Chem. *47*, 709 (1969).

79) Bartlett, P. D., Schaap, A. P.: J. Am. Chem. Soc. *92*, 3223 (1970).

80) Mazur, S. C., Foote, C. S.: J. Am. Chem. Soc. *92*, 3235 (1970).

81) Bollyky, L. J.: J. Am. Chem. Soc. *92*, 3230 (1970).

82) Wildes, P. D., White, E. H.: J. Am. Chem. Soc. *93*, 6286 (1971).

83) Filespescu, N., Mushrush, G. W.: J. Phys. Chem. *72*, 3516 (1968).

84) Wilson, T., Schaap, A. P.: J. Am. Chem. Soc. *93*, 4126 (1971).

85) Cordes, H. F., Richter, H. P., Heller, C. A.: J. Am. Chem. Soc. *91*, 7209 (1969).

86) De Corpo, J. J., Barovski, A., McDowell, M. V., Saalfeld, F. E.: J. Am. Chem. Soc. *94*, 2879 (1972).

87) Murray, R. W. Kaplan, M. L.: J. Am. Chem. Soc. *90*, 537 (1968); *91*, 5358.

88) Hoffmann, R.: J. Chem. Phys. *39*, 1397 (1963).

89) Adam, W., Liv, J. C.: J. Am. Chem. Soc. *94*, 2894 (1972).

90) McCapra, F., in: J. N. Bradley, R. D. Gillard, R. F. Hudson (eds.): Essays in chemistry, Vol. 3, p. 101. New York: Academic Press 1972. This very condensed article also deals with other chemiluminescent systems.

91) Lee, J., Seliger, H. H.: Photochem. Photobiol. *11*, 247 (1970).

92) Melhuish, W. H.: J. Phys. Chem. *65*, 229 (1961).

93) McCapra, F.: Pure Appl. Chem. *24*, 611 (1970).

94) Roswell, D. F.: Personal communication.

95) Klockenbring, G.: Dissertation, Techn. Univ. Clausthal 1969.

96) Wei, C. C., White, E. H.: Tetrahedron Letters *1971*, 3559.

97) Gundermann, K.-D., Horstmann, W., Bergmann, G.: Liebigs Ann. Chem. *684*, 127 (1965).

98) Lathia, D.: Dissertation, Techn. Univ. Clausthal 1969.

99) Gundermann, K.-D., Nolte, W.: unpublished results.

100) Gundermann, K.-D.: Chimia *25*, 261 (1971). — Spengler, H.: Dissertation, Techn. Univ. Clausthal 1969.

101) Gundermann, K.-D., Fiege, H., Klockenbring, G.: Liebigs Ann. Chem. *738*, 140 (1970); *743*, 200 (1971).

102) Gundermann, K.-D., Böttcher, C.: unpublished results. — Böttcher, C.: Dissertation, Techn. Univ. Clausthal 1972.

103) Spengler, H.: Dissertation, Techn. Univ. Clausthal 1971.

104) Rapaport, E., Cass, M. W., White, E. H.: J. Am. Chem. Soc. *94*, 3153 (1972).

105) Drew, H. D. K., Garwood, R. F.: J. Chem. Soc. *1939*, 836.

106) White, E. H., Bursey, M. M.: J. Am. Chem. Soc. *86*, 941 (1964).

107) White, E. H., Nash, E. G., Roberts, D. R., Zafiriou, O. C.: J. Am. Chem. Soc. *90*, 5932 (1968).

108) Bersis, D. S., Nikokavouras, J.: Nature (London) *217*, 451 (1968).

109) Gorsuch, J. D., Hercules, D. M.: Photochem. Photobiol. *15*, 567 (1972).

110) Omote, Y., Yamamoto, H., Sugiyama, M.: Chem. Commun. *1970*, 914.

111) Lee, J., Seliger, H. H.: Photochem. Photobiol. *11*, 247 (1970).

112) Förster, T.: Z. Elektrochem. *54*, 531 (1950).

113) Van Duuren, B. L.: Chem. Rev. *63*, 325 (1963).

114) White, E. H., Bursey, M. M.: J. Org. Chem. *31*, 1912 (1966).

115) Lee, J., Seliger, H. H.: Photochem. Photobiol. *15*, 227 (1972).
116) White, E. H.: In: W. D. McElroy and B. Glass, A Symposium on Light and Life, p. 183. Baltimore 1961.
117) White, E. H., Zafiriou, O. C., Kaegi, H. H., Hill, J. H. M.: J. Am. Chem. Soc. *86*, 940 (1964).
118) Matheson, I. B. C., Lee, J.: Photochem. Photobiol. *12*, 9 (1970).
119) Baxendale, J. H.: Chem. Commun. *1971*, 1489.
120) Rapaport, E., Cass, M. W., White, E. H.: J. Am. Chem. Soc. *94*, 3153 (1972).
121) Drew, H. D. K., Garwood, R. F.: J. Chem. Soc. *1939*, 836.
122) Cormier, M. J., Prichard, P. M.: J. Biol. Chem. *243*, 4706 (1968).
123) White, E. H., Roswell, D. F.: J. Am. Chem. Soc. *89*, 3944 (1967).
124) White, E. H., Roswell, D. F., Zafiriou, O. C.: J. Org. Chem. *34*, 2462 (1969).
125) McCapra, F., Hann, R. A.: Chem. Commun. *1969*, 442.
126) Roswell, D. F., Paul, V., White, E. H.: J. Am. Chem. Soc. *92*, 4855 (1970).
127) Roberts, D. R., White, E. H.: J. Am. Chem. Soc. *92*, 4861 (1970).
128) Albrecht, H. O.: J. Physik. Chem. *136*, 321 (1928).
129) Rauhut, M. M., Semsel, A. M., Roberts, B. G.: J. Org. Chem. *31*, 2431 (1966).
130) Eicke, H. F., Fiege, H., Gundermann, K.-D.: Z. Naturforsch. *25b*, 484 (1970).
131) Gundermann, K.-D., Unger, H.: unpublished results.
132) Kautsky, H., Kaiser, K. H.: Z. Naturforsch. *5b*, 353 (1950).
133) Omote, Y., Miyake, T., Sugiyama, N.: Bull. Chem. Soc. Japan *40*, 2446 (1967).
134) Gundermann, K.-D., Scholz, G.: unpublished results. — Scholz, G.: Dissertation, Techn. Univ. Clausthal 1971.
135) Totter, J. R.: Photochem. Photobiol. *3*, 231 (1964).
136) Janzen, E. G., Pickett, J. B., Happ, J. W., de Angelis, W.: J. Org. Chem. *35*, 88 (1970).
137) Happ, J. W., Janzen, E. G.: J. Org. Chem. *35*, 96 (1970).
138) Happ, J. W., Janzen, E. G., Rudy, B. C.: J. Org. Chem. *35*, 3382 (1970).
139) McCapra, F., Richardson, D. G.: Tetrahedron Letters *1964*, 3167.
140) Rauhut, M. M., Sheehan, D., Clarke, R. A., Semsel, A. M.: Photochem. Photobiol. *4*, 1097 (1965).
141) Cass, M. W., Rapaport, E., White, E. H.: J. Am. Chem. Soc. *94*, 3168 (1972).
142) Rapaport, E., Cass, M. W., White, E. H.: J. Am. Chem. Soc. *94*, 3160 (1972).
143) Hercules, D. M.: J. Phys. Chem. *74*, 2114 (1970).
144) Watne, B. M.: Thesis, Massachusetts Institute of Technology, Cambridge, Mass. 1967.
145) Maloy, J. T., Prater, K. B., Bard, A. V.: J. Am. Chem. Soc. *93*, 5959 (1971); J. Phys. Chem. *72*, 4348 (1968).
146) Maloy, J. T., Bard, A. V.: J. Am. Chem. Soc. *93*, 5968 (1971).
147) Zweig, A., Maurer, A. H., Roberts, B. G.: J. Org. Chem. *32*, 1322 (1967).
148) Faulkner, L. R., Bard, A. J.: J. Am. Chem. Soc. *91*, 209 (1969).
149) Hoytink, J.: Discussions Faraday Soc. *45*, 14 (1968).
150) Freed, D. J., Faulkner, L. R.: J. Am. Chem. Soc. *93*, 2097 (1971).
151) Chang, J., Hercules, D. M., Roe, D. K.: Electrochim. Acta *13*, 1197 (1969).
152) Faulkner, L. R., Bard, A. J.: J. Am. Chem. Soc. *91*, 6495, 9497 (1969).
153) Siegel, T. M., Mark, H. B.: J. Am. Chem. Soc. *93*, 6281 (1971).
154) Cruser, S. A., Bard, A. J.: J. Am. Chem. Soc. *91*, 267 (1969).
155) Haas, J. W., Baird, J. E.: Nature (London) *214*, 1006 (1966).
156) Keszthelyi, C. P., Tachikawa, H., Bard, A. J.: J. Am. Chem. Soc. *94*, 1522 (1972).
157) Weller, A., Zachariasse, K.: J. Chem. Phys. *46*, 4984 (1967).

158) Weller, A., Zachariasse, K. in: E. C. Lim, Molecular luminescence, p. 895. New York: Benjamin 1969.

159) Weller, A., Zachariasse, K.: Chem. Phys. Letters *10*, 197 (1971).

160) Zachariasse, Dissertation, Free University of Amsterdam 1972.

161) Lyttle, F. E., Hercules, D. M.: Photochem. Photobiol. *13*, 123 (1971).

162) Hammond, G. S., Saltiel, J., Lamola, A. A., Turro, N. J., Bradshaw, J. S., Cowan, D. O., Counsell, R. C., Vogt, V., Dalton, C.: J. Am. Chem. Soc. *86*, 3197 (1964).

163) Porter, G., Wilkinson, F.: Proc. Roy. Soc. (London), Ser. *A*, *264*, 1 (1961).

164) Higginson, W. C. E., Wright, P.: J. Chem. Soc. *1957*, 4685.

165) Rosinsky, D. R.: J. Chem. Soc. *1957*, 4685.

166) Hercules, D. M. in: E. D. Bransome (ed.), Current status of liquid scintillation counting, in press.

167) Tokel, N. E., Bard, A. J.: J. Am. Chem. Soc. *94*, 2862 (1972).

168) Johnson, F. H., Haneda, Y.: Bioluminescence in progress. Princeton: University Press 1966.

169) Goto, T., Kishi, Y.: Angew. Chem. *80*, 417 (1968).

170) Hopkins, T. A., Seliger, H. H., White, E. H., Cass, M. W.: J. Am. Chem. Soc. *89*, 7148 (1967).

171) De Luca, M., Dempsey, M. E.: Biochem. Biophys. Res. Commun. *40*, 117 (1970).

172) White, E. H., McCapra, F., Field, G. F.: J. Am. Chem. Soc. *85*, 337 (1963).

173) White, E. H., Worther, H., Field, G. F., McElroy, W. D.: J. Am. Chem. Soc. *88*, 2015 (1966) and previous papers cited there.

174) Plant, P. J., White, E. H., McElroy, W. D.: Biochem. Biophys. Res. Commun. *31*, 98 (1968).

175) White, E. H., Worther, H., Field, G. F., McElroy, W. D.: J. Org. Chem. *30*, 2344 (1965).

176) Suzuki, N., Sato, M., Nishikawa, K., Goto, T.: Tetrahedron Letters *53*, 4683 (1969).

177) McCapra, F., Chang, Y. C., Francois, V. P.: Chem. Commun. *1968*, 22.

178) McCapra, F., Wrigglesworth, R.: Chem. Commun. *1969*, 91.

179) White, E. H., Wei, C. C.: Biochem. Biophys. Res. Commun. *39*, 1219 (1970).

180) Omote, Y.: J. Soc. Org. Synthet. Chem. *29*, 1085 (1971).

181) McCapra, F., Wrigglesworth, R.: Chem. Commun. *1968*, 1256.

181a) Omote, Y., Yamamoto, H., Funasaki, K., Akutagawa, M., Sugiyama, N.: Bull. Chem. Soc. Japan *42*, 3014 (1969).

182) White, E. H., Wielko, J., Roswell, D. F.: J. Am. Chem. Soc. *91*, 5194 (1969).

183) Güsten, H., Ullman, E. F.: Chem. Commun. *1970*, 28.

184) Weber, K., Matkovich, J.: Arch. Toxikologie *21*, 38 (1965); and previous papers.

185) Babko, A. K., Kalinichonko, H. L.: Ukr. Khim. Zh. *31*, 1316 (1965).

186) Seitz, W. R., Suydam, W. W., Hercules, D. M.: Anal. Chem. , in press.

187) Roberts, B. G., Hirt, R. C.: Appl. Spectry. *21*, 250 (1967).

188) Bass, A. M., Kessler, K. G.: J. Opt. Soc. Am. *49*, 1223 (1959).

189) Bezman, R., Faulkner, L. R.: J. Am. Chem. Soc. *94*, 3699 (1972).

190) Omote, Y., Yamamoto, H., Tomioka, S., Sugiyama, N.: Bull. Chem. Soc. Japan *42*, 2090 (1969).

191) Kalbhen, D. A.: Internat. J. Appl. Radiation Isotopes *18*, 655 (1967).

192) Shevlin, P. B., Neufeld, H. A.: J. Org. Chem. *35*, 2178 (1970).

193) Stross, F. H., Branch, G. E.: J. Org. Chem. *3*, 385 (1938).

194) Taylor, K. J., Jarman, P. D.: J. Am. Chem. Soc. *93*, 257 (1971).

195) Negishi, K.: J. Phys. Soc. Japan *16*, 1450 (1961).

196) Stauff, J., Bergmann, U.: Z. Physik. Chem. (Frankfurt) *78*, 263 (1972).
197) Cormier, M. J., Hercules, D. M., Lee, J. (ed.): Chemiluminescence and bio-luminescence. New York: Plenum Press 1973.
198) Rauhut, M. M., Roberts, B. G., Semsel, A. M.: J. Am. Chem. Soc. *88*, 3604 (1966).
199) Stauff, J., Sander, U., Jaeschke, W.: *loc. zit.* [197], p. 131.
200) Heringa, J. H., Strating, J., Wynberg, H., Adam, W.: Tetrahedron Letters *1972*, 169.
201) Cia-Sen, D., Wilson, Th.: *loc. zit.* [197], p. 265.
202) De Luca, M., Dempsey, M.: *loc. zit.* [197], p. 345.
203) Saimomura, O., Johnson, F. H.: Biochem. Biophys. Res. Commun. *44*, 340 (1971); *loc. zit.* [197], p. 337.
204) Discussion remarks of White, E. H., McCapra, F.: *loc. zit.* [197], p. 358.
205) Adam, W., Rucktäschel, R.: J. Org. Chem. *37*, 4128 (1972).
206) White, E. H., Brundrett, R. B.: *loc. zit.* [197], p. 231.
207) Wildes, P. D., White, E. H.: J. Am. Chem. Soc. *95*, 2610 (1973); *94*, 6223 (1972).
208) Beck, M. T., Joo, F.: Photochem. Photobiol. *16*, 491 (1972).
209) Gundermann, K.-D., Röker, K. D.: Angew. Chem. Intern. Ed. Engl. *12*, 425 (1973).
210) Hoytink, G. J.: *loc. zit.* [197], p. 147.
211) Bezman, R., Faulkner, L. R.: J. Am. Chem. Soc. *94*, 6331 (1972).
212) Martin, J. E., Hart, E. J., Adamson, A. W., Gafney, H., Halpern, J.: J. Am. Chem. Soc. *94*, 9238 (1972).
213) Siegel, T. M., Mark, Jr., H. B.: J. Am. Chem. Soc. *94*, 9020 (1972).
214) Weller, A., Zachariasse, K.: *loc. zit.* [197], p. 181.
215) Weller, A., Zachariasse, K.: Chem. Phys. Letters *10*, 424 (1971).
216) Bard, A. J., Kreszthelyi, C. P., Tachikawa, H., Tokel, N. E.: *loc. zit.* [197], p. 193.
217) Maulding, D. R., Roberts, B. G.: J. Org. Chem. *37*, 1458 (1972).

Substituent Effects in Photochemical Cycloaddition Reactions

Professor William C. Herndon

Department of Chemistry, The University of Texas, El Paso, Texas, USA

Contents

I. Introduction

Photocycloaddition reactions occupy an important position in organic chemistry as a synthetic tool. Many different types of photocycloadditions have been discovered, and several hundreds of examples have been listed. [1-3] An estimate of the large size of the field can be obtained from the review articles and books, numbering over thirty, published in the decade 1960—1970. [4] Since, from the very beginning, organic photochemists have speculated about the mechanisms of their reactions, papers that describe experiments designed to test mechanistic schemes and theoretical results are also very numerous.

This article will only discuss two particular kinds of photocycloaddition reactions, the photodimerization or cross-cycloaddition of two olefins to yield a cyclobutane derivative, and the photoreaction of an olefin with a carbonyl compound to give an oxetane, Eq. 1 and Eq. 2. The inportance of substituent effects in reactions of these types is pointed

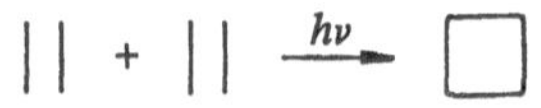

$$|| + || \xrightarrow{h\nu} \square \tag{1}$$

$$|| + \begin{matrix}O\\||\end{matrix} \xrightarrow{h\nu} \square^O \tag{2}$$

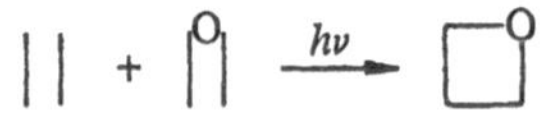

up by the facts that one of the parent reactions, ethylene dimerization, is difficult to carry out [5], and the second reaction, formaldehyde-ethylene cycloaddition, has not been observed. However, many synthetically useful photoaddition of substituted olefins and carbonyl compounds are known. [6-17]

The overall results of substituent effects are observed in the products of a reaction, their rates of formation, and their stereochemistries. The purpose of this article is to apply very simple theoretical techniques to correlations and predictions of the rate and stereoselectivity effects of substituents in [2+2] photocycloadditions. The theoretical methods that will be used are *perturbational molecular orbital (PMO) theory* and its pictorial representation, the interaction diagram. Only an outline of the theory will be given below, since several more detailed descriptions are available. [4,18-34]

The discussions of actual reactions will be qualitative, with numerical calculations giving way to semi-quantitative correlations obtainable without excessive numerical computations. The actual numerical results of quantum-mechanical calculations on large molecular systems are always obtained after many, sometimes drastic approximations. This problem is magnified when photochemical processes are involved, since excited states are most easily and incorrectly represented by populating the virtual orbitals of ground-state wave functions. Consequently, the discussion will center on those points that are not likely to be artifacts of computational methods. Even with this limitation, several facets of these reactions will be understandable on the basis of this theory. Perhaps this PMO method will answer Hammond's appeal [35] for a unifying theory to explain structure-reactivity relationships.

II. Theory

At some stage during a cycloaddition reactions, an interaction between two molecular species must occur. Let us imagine that two possible mutual orientations of the reacting molecules are possible that lead to

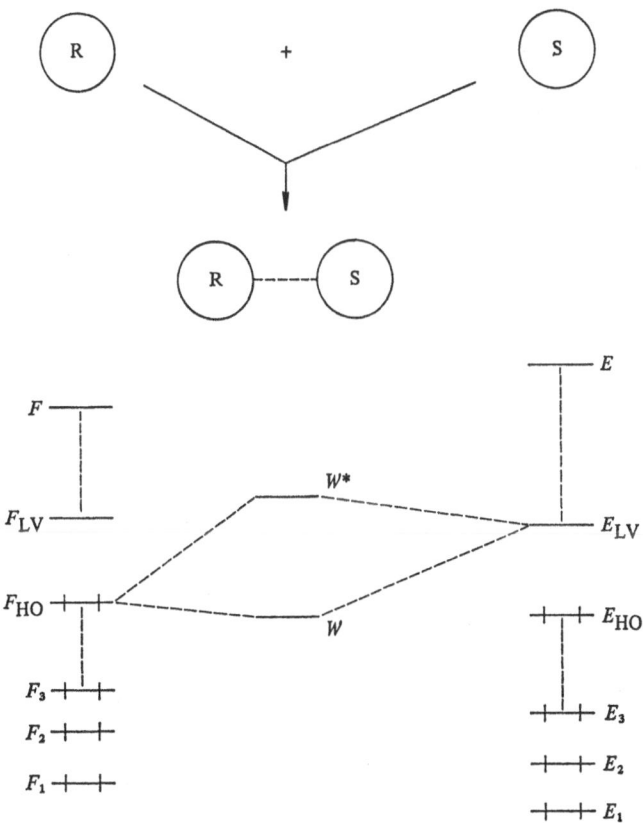

Fig. 1. Interaction diagram, reaction of R and S

two different products. The relative stabilities of these two orientations would then control the relative rates of formation of the different products. If the reactions were irreversible, the relative amounts of products could be predicted if the relative stabilities of the orientation complexes could be calculated. PMO theory can be used to carry out a calculation of this type.

To illustrate the method, consider an interaction diagram, Fig. 1, for two molecules, R and S, uniting to form a complex R......S. The energy levels of R are F_1, F_2, F_3, ... F, the highest occupied level (HOMO) being F_{HO}. The energy levels of S are E_1, E_2, E_3, ... E with the lowest vacant level (LVMO) being E_{LV}. The total intermolecular interaction is a sum of the interactions of each molecular energy level of R with each level of S, plus Coulombic terms that are appropriate to the reaction. The energies and coefficients of the MO's of R and S are assumed to be known, so that the orbital interactions and Coulombic terms can be calculated.

Considering only the interaction between HOMO of R and LVMO of S, elementary perturbation theory shows that the result of the orbital interaction is a repulsion of the levels, the occupied level becomes more stable, the unoccupied level less stable. The simplest Hückel-type formulation of PMO theory gives equations for the intermolecular perturbation energy change Δ that are quite simple in form, Eqs. 3—6 [18,20,22,26—28] Q is a first-order Coulombic energy that can be calculated in terms of

$$\Delta = Q + \Delta_1 + \Delta_2 + CT + P \tag{3}$$

$$Q = - Q_R Q_S \Omega / \varepsilon \tag{4}$$

$$\Delta_1 = N\, C_R C_S^\gamma{}_{RS} \tag{5}$$

$$\Delta_2 = N\, \frac{(C_R C_S^\gamma{}_{RS})^2}{F - E} \tag{6}$$

the net charges on the atoms of R and S. Q may be a stabilizing or destabilizing term. Both Δ_1 and Δ_2 are stabilizing energy terms, and they can be calculated in terms of the coefficients C_R and C_S of the interacting molecular wave functions, the energies of the interacting levels, and the interaction integrals γ_{RS}. N is an occupancy number, equal to unity if one or three electrons are involved in the interaction, and equal to two if two electrons are involved.

The first-order term Δ_1 arises if two partially occupied interacting orbitals are degenerate or nearly degenerate in energy. The second-order term Δ_2 results from stabilizing mixing of occupied orbitals of one molecule with vacant orbitals of the other molecule. In applications the interaction integral γ has been taken as proportional to overlap integrals or simply assigned arbitrary values depending upon the mechanism to be calculated. In Eqs. 2, 3, and 4 each term is understood to be correctly summed over all relevant interactions between the two molecules.

The charge-transfer and polarization terms, Eq. 1, CT and P, can be understood be referring to Fig. 2. These configurations may make a contribution to the perturbation energy of the composite state if one

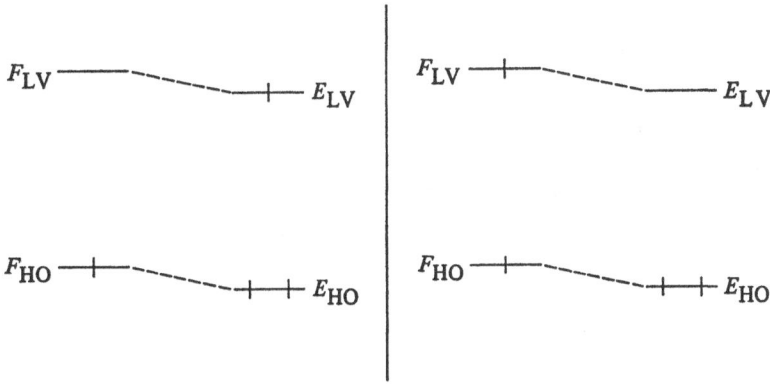

Fig. 2. Charge-transfer and polarization configurations

molecule is a good donor, and the other a good acceptor, or if both molecules are quite polar. The first-order parts of the contributions of these configurations, Eqs. 4 and 5, would generally dominate over second-order terms. The extents of the contributions would be calculated according to configuration-interaction methods. Large effects could be expected in photochemical reactions.

An advantage of PMO theory is that it is capable of elaboration in many different ways. A simple theory, capable of elaboration, is desirable for many reasons, the most important being that simple theory is simply applied, and more elaborate theory is more easily dissected into parts, to each of which theoretical or physical interpretation may be given.

The first photochemical reactions to be correlated with PMO theory were the dimerizations of anthracene, tetracene, pentacene, and acenaphthylene. [36] More detailed energy surfaces for the photodimerization reactions of butadiene have also been calculated. [30] In the category of simplified calculations lie studies of the regiospecificity of Diels-Alder reactions [37], and reactivity in oxetane-forming reactions. [38,39] In these studies, Hückel MO calculations provided the starting wave functions and the intermolecular perturbation integral was assigned a finite value of 1/2 of the Hückel exchange integral β at only the sites where bond-formation takes place. No contributions of Q, CT or P terms were included in the above calculations. Calculations on Diels-Alder reactions using Extended Hückel [40] wave functions, with considerations of all valence orbital interactions, and with interaction integrals proportional to overlap integrals were the first of this type to appear. [41] Some similar results for charge-transfer complexes using CNDO/2 [42] wave-functions as starting orbitals have also been reported. [43,44] Other calculations of interest to photochemists are on the cycloaddition reactions of quinones with olefins, [45] and the excited state addition of maleic anhydride to benzene. [46]

In the category of more elaborate calculations are an SCF—PMO treatment of the reaction of cyclopentadiene with cyclopropene [33], and calculations including overlap on the photodimerization of linear steroidal dienones. [31] A very fine, detailed theoretical study of thymine photodimerization has been recently presented. [47] Other PMO calculations on photocycloadditions of biochemical interest have also been published. [48,49]

III. Interaction Diagrams

The molecular interaction diagram is an excellent starting point for qualitative considerations of mechanism in cycloaddition reactions. Fig. 3, Section VI, illustrates the interaction diagram for the cycloaddition of an $n-\pi^*$ excited state of formaldehyde reacting with ethylene to give oxetane, Eq. 2. [38,39] The interaction diagram is a combination of two molecular diagrams, based on the results of MO calculations and/or experimental results. The energy level occupancies for the two reacting species are indicated by vertical bars, and the dominating orbital interactions that contribute to the stabilizing perturbation energy are pointed out by connecting dashed lines. The phase of the wave functions, and the coefficient values would be obtained from calculations.

The bonding levels, Fig. 3, are placed in accordance with photoionization spectroscopy experiments [50-52], and the higher energy levels are positioned on the basis of known excitation values to give the singlet excited states. [53-55] The energy levels of singlets are used rather than those of triplet excited states because in these simple systems the triplet states have very different geometries from ground states, and consequently relatively large singlet-triplet splittings are observed. [56] With larger molecules the singlet-triplet energy differences are much smaller, and the geometry of the excited state is more likely to be similar to that of the ground state.

Particular interactions between energy levels could stabilize suprafacial-suprafacial $[2+2]$ concerted reactions, $[_2\pi_s+_2\pi_s]$, [57] or suprafacial-antarafacial modes, $[_2\pi_s+_2\pi_a]$. Other interactions might be favorable toward biradical reactions. These diagrams will be discussed in detail in Section VI.

IV. Mechanisms and Examples

A. General Considerations

In broad outline, photochemical cycloaddition reactions involve the following steps; vertical excitation of one reactant, vibrational decay of the excited molecule toward its lowest excited state configuration (possibly intersystem crossing), interaction of excited reactant with second reactant, deexcitation from the potential energy surface of excited state to the ground state potential energy surface of a product. From the standpoint of PMO theory, each one of these stages may control a different aspect of the calculation. For example, the actual reactive excited state must be known so that the correct level occupancies may be assigned to the photoexcited molecule. And the nature of the interaction must be guessed so that realistic interaction integrals may be chosen.

Each one of the two types of photoreactions to be discussed has several other possible complicating features, and these will be examined in turn in this section.

B. Olefin-Carbonyl Cycloadditions

The photocycloaddition of an aldehyde or ketone with an olefin to yield an oxetane was reported by Paterno and Chieffi in 1909. [58] Contemporary studies on the synthetic utility and mechanistic features were initiated nearly 50 years later by Büchi et al. [59] Two review articles summarizing synthetic aspects of Paterno-Büchi reactions have been published [6,12], and mechanistic studies have been reviewed several times. [6,38,60-62] The reaction involves the addition to olefin of a photo-excited carbonyl moiety. This circumstance makes it advantageous to review this reaction before a discussion of olefin-olefin additions, because the solution photochemistry of carbonyl compounds is probably better understood than any other aspect of organic photochemistry. Many of the reactions of carbonyl compounds have been elucidated during studies of the important phenomena of energy transfer and photosensitization. [63-65]

A clear division of Paterno-Büchi reactions into several distinct categories is possible on the basis of the type of reacting carbonyl compound (alkyl or aromatic), the excited state responsible for reaction (n—π^* or π—π^*, singlet or triplet), and the type of olefin (electron deficient or electron-rich). Some examples of these reactions are given in Eqs. 7—11, where only the oxetane products are shown.

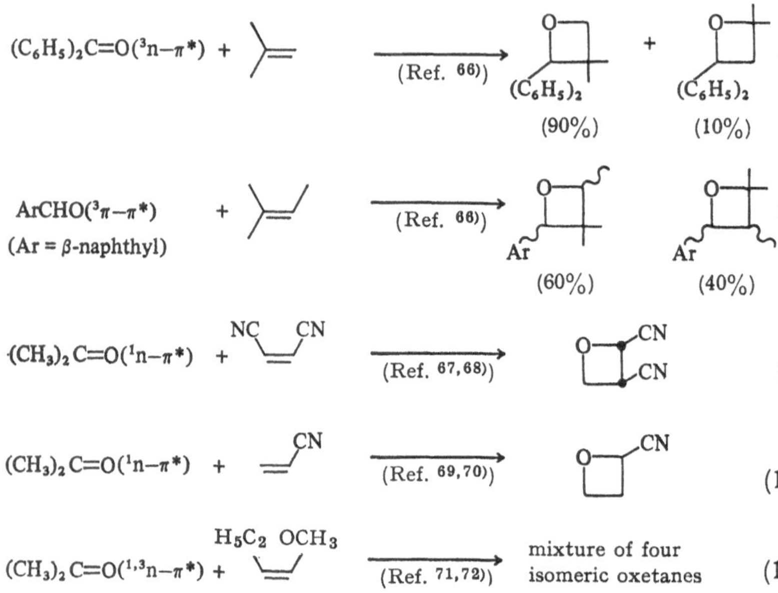

$$(C_6H_5)_2C=O(^3n-\pi^*) + \ \xrightarrow{\text{(Ref. 66)}} \quad \underset{(C_6H_5)_2}{\square} \ + \ \underset{(C_6H_5)_2}{\square} \quad (7)$$

$$\underset{(90\%)}{} \qquad \underset{(10\%)}{}$$

$$ArCHO(^3\pi-\pi^*) + \ \xrightarrow{\text{(Ref. 66)}} \quad \underset{Ar}{\square} \ + \ \underset{Ar}{\square} \quad (8)$$

(Ar = β-naphthyl)

$$\underset{(60\%)}{} \qquad \underset{(40\%)}{}$$

$$(CH_3)_2C=O(^1n-\pi^*) + \underset{NC\ \ \ \ CN}{} \ \xrightarrow{\text{(Ref. 67,68)}} \quad \underset{}{\square}\text{CN} \quad (9)$$

$$(CH_3)_2C=O(^1n-\pi^*) + \underset{CN}{} \ \xrightarrow{\text{(Ref. 69,70)}} \quad \underset{}{\square}\text{CN} \quad (10)$$

$$(CH_3)_2C=O(^{1,3}n-\pi^*) + \underset{H_5C_2\ \ OCH_3}{} \ \xrightarrow{\text{(Ref. 71,72)}} \quad \begin{array}{l}\text{mixture of four}\\\text{isomeric oxetanes}\end{array} \quad (11)$$

All of the elements of stereo- and regioselectivity and reactivity that theory must explain are found in the above reactions. The triplet excited states of the aryl carbonyl compounds demonstrate regioselectivity that has been previously explained on the basis of the relative stabilities of the two possible biradical intermediates, 1 and 2. [65,66] The selectivity

$$\underset{Ar}{\overset{O}{\square}} \qquad \underset{Ar}{\overset{O}{\square}}$$

1   2

observed does seem rather low compared to other free radical addition reactions. [73] No high stereoselectivity for the reactions, Eqs. 7 and 8, is observed, in contrast to the stereospecificity exhibited by the reactions of ^1n—π^* alkyl ketones with electron-dificient olefins. [67-70] These reactions are also regiospecific, cf. Eq. 10. Finally, the reactions of alkyl ketones with electron-rich olefins seems to involve both excited states of the ketone, Eq. 11, with about equal regioselectivities. The ratio of 2-alkoxy olefins to 3-alkoxy oxetanes, *3/4*, was not determined in the given reaction [71,72], but the reaction of propionaldehyde with ethyl

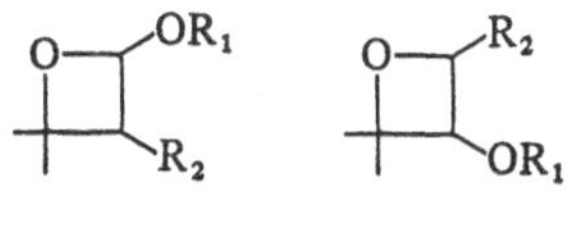

3   *4*

vinyl ether gave almost exclusively compounds of structure *3*, [74] or in a separate report 75% *3*, 25% *4*. [75]

Benzaldehyde and ethyl vinyl ether also gave a 3 to 1 ratio of *3* and *4*. [75] The reaction with triplet ketone was found to be less stereoselective than the reaction with singlet ketone.

The quantum yields for oxetane formation have not been determined in every case, and only a few relative rate constants are known. The reactivities of singlet and triplet states of alkyl ketones are very nearly equal in attack on electron rich olefins. [72] However, acetone singlets are about an order of magnitude more reactive in nucleophilic attack on electron-deficient olefins. [61] Oxetane formation is competitive with α-cleavage, hydrogen abstraction and energy-transfer reactions [60–64] so the absolute rates must be reasonably high. Aryl aldehydes and ketones add to olefins with lower quantum yields, [66] and $^3\pi$-π^* states are particularly unreactive. [76]

The mechanism given by Eqs. 12—18 is a minimum mechanistic scheme for oxetane formation derived mainly by combining suggestions

$$K \longrightarrow {}^1K^* \qquad \text{excitation} \qquad (12)$$

$$^1K^* \longrightarrow {}^3K^* \qquad \text{intersystem crossing} \qquad (13)$$

$$^1K^* + 0 \longrightarrow {}^1[K \ldots 0]^* \qquad \text{complex formation} \qquad (14)$$

$$^3K^* + 0 \longrightarrow {}^3[K \ldots 0]^* \qquad \text{complex formation} \qquad (15)$$

$$^{1 \text{ or } 3}[K \ldots 0]^* \longrightarrow BR \qquad \text{biradical formation} \qquad (16)$$

$$^{1 \text{ or } 3}[K \ldots 0]^* \longrightarrow \text{oxetanes} \qquad \text{product formation} \qquad (17)$$

$$BR \longrightarrow \text{oxetanes} \qquad \text{product formation} \qquad (18)$$

149

of investigators in the field [65-72] with some recent results on energy transfer. [77,78] The excited ketone K might also undergo fluorescence, phosphorescence, radiationless decay or energy transfer to olefin 0 (via complex ?) [78] but these reactions have been omitted to simplify the scheme. Formation of biradical BR or complex could be reversible, and these complications are also neglected.

Eq. 17 is meant to represent the possibility for a concerted formation of oxetane product. A problem that always exist in cycloadditions is the question of whether the reaction takes place by a two-step biradical reaction pathway or through a concerted mechanism. Such questions have not even been resolved for purely thermal reactions. [4] A recent speculation on this point proposes almost universal concertedness for all cycloaddition reactions. [79] In that work, mixed stereochemistry in the products of [2+2] cycloaddition reactions is generally attributed to a mixture of two concerted reactions, suprafacial-suprafacial, and suprafacial-antarafacial. It will be seen later that the PMO calculations generally do not support this idea. A mixture of biradical and concerted reactions is in better agreement with experimental facts.

C. Olefin Dimerizations and Cycloadditions

There are so many different examples of photochemical dimerizations and cross-cycloadditions [8-11,13-17] of olefinic compounds that one is not surprised to find several variations of mechanistic patterns. Simple olefins, including dienes and strained small ring, bicyclic olefins and styrene derivatives form a class of compounds that undergo such reactions sensitized by triplet energy donors. Some examples are given in Eqs. 19—23, where only cyclobutane products are depicted. Theory

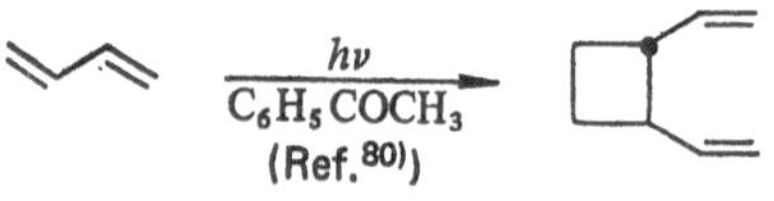

$$\text{(19)}$$

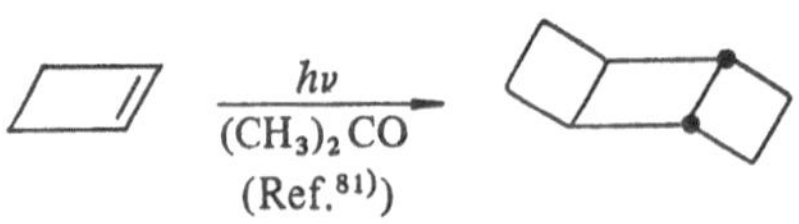

$$\text{(20)}$$

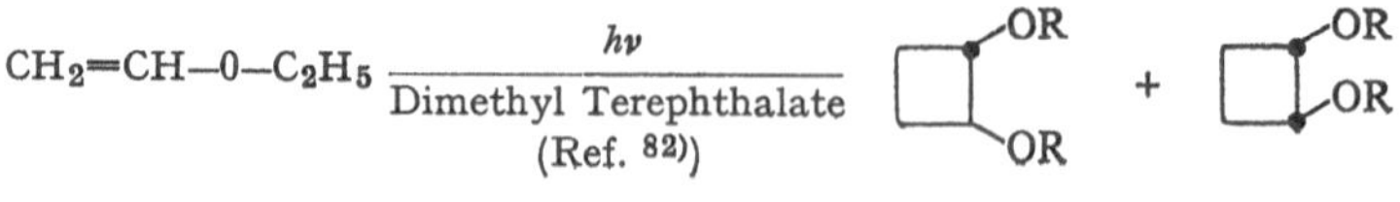

$$CH_2=CH-O-C_2H_5 \xrightarrow[\text{Dimethyl Terephthalate}]{h\nu} \text{[Ref. 82]}$$

$$\text{(21)}$$

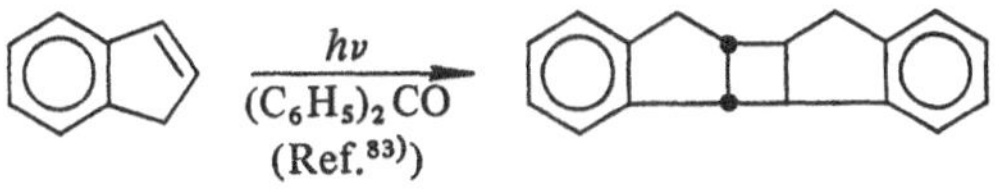

$$\text{(22)}$$

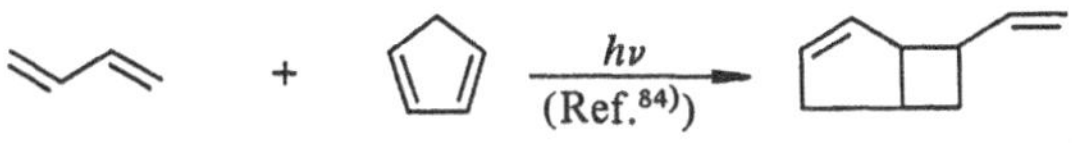

$$\text{(23)}$$

must account for the high regiospecificity (head to head dimers are formed), accompanied by the lack of stereospecificity. These results have been rationalized previously by assuming biradical intermediates. [80]

Sensitized cross-dimerizations form a second group with a few examples shown in Eqs. 24—27. Most examples involve electron-deficient olefins as one addend. The reactions are again highly regioselective

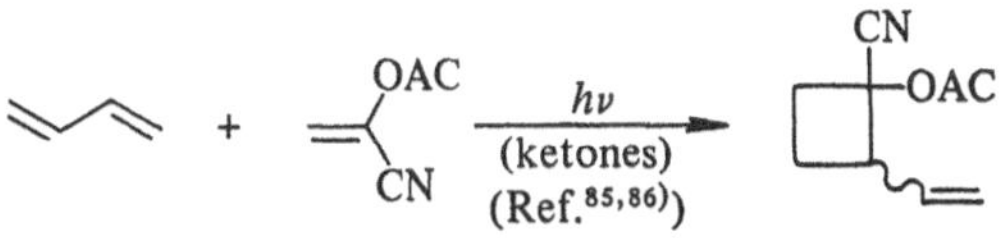

$$\text{(24)}$$

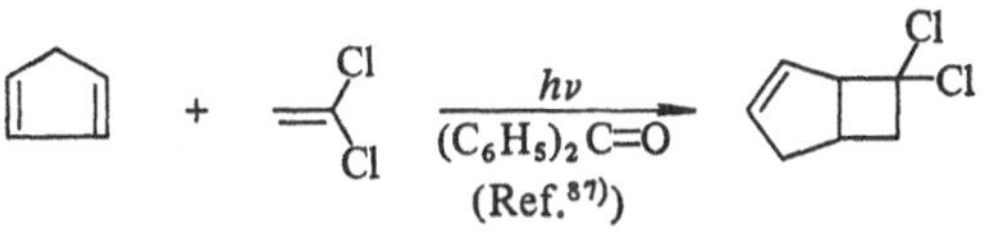

$$\text{(25)}$$

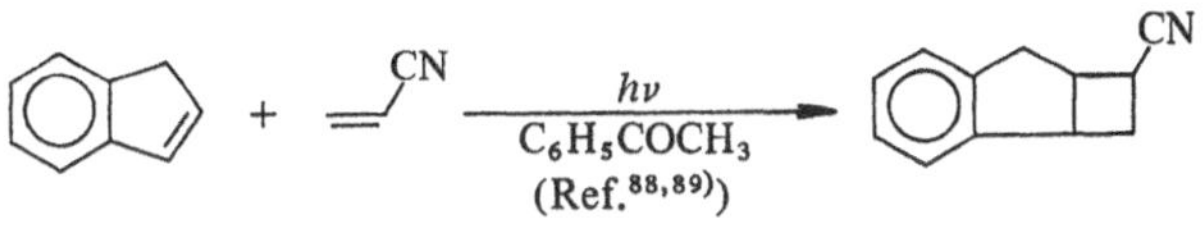

$$\text{(26)}$$

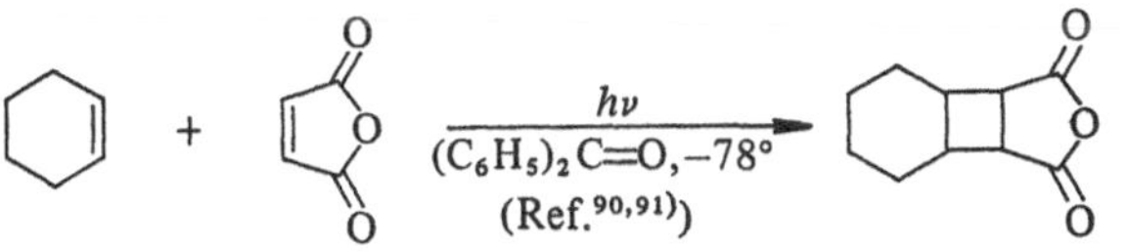

$$\text{(27)}$$

and non-stereoselective. Biradical intermediates have been proposed, and estimates of competition between bond-rotation and bond-closure convincingly account for the proportions of stereoisomers formed. [86,90] In the reaction, Eq. 24, closure of biradical to some cyclohexene derivatives is expected and experimentally confirmed. [85,86] The numerous photoadditions of maleic anhydride and derivatives [60] are efficiently photosensitized, and a complex of substrate-anhydride has been proposed for aromatic substrates as the energy acceptor. [93]

A third important group of reactions that may be discussed with the first two groups is not necessarily photosensitized. The dimerizations and additions of cyclic α,β-unsaturated ketones can be initiated by direct n-π^* excitation of the ketone, followed by addition reactions. However, the reactions are efficiently photosensitized by triplet sensitizers, and it is reasonable to propose that the unsensitized cycloaddition reactions also proceed via triplet states. [8,63,94] Examples are given in Eqs. 28–31.

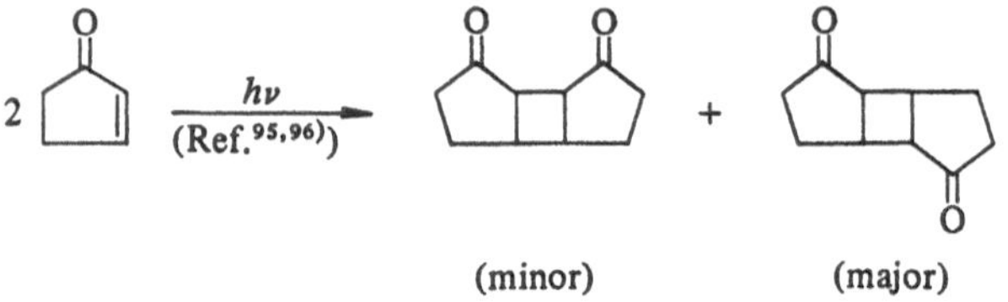

(28)

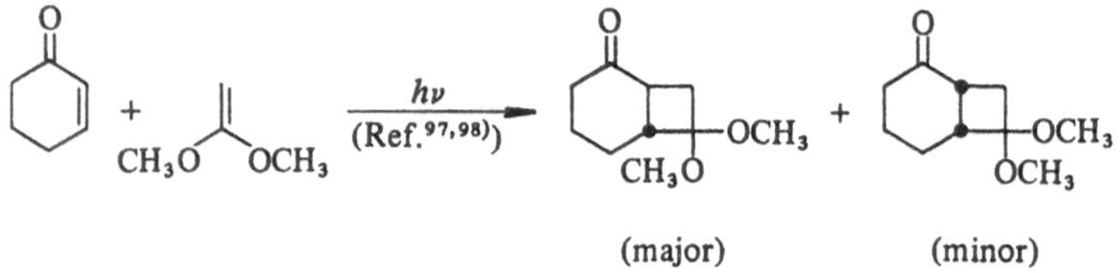

(minor)   (major)

(29)

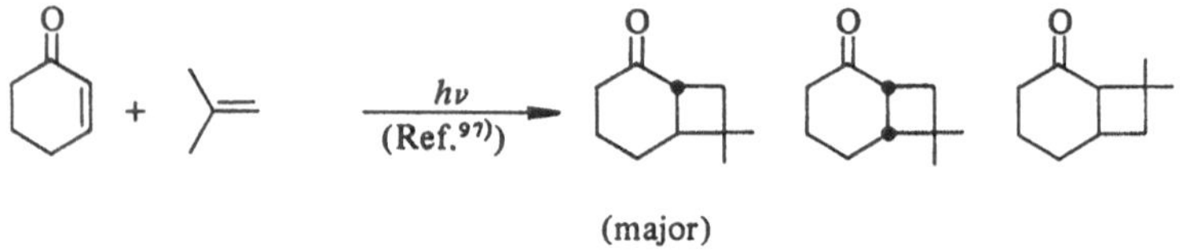

(major)   (minor)

(30)

(major)

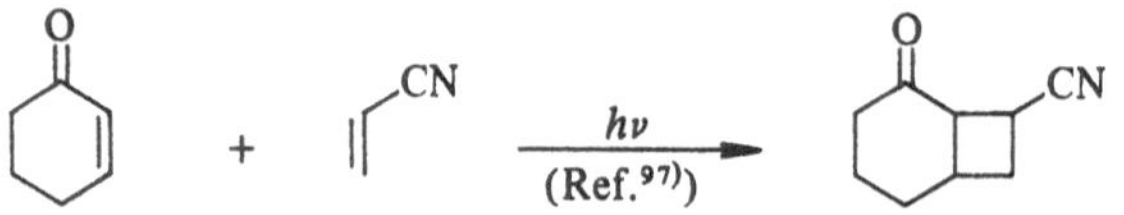

(31)

Complicating aspects of the reactions include the solvent dependence of the dimerization of cyclopentenone, the head-to-head ratio increasing with polarity of solvent [99,100], and the plethora of products from the reaction cyclohexenone-isobutylene, where several olefinic products are not shown in Eq. 30. [97] The complicated features of these reactions are so well-described in a recent review article [94] that no furhter outline will be provided here. Biradical mechanisms can account for a great deal of the experimental observations, particularly the observed regio-selectivity and the lack of stereospecificity.

For the foregoing series of reactions, a minimal mechanism beginning after energy transfer to olefin is summarized in Eqs. 32—35.

$$^{3}0^{*} \quad + \; 0 \quad \longrightarrow \quad ^{3}[0\ldots.0]^{*} \qquad \text{complex formation} \qquad (32)$$

$$^{3}[0\ldots.0]^{*} \quad \longrightarrow \quad \text{BR} \qquad\qquad \text{biradical formation} \qquad (33)$$

$$^{3}[0\ldots.0]^{*} \quad \longrightarrow \quad \text{products} \qquad\qquad\qquad\qquad (34)$$

$$\text{BR} \qquad\quad \longrightarrow \quad \text{products} \qquad\qquad\qquad\qquad (35)$$

Eq. 34 is included so that the possibility of synchronous concerted formation of cyclobutanes may be considered. This step is formally a forbidden triplet to singlet process, but several workers have taken the position that this may not be an inhibiting factor for concerted reactions. [31,38,99]

A remaining class of dimerizations is represented by the reactions shown in Eqs. 36—39. These are singlet state reactions, and they are all characterized by stereospecificity and regiospecificity. Production of the *trans* isomer from the steroidal dienone in Eq. 39 is anomalous and poses a

$$(36)$$

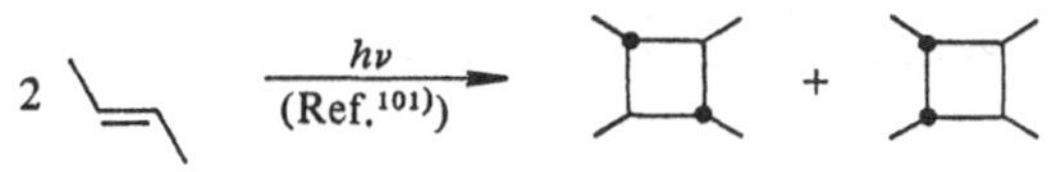

$$(37)$$

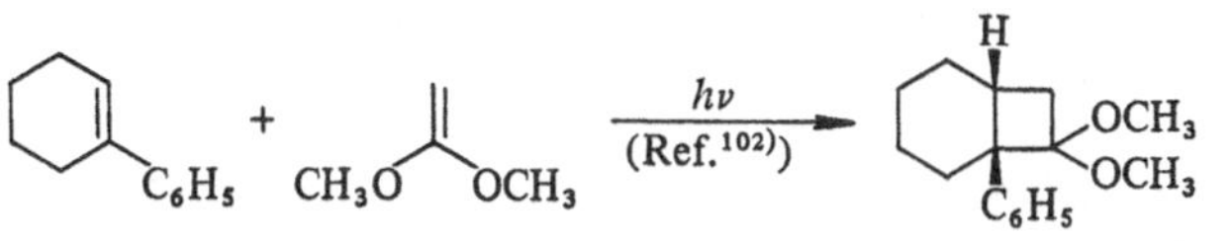

$$(38)$$

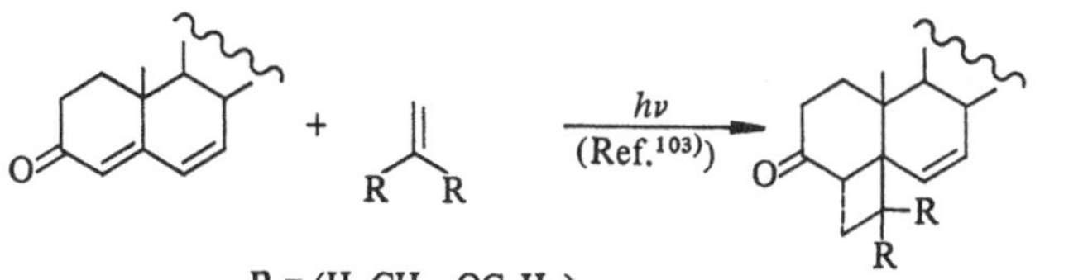

$$R = (H, CH_3, OC_2H_5)$$

$$(39)$$

W. C. Herndon

fascinating challenge to theory. The other reactions give *cis* addition exclusively, and a logical conclusion is that concerted cycloaddition reactions have taken place. A minimal mechanism would involve Eqs. 32—34 with singlets replacing triplets.

V. Assumptions of PMO Theory

A very complex group of observations and speculations has been presented in Section IV. It might almost seem illogical to apply a single theoretical approach to so diverse a set of reactions, but the utility of PMO theory for correlating the several aspects of Paterno-Büchi reactions has already been demonstrated. [38,39] The newer results to be presented below will help to confirm the idea that PMO theory gives a unified useful theoretical picture for the majority of [2+2] photocycloaddition reactions. First, however, the many approximations and postulates inherent in this work should be made clear.

In calculations and interaction diagrams, only the most simplistic MO models will be chosen to represent ground and excited states of reactants. An olefin then has a σ-bond framework largely neglected in discussing the reactivity of the molecule. The bonding level will be characterized by a π-electron wave function with no nodes between the two basis p orbitals of the π-bond. The first π-antibonding level has one node in the wave function, and a first excited state has electron-occupancy of unity in each level.

The description of the carbon-oxygen double bond is analogous, but in addition to the σ-bonds there are unshared pairs of electrons on oxygen so that two excited states are possible, π-π^* and n-π^*. For n-π^* excitation the resultant half-vacant orbital on oxygen should possess electrophilic reactivity, and the electron rich π-system should have nucleophilic characteristics. [62]

Reactivity differences might arise as a consequence of the inherent difference between a singlet and triplet state of a molecule. Since most of our molecular level parameters will be the result of HMO calculations, calibrated by spectroscopic values when possible, triplet and singlet reactivities are indistinguishable. This point has been discussed previously [38], the conclusion is that approximate HMO wave functions are equally good or equally bad for both states. We do expect that if reactivity and stereochemistry are determined at a biradical stage, singlet and triplet reactions should be similar in rate. There schould be no intrinsic difference between the rate of a bimolecular reaction involving a singlet excited molecule as compared with the same reaction involving the same excited molecule, identical in all respects except that of

154

spin multiplicity. However, there is no doubt that the concerted reaction of a triplet compound to yield ground-state product is a spin-forbidden process. In the absence of experimental facts or theoretical treatments of this problem, we simply guess that spin-forbiddeness inhibits a reaction in comparison with spin-allowed processes.

If one examines the minimal sequences of reaction steps for [2 + 2] cycloadditions, Eqs. 12—18, 32—35, one concludes that stereochemistry of addition, and perhaps relative reactivities might be calculable at several points. Oriented complexes could control regiospecificity, or the transition state leading to a biradical could be the important stage. Relative rates of product formation would be derived from relative perturbation stabilization energies for different configurations of the two reactants.

Ideally, one would like to compare observed selectivities with calculated energies of transition states leading to various products. The PMO method in the simplified form used here is not capable of application at that level, so assumptions about the important interactions between the reacting molecules are necessary before calculations. This is handled by fixing the value of the perturbation exchange integral as equal to 0.5β, where β is the usual HMO resonance integral, only at the positions where bonding to form product actually takes place. A concerted reaction would have two bonding points, and a biradical reaction would only have one. The value of 0.5β is justified by adducing that an incipient bond is half-formed in a transition state.

A serious assumption in applying PMO theory to the cycloaddition reactions of photoexcited molecules regards the interpretation of calculated perturbation energies. For ground-state reactions a perturbation calculation approximates the degree of stabilization of the transition state. However, the process of reaction for a photoexcited molecule involves transferance from an excited potential-energy surface to a less energetic ground-state potential energy surface.[104,105] A reaction surface involving an excited molecule probably has a negative slope. With this in mind, PMO energies can be reinterpreted as a measure of the curvature of the slope of the energy surface at the beginning of a reaction.[106] With realistic excited state wave functions, a PMO approach to reactivities of excited molecules might be more successful than the same approach to ground-state chemistry.[107,108]

VI. Applications of PMO Theory to Paterno-Büchi Reactions

Let us consider the reaction of ethylene with formaldehyde to yield oxetane as the prototype for this kind of reaction. Fig. 3 is the interaction diagram, assuming that the aldehyde is in an n-π^* excited state. The dominant orbital interactions are

$$F(\pi^*) \longrightarrow E(\pi^*) \quad \text{and} \quad E(\pi) \longrightarrow F(n).$$

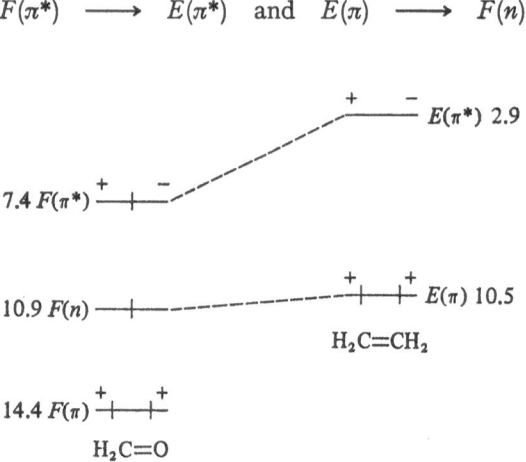

Fig. 3. Interactions diagram. Formaldehyde (n-π^*) and ethylene (energies in eV)

The first interaction has favorable orbital phase overlap for a concerted $(2\pi_s + 2\pi_s)$ reaction. The interaction integral γ, Eqs. 3—6, for a concerted process would have a maximum value if the two molecules approached each other so that the reacting orbitals could overlap in the most efficient manner. The best geometry would involve a face-to-face reaction of the two reactant species. The stereochemical consequence of such a reaction would be specific retention of substituent relative geometries.

The $E(\pi) \rightarrow F(n)$ interaction would have a large interaction integral if the half-vacant non-bonding orbital of the formaldehyde and one of the π orbitals of ethylene overlaped strongly. This can occur if the ethylene π orbital approaches the formaldehyde molecule. in roughly the plane of the molecule which is the nodal plane of the π system of the formaldehyde. Slight geometric distortions of the excited formaldehyde from planarity would not significantly affect this argument. Although geometries of the two reacting species can be chosen in which both new bonds are formed in concert, the most likely result of this latter interaction is to facilitate the formation of a biradical intermediate. If biradicals were formed, some loss of stereochemical specificity should be observed, the amount dependent upon the ratio of bond-rotation processes to bond closure processes.

Calculations based on the energy levels in Fig. 3 are very simple to carry out, and are shown in Eqs. 40 and 41. The almost degenerate interaction is calculated according to first-order theory, Eq. 5, and the

$$E(\pi) \rightarrow F(n), \quad \Delta_1 = (1) \left(\frac{\sqrt{2}}{2}\right) \gamma = 0.35\,\beta \tag{40}$$

$$F(\pi^*) \rightarrow E(\pi^*) \quad \Delta_2 = \frac{\left[(.85)\left(\frac{\sqrt{2}}{2}\right) + (.53)\left(\frac{\sqrt{2}}{2}\right)\right]^2 \gamma^2}{4.5/3.8} = .80\frac{\gamma^2}{\beta} = 0.20\,\beta \tag{41}$$

π-π interaction is computed by second-order theory, Eq. 6. The value of $\beta = 3.8$ eV is obtained from the energy difference between π and π^* levels of ethylene (2 β by an HMO calculation). The very crude calculation gives both interactions as sizable, and one could guess that the two types of reactions could be competing processes in the formaldehyde-ethylene case. Unfortunately, this particular experiment has not been carried out, but several examples of reactions with substituted carbonyl compounds and olefins, have already been cited, results of which have shown that the two mechanisms outlined above are quite likely to be correct descriptions of the actual molecular processes. [66-71]

The effect of substituents upon mechanism preference can be directly deduced from the interaction diagram, after tracing the effects of substituents on the molecular energy levels of the addends. Electron-attracting substituents (electronegative groups) lower all levels of a molecule, the filled levels being affected more than the vacant levels. Electron-donating groups raise both filled and vacant levels. The known values of ionization potentials for substituted aldehydes and ketones are similar to that of formaldehyde [50-52], so the energy levels of non-bonding electrons are affected only by a second-order inductive effect. Conjugative substituents lead to bands of π energy levels, the highest occupied level raised and the lowest vacant level lowered in energy. The conjugative substituent also spreads the electronic density, so that the eigenvectors at reactive sites are lowered in either π or π^* level.

The reaction of cis or trans-dicyanoethylene with acetone, Eq. 9. [67,68] illustrates the use of the theory very nicely. The electronic energy levels of acetone are all raised by the electron-donating methyl groups. At the same time the energy levels of the olefin are both lowered by the electron-attracting cyano substituents, Fig. 4. The interaction $E(\pi) \rightarrow F(n)$ is weakened as the interaction $F(\pi^*) \rightarrow E(\pi^*)$ is strengthened. Roughly, the ratio of concerted reaction to biradical process should be increased by some large factor. A concerted $[_\pi 2_s + _\pi 2_s]$ process would be predicted giving stereospecific retention of the substituents on the ethylenic moiety. Quenching experiments show that ^1n-π^* state of acetone is the reactive species and 100% retention of configuration of the cyano groups is observed. [67,69]

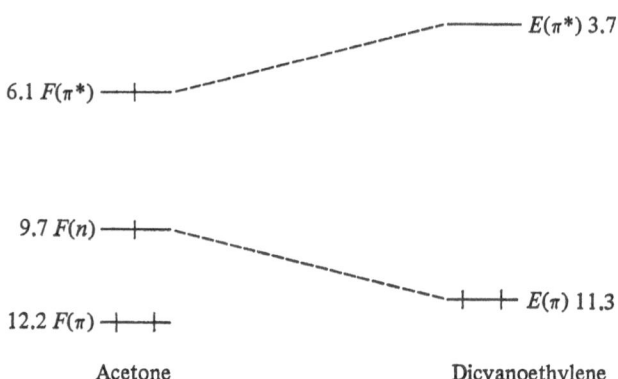

Fig. 4. Interaction diagram. Acetone and dicyanoethylene (energies in eV)

In a similar reaction, acrylonitrile plus acetone, [69] Eq. 10, the sole addition product has the orientation shown in 5. Coefficients from an HMO calculation of the LVMO of acrylonitrile are also shown. The

$$\text{H}_2\text{C} \text{----- CH ----- C ----- N}$$
$$+.66 \quad -.23 \quad -.58 \quad +.43$$
$$h(\text{N}) = 1.0, \quad k(\text{CN}) = 1.0$$

5

numerator part of the second-order energy for concerted reaction would be much larger for the orientation of the observed product than for the inverted orientation. Biradical intermediates would predict the non-observed product.

Charge-transfer from ketone to olefin would only increase the propensity toward a concerted reaction, since the interaction of a positively charged carbonyl compound for a negatively-charged olefin would give rise to an attractive Coulombic force. For the orientation problem, total π-charges after CT are shown in structure 6, with the resultant molecular

$$-.712$$
$$\text{O} \quad +.553 \qquad -.086 \quad ^{\text{CN}}$$
$$+.447 \qquad -.202$$

6

dipoles indicated by arrows. Again CT would increase the propensity toward the observed result. The charge interactions at the actual bonding sites may not be as important as the overall dipolar force in establishing the orientational preference. Generally, one will find that inclusion of CT state interactions in photochemical reactions will reinforce the conclusions obtained by neglect of such interactions. Note that CT from $E(\pi)$ to $F(n)$ would stabilize an alignment of the two reactants which would lead to the incorrect regiospecificity. Therefore, the observed regiospecificity infers that a concerted reaction is taking place as well as indicating that a biradical intermediate is not a correct postulate.

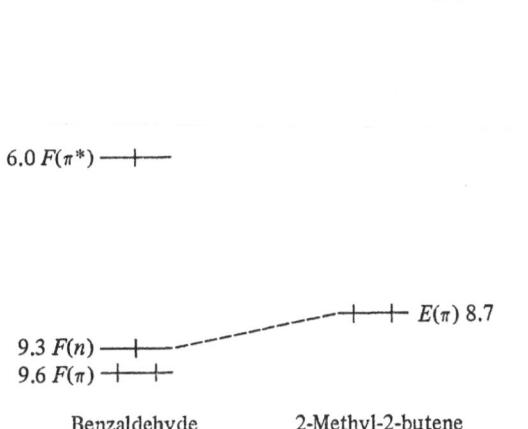

$$C_6H_5CHO + \quad \xrightarrow{h\nu} \quad \qquad + \qquad \qquad \qquad \qquad (42)$$

$$7 \ (40\%) \qquad 8 \ (25\%)$$

Photolysis of benzaldehyde and trimethylethylene yields a mixture of *cis* and *trans* oxetanes with the two orientations shown in Eq. 42. Orientation 7 predominates and biradical intermediates generated after formation of a bond involving the lone nonbonding electron of an n, π^* excited benzaldehyde have been postulated. [66] Fig. 5 is the interaction diagram, the molecular parameters being based on HMO calculations, and spectroscopic experiments. [55,56,109] The orbital interaction $E(\pi) \rightarrow F(n)$ is obviously dominant since the energy gap between $F(\pi^*)$ and $E(\pi^*)$ is over 4 eV. Therefore, a biradical mechanism should be postulated. The dominant orbital interaction is largest for attack of the

$$——— E(\pi^*) \ 1.9$$

$$6.0 \ F(\pi^*) ——+——$$

$$——+—+ \ E(\pi) \ 8.7$$
$$9.3 \ F(n) ——+——$$
$$9.6 \ F(\pi) —+—+—$$

Benzaldehyde 2-Methyl-2-butene

Fig. 5. Interaction diagram. Benzaldehyde and 2-methyl-2-butene

ethylene carbon with one methyl substituent on the carbonyl non-bonding orbital, since the ground state of the olefin is polarized $(CH_3)_2C^{\oplus}$ —$^{\ominus}$ CHCH$_3$. This gives a larger perturbation energy, and infers a more stable transition state leading to the biradical which could later close to yield the major products. This argument, based on relative stabilities of transition states, gives the same results as an argument based on deduced relative stabilities of biradical intermediates. The *cis-trans* isomerized mixture is of course a consequence of rotation around single bonds in the biradicals.

An additional factor supporting biradical intermediates in cases involving aromatic aldehydes is that the experimental facts point to the ^3n-π^* state as the reactive moiety. The previously discussed spin-conversion factor favoring a two-step process should come into play, with inter-system crossing occuring after the biradical is formed. One interesting aspect of this theoretical method is that the calculations themselves, in conjunction with stereochemistry of the reaction, can lead to a decision of the most probable reactive state. If one carries out the PMO calculations assuming a π-π^* excited state [38], 4-center cycloaddition reactions are favored and the major product is predicted to be **8**. Only assumption of an n-π^* reaction gives agreement with the experimental facts.

In reactions of vinyl ethers with alkyl ketones (electron-rich olefins plus electron-rich ketones), Eq. 11, both ^1n-π^* and ^3n-π^* triplet states of the ketones have been found to be reactive[71,74,75], the reaction of the singlet having a low regioselectivity and a high stereoselectivity. The triplet reaction is nonstereo and regioselective. [72] The mechanistic analyses were carried out in terms of biradical intermediates that result from nucleophilic attack on the half-vacant orbital of oxygen for both singlet and triplet reactions. The different stereochemical results were attributed to the greater tendency of a triplet biradical to undergo internal rotational motions because of the necessity for spin inversion before closure can occur.

Qualitatively, the interaction diagram would closely resemble that in Fig. 3, since electron-donating substituents in both addends would raise the molecular levels of both the carbonyl compound and the olefin. Only the energy gap, $E(\pi) \rightarrow F(n)$, would increase, the net result being that the calculated ratio of concerted to biradical reaction, Eqs. 40 and 41, should be even closer to unity than in the formaldehyde-ethylene case. Detailed calculations [38] support this conclusion, so PMO theory predicts that the overall stereochemical results are due to a combination of concerted (singlet) and biradical (triplet) mechanisms. This explanation agrees with the experimental facts, and it bypasses the necessity to postulate differential rates of rotation and closure for different kinds of biradical intermediates.

Alkenals also add to simple olefins to yield oxetanes, Eq. 43. Using acetaldehyde and the isomeric 2-butenes the quantum yield is low, and

$$
\begin{array}{c}
\text{H} \\
\underset{|}{\text{CH}_3\text{C}}\text{=O} + \text{CH}_3\text{CH}\text{=CHCH}_3 \xrightarrow{h\nu} \\
(\textit{cis or trans})
\end{array}
\qquad \underset{9}{\square} \quad \underset{10}{\square} \quad \underset{11}{\square} \qquad (43)
$$

the reaction is highly stereoselective. [110a] With *cis*-2-butene the ratio of $(9+11)$: *10* is 10:1, and with *trans*-2-butene the ratio $(9+10)$:*11* is 25:1. The prior mechanistic suggestion of singlet biradical intermediates is not untenable, but the stereochemical results and PMO theory suggest a concerted reaction. This explanation also accounts for regioselective effects that have been found in these types of reactions. [110b]

Excitation of carbonyl compound to π-π^* excited states occurs with naphthaldehydes and electron rich acetophenones as in the case of 4-methoxy-acetophenone. [111] An example of a Paterno-Büchi reaction involving a $^3\pi$-π^* state is shown in Eq. 8, and a second example in Eq. 44. PMO theory at present cannot make a definite prediction about the

$$(44)$$

mechanism of such a reaction because of the triplet-singlet spin-conversion problem. Two interactions both stabilizing a concerted $[_\pi 2_s +_\pi 2_s]$ state are present. These are (Fig. 6) $E(\pi) \rightarrow F(\pi)$ and $F(\pi^*) \rightarrow E(\pi^*)$.

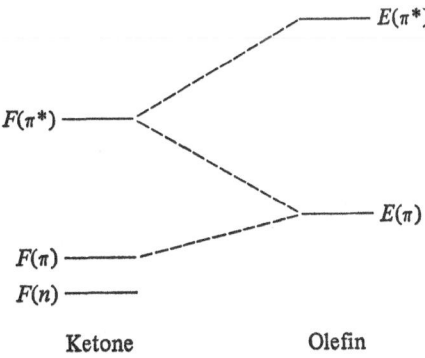

Fig. 6. Qualitative interaction diagram, aromatic ketone and electron-rich olefin

An interaction, $E(\pi) \rightarrow F(\pi^*)$, stabilizing a concerted $[_2\pi_s + _2\pi_a]$ reaction [79] could not be competitive because the magnitude of the interaction integral γ would be too low. Experiments with substituted olefins in which the extent of stereoselectivity could be determined would be useful in shedding more light on this point.

In the reactions of aliphatic carbonyl compounds with conjugated olefins a very clear distinction of mechanism is possible after comparing calculations with experimental results. Examples are shown in Eqs. 45 [112,113] and 46. [114] After n-π^* excitation of the aldehyde the domi-

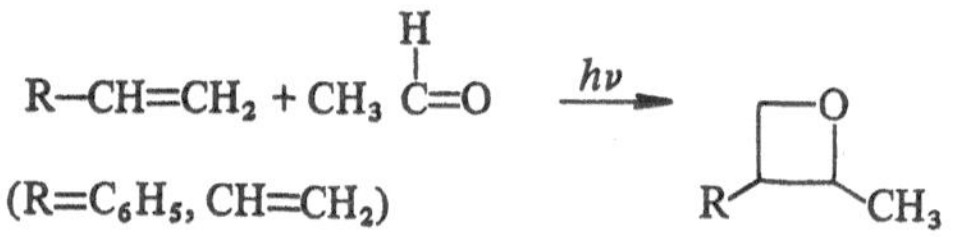

$$R-CH=CH_2 + CH_3 \overset{H}{\underset{|}{C}}=O \xrightarrow{h\nu} \qquad (45)$$

$$(R=C_6H_5, CH=CH_2)$$

$$+ \qquad \xrightarrow{h\nu} \qquad (46)$$

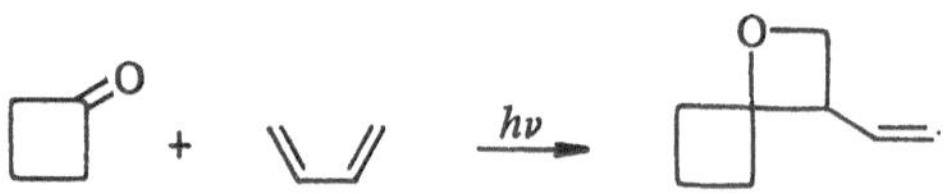

nant orbital interactions must be $E(\pi) \rightarrow F(n)$ and $F(\pi^*) \rightarrow E(\pi^*)$, the first interaction as usual leading to biradical, the second stabilizing a concerted transition state. The coefficients of $F(\pi^*)$ and $E(\pi^*)$ are shown below in diagram 12, l signifying large relative value and s small relative

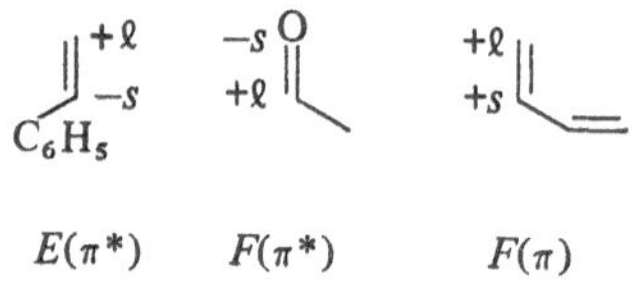

$$E(\pi^*) \qquad F(\pi^*) \qquad F(\pi)$$

12

value. A concerted reaction would give the opposite orientation to that of the observed product ($l^2 + s^2 > 2\,ls$, Schwarz inequality). Here, the regiospecificity is direct evidence against a concerted reaction.

Several classes of reactions have now been distinguished and the results are summarized in Table 1 [39], where BR stands for biradical and C stand for concerted. The table illustrates the fact that substituents must be considered in tandem, and that biradical mechanisms are more common for these reactions than concerted reactions. The table is not recommended as a device for predicting mechanism and stereochemistry. Instead, a careful interaction diagram using experimental ionization

162

Table 1. Substituent effects in photoformation of oxetanes

Olefin	n-π^* (C=O) Substituents			$^3\pi$-π^* (C=O)
Substituents	Attracting	Donating	Conjugating	Conjugating
Attracting	BR + C	C	C + BR	BR
Donating	BR	BR + C	BR	BR
Conjugating	C + BR	BR	C + BR	BR

potentials and spectroscopically measured excited state energies should be constructed. Identification of the important orbital interactions should distinguish the mode of reaction.

VII. Applications of PMO Theory to Olefin Photocycloaddition Reactions

A. Dimerizations

The photocycloaddition of two identical olefins is allowed to be a concerted $[_2\pi_s + _2\pi_s]$ reaction according to the Woodward-Hoffman [57] rules if the factor of excited state spin multiplicity is ignored. The construction of a simple interaction diagram, Fig. 7, with arbitrary

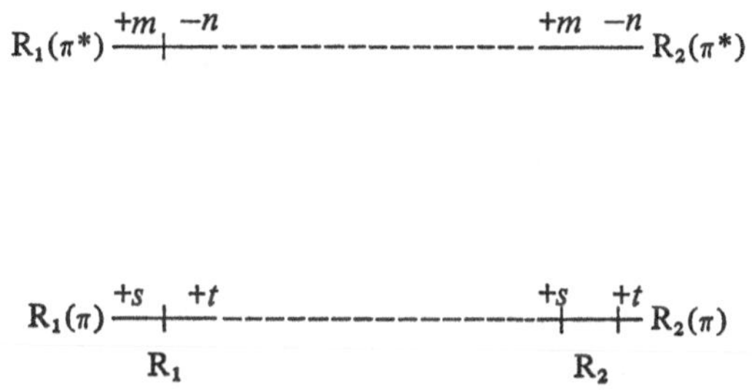

$$R_1(\pi^*) \xrightarrow{\ +m\ \ -n\ } \text{-----------------------} \xrightarrow{\ +m\ \ -n\ } R_2(\pi^*)$$

$$R_1(\pi) \xrightarrow{\ +s\ \ +t\ } \text{---------------------} \xrightarrow{\ +s\ \ +t\ } R_2(\pi)$$

$$\qquad\qquad R_1 \qquad\qquad\qquad\qquad R_2$$

Fig. 7. Interaction diagram. Photocycloaddition of two identical olefins

values of the orbital coefficients shows that the dominant interactions are both first-order and are in phase for a potential concerted reaction. Such reactions would be stereospecific, but the calculated first-order PMO energies, Eq. 5, shows that the reactions should also be highly regioselective. Dimerization should occur preferentially in a head-to-head orientation. For example this follows from the interaction $R_1(\pi^*) \rightarrow R_2(\pi^*)$ because (Schwarz inequality) $(m^2 + n^2)\gamma$ is always larger than $(2\,mn)\gamma$.

Very few examples of singlet π-π^* dimerizations are reported, the reactions of 2-butenes in Eqs. 36 and 37 being examples with observable stereochemistry. The reactions are stereospecific in agreement with theory. The unsensitized reactions of styrene [115] and some stilbene derivatives [116] may also proceed via excited $^1\pi$-π^* states, Eq. 47 and 48. Both reactions show the predicted regiospecificity. Of course, a

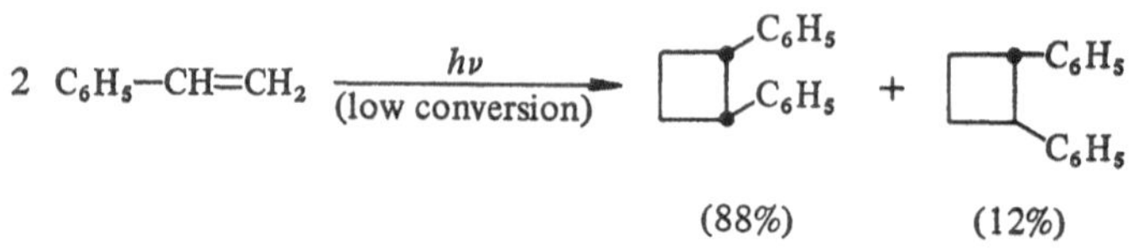

2 C6H5—CH=CH2
hν
(low conversion)
C6H5
C6H5
+
C6H5
C6H5
(88%)
(12%)

(47)

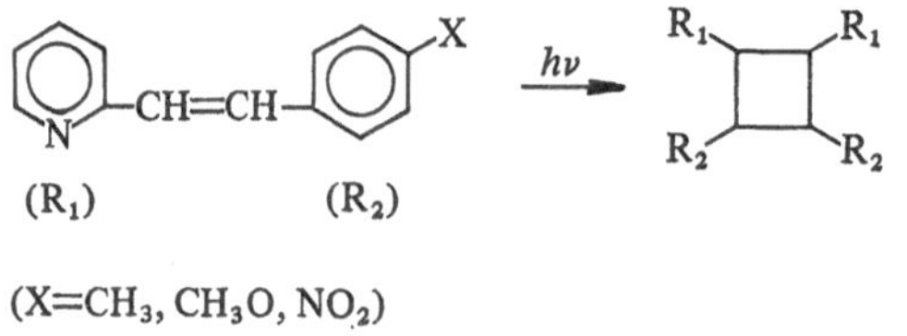

X
CH=CH
N
(R1)
(R2)
(X=CH3, CH3O, NO2)
hν
R1
R1
R2
R2

(48)

biradical intermediate could account for the head-to-head orientation in Eq. 47, but sensitized photolysis of styrene [115] and thermal dimerization [115] give mostly *trans*-head-to-head dimer instead of the *cis* compound. Since both of these latter reactions probably occur by a biradical reaction, the unsensitized reaction is likely to be concerted.

Another illustrative example concerns the photodimerization of coumarin. The direct irradiation in ethanol, Eq. 49,

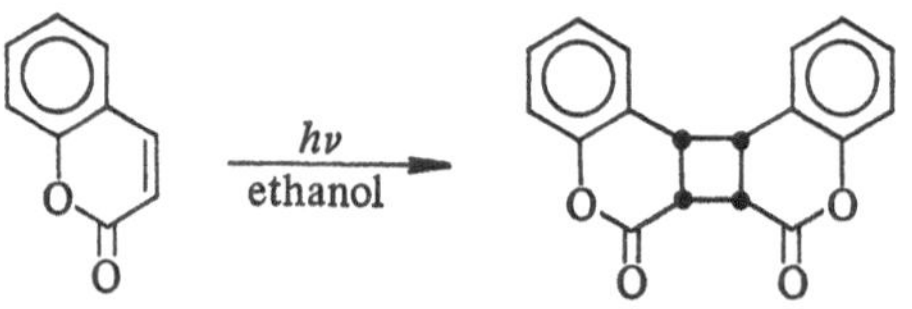

hν
ethanol
O
O

(49)

gives the *cis* head-to-head dimer. [117] An addition of excited-singlet coumarin to a ground-state courmarin with singlet excimer intermediate has been proposed. [118,119] The head-to-head geometry and the *cis* face-to-face orientation of the molecules would be stabilized by the dominant perturbations, Fig. 7 [120–122],in harmony with the isolation of the observed product.

However, there is a dichotomy. Is the product orientation determined at the excimer stage, or do the PMO theory calculations refer to transition states? Efficient merging of excimer geometry into transition state geometry would cause this question to be of little importance for predictive purposes. The same problem arises when considering cross-photocycloadditions. Complex-formation (exciplex) [77,78] has been included in the mechanistic Eqs. 12—18, 32—35, that are given in Section IV. We will not consider the problem further in this paper, but will simply assume that the calculations refer to a product-determining step, whatever its nature. It is clear that PMO theory predicts that singlet state photodimerizations should generally give head-to-head *cis* dimers.

The dimerization of coumarin can also be photosensitized [123], as can nearly all photodimerizations. Triplet-triplet sensitization is the usual type of sensitization [63—65] employed in photocycloadditions and there are numerous examples. Some have been given in Eqs. 19—22, and 28, and additional results will be shown below. The regioselectivity should be calculable from a PMO calculation, and referring to Fig. 7, one again deduces a general rule of head-to-head orientation. However, since the close face-to-face association in a singlet excimer is not possible in a triplet complex, [120—122] and since biradials are logical intermediates from the additions of triplet-state molecules to their ground state counterparts, several factors, including competition between bond-rotation and bond-closure, will determine the stereoselectivity of any particular reaction.

Triplet photoaddition of simple non-cyclic monoolefins is unknown. The sensitized dimerization of ethyl vinyl ether gives exclusively head-to-head adducts, Eq. 21, and probably should not be classed as an example of simple acyclic olefin. Usually the triplets have high energies and are severly twisted. [55] Some cyclic rigid molecules, Eq. 20, that do dimerize [63] do not incorporate substituents that allow regioselectivity to be determined. Butadiene gives principally head-to-head dimerization, Eq. 19, concordant with the PMO prediction, and so does indene, Eq. 22. The *anti* dimer that is formed would not be expected from a singlet excimer reaction.

Dimerizations of α,β-unsaturated carbonyl compounds are perhaps the most interesting reactions, and certainly are the subjects of the most wide-spread investigations. Many are photosensitized, including that of coumarin, Eq. 50. [123] Other reactions of simple enones also involve

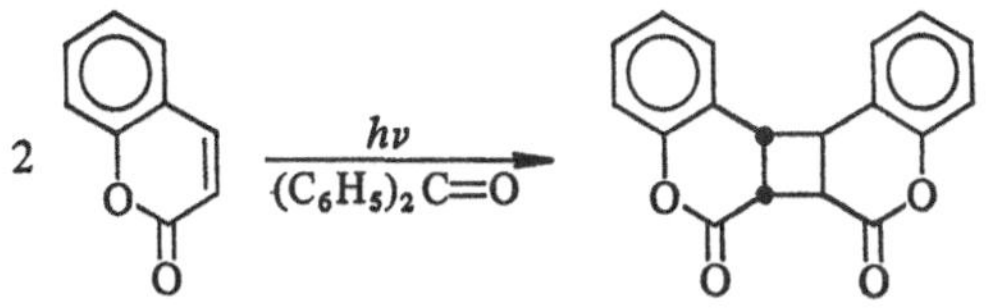

$$2 \quad \text{(structure)} \quad \xrightarrow[\text{(C}_6\text{H}_5\text{)}_2\text{C=O}]{h\nu} \quad \text{(structure)} \qquad (50)$$

triplet states [94,100], even though no external sensitizer may be present. Eqs. 28 and 51. [95,96] A wide variety of pyrimidine derivatives that undergo similar dimerizations will not be discussed here. [125]

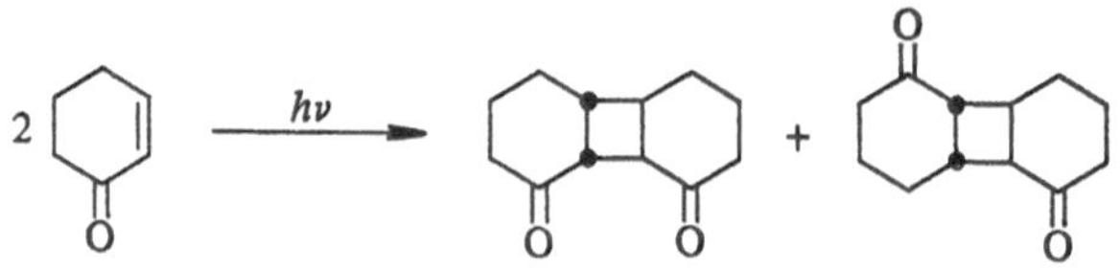

$$(51)$$

The reaction of coumarin is regiospecific in agreement with the general rule of predicted head-to-head orientation, but the reactions of cyclopentenone and cyclohexenone both yield more head-to-tail dimer then the predicted dimer. This is explained in the following way. The involved excited states are $^3\pi\text{-}\pi^*$ states [100,124], and both compounds are relatively small and highly polar. Recent calculations [126], including all valence electrons, have shown that the excited states are polarized toward the oxygen in the same direction as the ground-state, the principle polar effects being due to the sigma electrons. Even the $^1n\text{-}\pi^*$ state of formaldehyde is polarized toward oxygen [126,127], as has been shown experimentally.

An important factor that stabilizes the complex or transition state leading to biradical intermediate is the dipole-dipole interaction between the addends. The orientation given in *13* would be more stable than *14* from this factor, but the orientation in *14* is preferred by the orbital

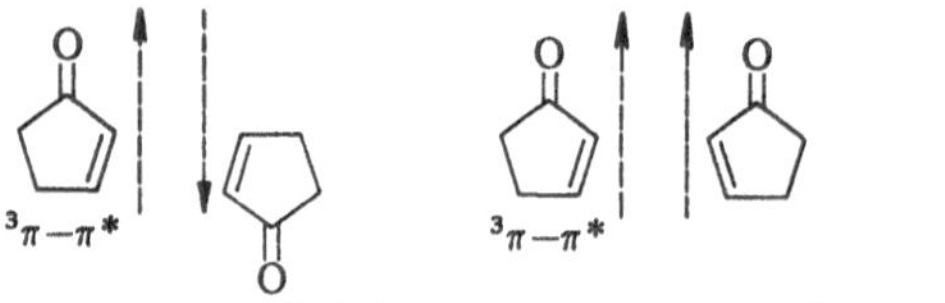

perturbations of Fig. 7. In highly polar solvents, the extent of head-to-head dimerization should increase since the stability of orientation *14* relative to that of *13* increases as solvent becomes more polar. Solvent effects in complete agreement have been reported. In acetonitrile the ratio of *13:14* is 5:4 (cyclopentenone), 1:2 (cyclohexenone). In dilute benzene the ratios are 5:1 for both ketones. [100] The PMO prediction and the unagreeable experimental results are therefore rationalized.

Incidently, the observed solvent effect may be supportive evidence against incursion in nonpolar solvents of a second reactive $^3n\text{-}\pi^*$ state. [129, 130] The $^3n\text{-}\pi^*$ state would be expected to be less selective for the head-to head dimer since one of the principle stabilizing interactions is missing.

Reactions of n-π^* excited molecules have been termed "half-allowed" reactions on this basis. [131] The ^3n-π^* state should be stabilized by the polar solvent [132,133], and one would then expect less of the head-to-head dimer. This argument ignores the dipole-dipole effect and could be invalid on those grounds.

Methyl sorbate gives three dimeric products from sensitized irradiation, Eq. 52. The *trans* configuration of the reactant double bonds is retained in all of the products. [134] Over 80% of the product arises from

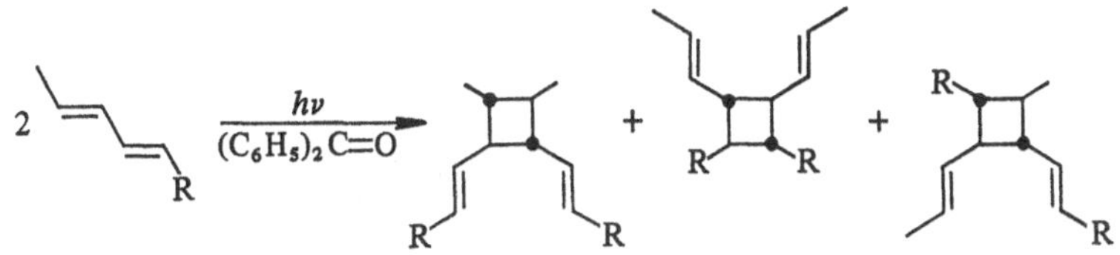

$$\text{(52)}$$

(R = CO$_2$CH$_3$)   (35%)   (47%)   (18%)

head-to-head or tail-to-tail dimerization in accordance with expectation. The production of the head-to-tail dimer in 18% yield is not anomalous from the standpoint of PMO theory. Using the HMO parameters, h (inductive methyl group) $= -0.5$, h (inductive carbonyl group) $= +1.0$, the first-order stabilizing perturbation energies are 0.94γ, 1.14γ, and 0.68γ for the order of products in Eq. 52.

The calculated results in the preceeding paragraph account for the relative amounts of products but the PMO energy differences seem too large. Using a value of $\beta = 3.0$ eV the respective PMO energies are 32.5, 39.4, and 23.5 kcal/mole. Complete specificity to yield the center product would be expected. The calculations assume a four-center mechanism, so it is of interest to see what an assumption of two-center reactions giving biradicals would predict. The calculated PMO energies for the three best orientations to give the observed products are 0.75γ, 0.77γ, and 0.68γ respectively, i.e. 26.0, 26.6, and 23.5 kcal/mole. These energy quantities account much more reasonably for the experimental results, and a reasonable prediction of biradical mechanism can be made. The biradical choice also is in consonance with the fact that the reaction is sensitized.

A remaining point to interpret is the complete retention of stereo-chemistry in the cyclobutane products. Models that have previously been proposed for excited diene triplets [80], *15* and *16*, have a full bond between atom 2 and 3 of the diene, and cannot adequately represent the triplet dienes formed in this reaction. A great deal of double bond character at the terminal bonds is required to prevent rotation. This

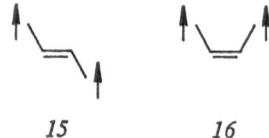

15   16

indicates that the full double bond between atoms 2 and 3 is not a correct representation.

Many derivatives of quinones, cinnamic acids, and mucconic acids photodimerize in solid phases to give results [16] that in many cases are not in agreement with the general PMO rule of head-to-head reaction. However, it is clear that those reactions are controlled by topochemical effects, *i.e.* the geometry and proximity of the reactants in the solid phase. [135] Consequently, PMO theory will not be useful for calculating reactions of that type.

B. Cross-Cycloadditions

Only a few of the many interesting possibilities will be discussed in this section. Examples of some of these have already been given in Eqs. 23—27, 29—31, 38, and 39. The unsensitized, probably singlet reactions in Eqs. 38 and 39 will be covered first.

Reaction 38 is relatively easy to understand with reference to the interaction diagram Fig. 8. The molecular parameters were obtained from HMO calculations.

$$-1.35\,\beta\ \underline{\quad.60\quad\quad-.70\quad}$$

$$\underline{\quad.60\quad\quad-.39\quad}-0.66\,\beta$$

$$0.60\,\beta\ \underline{\quad.75\quad\quad.41\quad}\ \text{-----------------}\ \underline{\quad.60\quad\quad.39\quad}\,0.66\,\beta$$

$$H_2C{=}C(OCH_3)_2 \qquad\qquad\qquad H_2C{=}CH{-}C_6H_5$$

Fig. 8. Interaction diagram. Phenylcyclohexene and dimethoxyethylene

The first-order interaction of the two bonding levels should be the controlling interaction. The calculated PMO energies for concerted reactions are 0.61γ for the observed orientation and 0.53γ for the other orientation. Calculated energies for biradical reactions are much smaller.

The difference in concerted reactions is large (5.5 kcal/mole) and no additional factor is needed to explain the high selectivity of the reaction. However, charge-transfer from olefin to aromatic should theoretically be an important stabilizing configuration in this reaction, and one should see if inclusion of CT will modify the conclusion. The charges on the atoms are shown in **17** after transfer of one electron. A contribution of a CT configuration would help to stabilize the preferred orientation relative

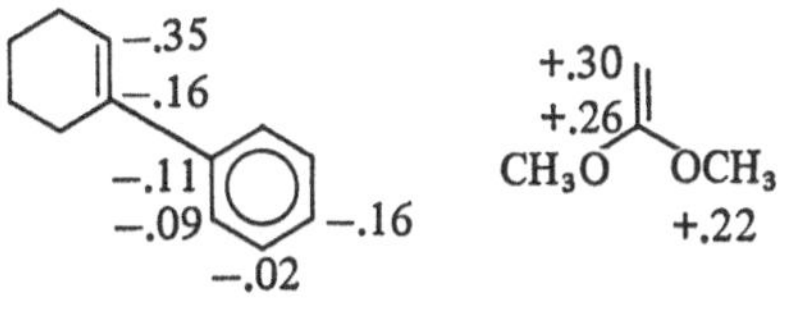

17

to the opposite geometry. It would be interesting to see if a change to more polar solvent would reverse the orientation. This could be an indication as to whether or not the right magnitudes of values for interactions integrals have been chosen.

The reaction in Eq. 39 [103)] is really quite easy to understand, although the result seems geometrically impossible if a molecular model is examined. The reaction involves an $^1 n\text{-}\pi^*$ state since dienic quenchers are noneffective, and the stereochemistry of the product is well-established. [103)] The reaction diagram, Fig. 9, provides the key to under-

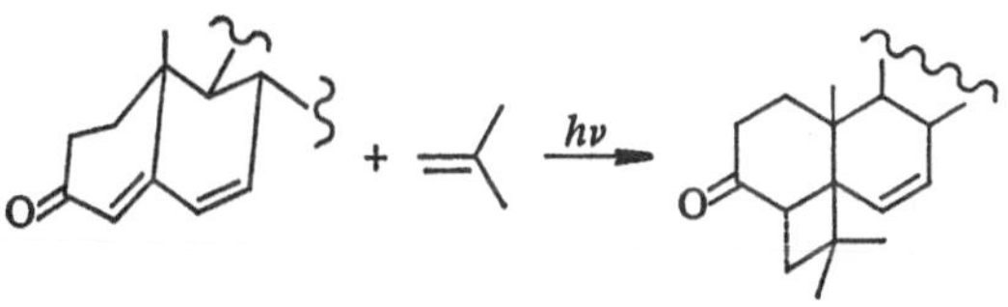

$$\text{(39)}$$

standing the problem. In Fig. 9, the level locations should be relatively accurate since their positions are based mainly on experimental data. The nonbonding n-electron of the dienone are chosen to be at the same energy as those in cyclopentenone from photoelectron spectroscopy [136)], and the excitation value of 2.82 eV is for the O-O band of the phosphorescence emission spectrum (2.66 eV) [137)] with an assumed singlet-triplet splitting of 0.16 eV. [138,139)]

The predominant interactions are all second-order. Interaction $K(\pi^*) \rightarrow O(\pi^*)$ stabilizes an all-suprafacial concerted cycloaddition,

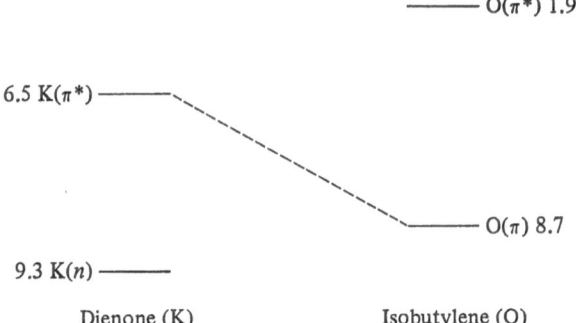

Fig. 9. Intermolecular diagram, isobutylene and a 3-keto-4, 6-dienic steroid. Energies in eV

and interaction $O(\pi) \rightarrow K(\pi^*)$ stabilizes a transition state for a concerted reaction that must be suprafacial-antarafacial and must yield a *trans* product. The energy gap for the former interaction is 4.6 eV while that for the latter is only 2.2 eV. One must conclude that formation of *trans* adduct should be favored, in agreement with the experimental facts.

A recent theoretical treatment of cycloaddition reactions using interaction diagrams [79] discusses diagrams like Fig. 9, and predicts [2+2] photocycloadditions of electron-rich olefins with electron-poor molecules to proceed in a suprafacial-antarafacial manner. The antarafacial component of the reaction is also predicted to be the excited addend in complete agreement with the facts cited above. However, the generalization of concertedness to all [2+2] cycloadditions of this type [79] is too broad. Sensitized and other triplet reactions have been postulated for a number of years to proceed via biradical intermediates. Many detailed correlations of experimental data are known. Furthermore, the problem of inhibition of a triplet concerted reaction due to its spin multiplicity has not been circumvented satisfactorily. If the concerted reaction has a very large gain in stabilization that can be obtained by forming bonds in concert, then the spin problem may be overcome. However, it is difficult to see how an antarafacial-suprafacial reaction $(^3n, \pi^*)$ could be stabilized relative to the always-allowed two-step biradical reaction to the extent required if concertedness is to predominate.

For example, the sensitized additions of dimethyl maleate and fumarate to cyclohexene, [140] Eqs. 53 and 54, are quoted [79] as examples of concerted $[_2\pi_a + _2\pi_s]$ and $[_2\pi_s + _2\pi_s]$ reactions respectively. Since three other cycloadducts with different stereochemistry are formed, comprising about one-third of the product mixture, and since both

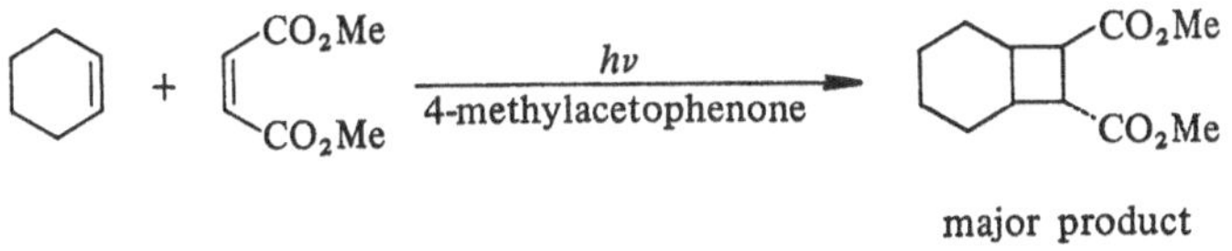

(53)

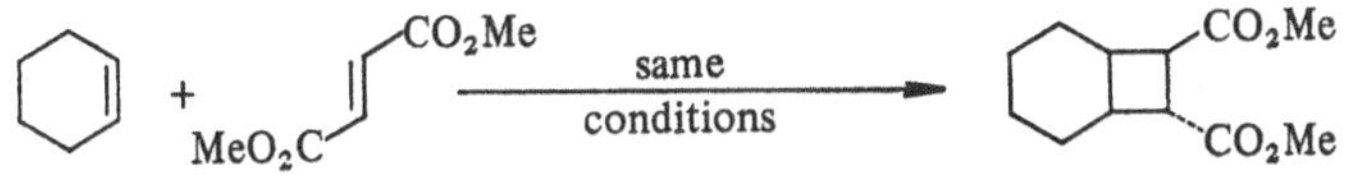

major product

(54)

reactions produce an almost identical mixture of products, one must suspect that the assumption of concertedness is not correct. The experimentalists did suggest that at least part of the major product could arise from a singlet excited state. [140] This fraction of reaction could be concerted, as demonstrated in previous arguments.

The question of the identity of the reactive excited state was left open, although π-π^* excited state was considered to be more probable.[140] The interaction diagram, Fig. 7, shows that a π-π^* state would have an additional stabilizing interaction $R_1(\pi) \rightarrow R_2(\pi)$ with orbital coefficients in phase for an all-suprafacial concerted reaction. The dominant reaction would theoretically depend upon the relative placements of the several levels, and since no experimental information for maleic and fumarate esters is presently available, a clear choice cannot be made. It is interesting that the reactive excited state could be inferred if both stereochemistry and a good molecular diagram were available.

A good interaction diagram can be drawn, Fig. 10, for the reaction of cyclopentadiene with 1,1-dichloroethylene, Eq. 25. [87]

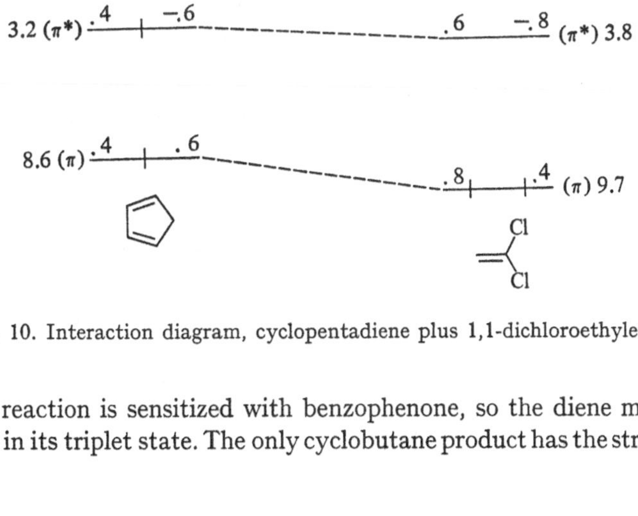

Fig. 10. Interaction diagram, cyclopentadiene plus 1,1-dichloroethylene

The reaction is sensitized with benzophenone, so the diene must be reactive in its triplet state. The only cyclobutane product has the structure shown in *18*. HMO calculations ($h_{Cl} = 2$, $k_{C-Cl} = 1$, inductive methyl parameter $= -0.5$) give the coefficients shown in Fig. 10. Calculated

18

PMO energies are 1.28γ (concerted) and 0.77γ (biradical) for the correct product, 1.26γ (concerted) and 0.66γ (biradical) for the nonobserved product. The energy difference of 0.11γ (3.8 kcal/mole) for a biradical mechanism accounts much more satisfactorily for the specificity of the reaction than the value obtained assuming a concerted reaction. This conclusion is somewhat mitigable since the permanent dipoles of the two reactants and CT states would all stabilize the preferred configuration. A general conclusion is that triplet conjugated dienes should react with electron-poor olefins with high regiospecificity and low stereospecificity.

The final type of reaction that will be discussed is the highly interesting cross photocycloaddition of cyclic α, β- unsaturated ketones with olefins. Examples were given in Eqs. 28—31. A general mechanism [94], to which there may be exceptions to be discussed later, would involve a triplet state of the enone and the reactions steps given in Eqs. 32, 33, and 35, complex formation, biradical formation, and product formation. An earlier idea that two different excited triplet states were reacting has been discounted. [100,141,142] The inefficiency of the reaction is attributed to an alternate decay of complex [77,78,100,142], and the excited state has a π-π^* configuration. [100,142]

A generalized interaction diagram for enone — olefin photoadditions is given in Fig. 11. The bonding levels of cyclopentenone are from photoelectron spectra [136], and the excited levels are phosphorescence emission values. [143] Olefin energy levels are from the prior figures, and the relative sizes of the eigenvectors are inferred from the ground-state polarities. Cyclohexenone systems would have very similar molecular diagrams to the cyclopentenone diagram, the excited states being only a few kilocalories lower in energy. [144]

The important stabilizing perturbations can be discerned for each case by examining the molecular diagram. Since a $^3\pi$-π^* state is involved the principle contributions are $O(\pi) \rightarrow K(\pi)$ and $K(\pi^*) \rightarrow O(\pi^*)$. The interaction of the antibonding energy levels are relatively less important because of the generally larger energy gaps. The reaction of the electron-rich isobutylene molecular may be exceptional because the $O(\pi) \rightarrow K(\pi)$ interaction is of smaller magnitude, and in addition, the $O(\pi) \rightarrow K(\pi^*)$ interaction must be considered.

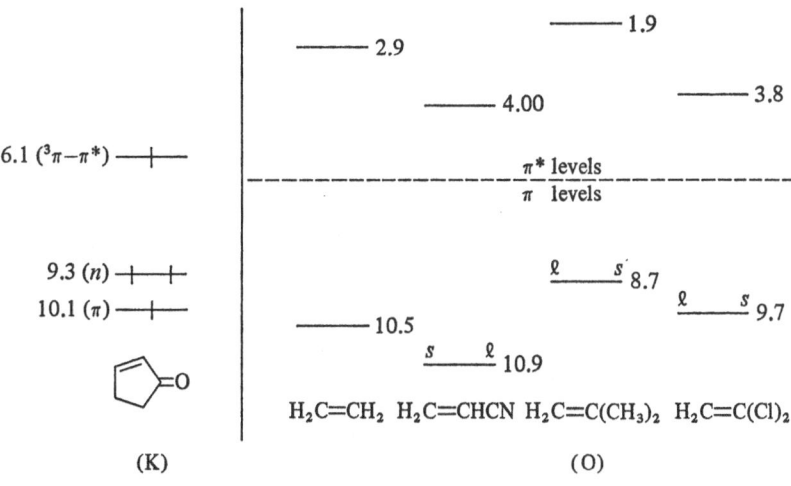

Fig. 11. Interaction diagram. α,β-Unsaturated ketone, and olefins (l and s stand for large and small respectively)

The reaction of ethylene with steroidal α,β-unsaturated ketones yields both *cis* and *trans* cyclobutane derivatives as illustrated in Eq. 55. [145)]

$$\text{(55)}$$

The *trans* compound cannot be formed by a concerted reaction from the $^3\pi\text{-}\pi^*$ state because the predominant level perturbation is the almost degenerate interaction of $K(\pi)$ and $O(\pi)$, that would give *cis* geometry of product. Since the *trans* adduct must be formed via biradical intermediate (the only other possibility), it has usually been convenient to suppose that the *cis* adduct is also formed from the same biradical intermediate. Another choice of mechanism is that *cis* compound is formed via a concerted reaction, and the *trans* compound arises from a biradical pathway. In this case, the spin prohibition could be outweighed by two factors, the favorable geometry and the stabilizing first-order perturbation.

The biradical possesses a large amount of excess energy, some of which can be used in the formation of the highly strained *trans*-fused cyclobutane ring. Of course, some biradical must necessarily form *cis* compound also. So a final possibility is that *cis* compound is formed by two reactions, biradical and concerted, and *trans* is produced only from the

173

biradical. The effect of temperature variation on the product distribution might help to clarify this aspect of the photochemistry.

Temperature changes could also have another effect. Steroidal ketones have been demonstrated to have nearly degenerate ^3n-π^* and $^3\pi$-π^* states, and the possibility of a thermal equilibrium between those states has been pointed out. [146] In fact, testosterone acetate and 10-methyl-$\Delta^{1,9}$-octalone, both very similar molecules to the reactant in Eq. 55, undergo photoreactions from two different triplet states. [144] So we ought to consider what the reactivity of the n-π^* state would be, and whether product distribution would differ from the two states. If so, temperature variation could change product distribution due to a change in the relative proportions of triplet states.

From examination of Fig. 11, it is inferred that the ^3n-π^* state is less reactive, and a biradical mechanism should be the major reaction pathway. The degenerate stabilizing perturbation of the bonding levels is missing, and concerted pathways are not likely if stabilized only by much smaller secondary interactions. If the ^1n-π^* singlet state could be intercepted in some way, the all-suprafacial concerted mechanism would be favored $[\mathrm{K}(\pi^*) \rightarrow \mathrm{O}(\pi^*)]$ relative to the suprafacial-antarafacial mechanism $[\mathrm{O}(\pi) \rightarrow \mathrm{K}(\pi^*)]$.

The regioselectivity of each one of the previously cited reactions, Eqs. 29—31, is well-correlated by the interaction diagram. The degenerate interaction of the bonding levels is controlling, and whether the reaction is concerted or biradical the major orientation should be as shown in 19. The olefin 1,1-dichloroethylene was taken as the model for 1,1-dimethoxy-ethylene.

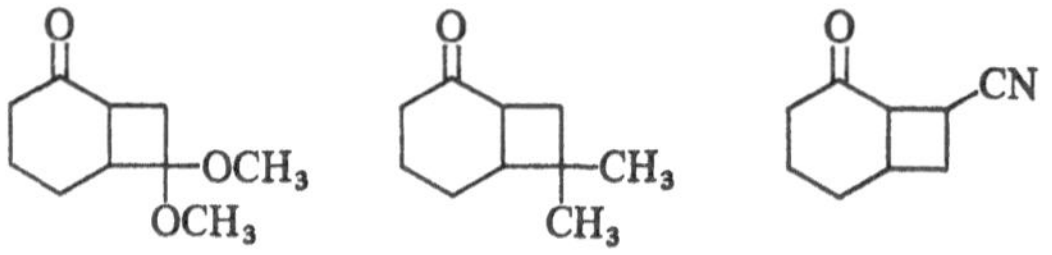

19

Considerations of mixed mechanisms and mixed reactive states might apply to these cases also, but there is no direct evidence that cyclo-hexenone itself reacts from other than the lowest $^3\pi$-π^* state. Both *cis* and *trans* products are produced, except possibly in the acrylonitrile reaction [97], and *cis* and *trans*-2-butene give the same mixture of products. A predominant biradical mechanism would seem the logical choice.

VIII. Concluding Remarks

Many workers may feel that the simple theoretical methods and calculations that we espouse in this essay are not suitable for the treatment of such complicated reactions. In fact, much of this material was submitted for publication in several journals during 1968 and 1969 and rejected, in part for that reason. Our object, in describing so many reactions and so varied a set of experimental results, was to demonstrate that simple perturbational methods can provide semiquantitative insights into several aspects of the mechanisms of reaction, and hence into the effects of substituents.

A great advantage of perturbation theory is that it can be elaborated to the degree thàt MO theory itself can attain. Another advantage is that experimental results can supply the starting molecular parameters. Even electron densities and eigenvectors might be available from NMR coupling constants and chemical shifts. A last advantage is that it is an easy method to apply and it seems to yield results highly congruent with experimental data.

Acknowledgement. The financial support of The Robert A. Welch Foundation is gratefully acknowledged. The author is also grateful for the opportunity of many stimulating discussions with colleagues, William B. Giles, Ernesto Silber, Jerold Feuer, Dale Otteson, Joseph Hull, and John N. Marx.

References

1) Schönberg, A., Schenck, G. O., Neumüller, O.-A.: Preparative organic photochemistry, Second Ed., Springer-Verlag New York, Inc. Berlin-Heidelberg-New York: Springer 1968.
2) Annual survey of photochemistry, Wiley-Interscience, New York, N. Y. Three volumes of this annual review have been published. The latest, Vol. 3 (1971) covers the literature of 1969.
3) Bryce-Smith, D. (Senior Reporter): Photochemistry, Specialist Periodical Report, The Chemical Society of London, Vol. 1, 1970; Vol. 2, 1971. The last volume covers the literature, July 1969 to June 1970.
4) Listed in Herndon, W. C.: Chem. Rev. *72*, 157 (1972).
5) Chesick, J. P.: J. Am. Chem. Soc. *85*, 3718 (1963).
6) Arnold, D. R.: Advan. Photochem. *6*, 301 (1968).
7) Burr, J. G.: Advan. Photochem. *6*, 193 (1968).
8) Chapman, O. L., Lenz, G.: Organic photochemistry (O. L. Chapman, ed.), Vol. 1, pp. 283—321. New York: Marcel Dekker, Inc. 1967.
9) Dilling, W. L.: Chem. Rev. *69*, 845 (1969).
10) Eaton, P. E.: Accounts Chem. Res. *1*, 50 (1968).
11) Fonken, G. L.: In: Organic photochemistry (O. L. Chapman, ed.), Vol. 1, pp. 197—246. New York: Marcel Dekker, Inc. 1967).
12) Muller, L. L., Hamer, J.: 1,2-Cycloaddition reactions, pp. 111—139. New York-Interscience Publishers 1967.
13) Mustafa, A.: Chem. Rev. *52*, 1 (1952).
14) Schaffner, K.: Advan. Photochem. *4*, 81 (1966).
15) Steinmetz, R.: Fortschr. Chem. Forsch., Topics Curr. Chem., *7*, 445 (1967).

16) Trecker, D. J., in: Organic photochemistry (O. L. Chapman, ed.), Vol. 2, pp. 63—111. New York: Marcel Dekker, Inc. 1969.

17) Warrener, R. N., Bremmer, J. B.: Rev. Pure Appl. Chem. 16, 117 (1966).

18) Dewar, M. J. S.: J. Am. Chem. Soc. 74, 3341, 3345, 3350, 3353, 3355, 3357 (1952).

19) Dewar, M. J. S.: Tetrahedron Suppl. 8, Part I, 75. (1966).

20) Dewar, M. J. S.: The molecular orbital theory of organic chemistry, Chapt. 6 and 8. New York: McGraw-Hill Bank Co. 1969.

21) Fukui, K., Fujimoto, H., in: Mechanisms of molecular migrations. (B. S. Thyagarajan, ed.), Vol. 2, pp. 117—190. New York: Interscience Publishers, Inc. 1969.

22) Fukui, K., Fujimoto, H.: Bull. Chem. Soc. Japan 41, 1989 (1968).

23) Fukui, K., Hao, H., Fujimoto, H.: Bull Chem. Soc. Japan 42, 348 (1969).

24) Fukui, K., Fujimoto, H.: Bull. Chem. Soc. Japan 42, 3399 (1969).

25) Fukui, K.: Fortschr. Chem. Forsch. Topics Curr. Chem. 15, 1, (1970).

26) Hudson, R. F., Klopman, G.: Tetrahedron Letters 1967, 1103.

27) Klopman, G., Hudson, R. F.: Theoret. Chim. Acta (Berl.) 8, 165 (1967).

28) Klopman, G.: J. Am. Chem. Soc. 90, 223 (1968).

29) Salem, L.: J. Am. Chem. Soc. 90, 543 (1968).

30) Salem, L.: J. Am. Chem. Soc. 90, 553 (1968).

31) Devaquet, A., Salem, L.: J. Am. Chem. Soc. 91, 3793 (1969).

32) Devaquet, A.: Mol. Phys. 18, 233 (1970).

33) Sustmann, R., Binsch, G.: Mol. Phys. 20, 1 (1971).

34) Sustmann, R., Binsch, G.: Mol. Phys. 20, 9 (1971).

35) Hammond, G. S.: Advan. Photochem. 7, 373 (1969).

36) Fukui, K., Yonezawa, T., Nagat, C.: Bull. Chem. Soc. Japan. 34, 37 (1961).

37) Feuer, J., Herndon, W. C., Hall, L. H.: Tetrahedron 24, 2575 (1968).

38) Herndon, W. C., Giles, W. B.: Mol. Photochem. 2, 277 (1970).

39) Herndon, W. C.: Tetrahedron Letters 1971, 125.

40) Hoffmann, R.: J. Chem. Phys. 39, 1397 (1963).

41) Herndon, W. C., Hall, L. H.: Theoret. Chim. Acta (Berl.) 7, 4 (1967).

42) Pople, J. A.: Accounts Chem. Res. 3, 217 (1970).

43) Herndon, W. C., Feuer, J.: J. Am. Chem. Soc. 90, 5914 (1968).

44) Mantione, M. J.: Theoret. Chim. Acta (Berl.) 11, 119 (1968); 15, 141 (1969).

45) Herndon, W. C., Giles, W. B.: Chem. Commun. 1969, 497.

46) Whangbo, M.-H., Lee, I.: J. Korean Chem. Soc. 13, 273 (1969).

47) Sayre, R., Harlos, J. P., Rein, R., in: Molecular orbital studies in chemical pharmacology (L. B. Kier, ed.), pp. 207—237, Springer-Verlag New York, Inc. Berlin—Heidelberg—New York: Springer 1970.

48) Nagata, C., Imumura, A., Tagashira, Y., Kodama, M.: J. Theoret. Biol. 9, 357 (1965).

49) Song, P.-S., Harter, M. L., Moore, T. A., Herndon, W. C.: Photochem. Photobiol. 13, 521 (1971).

50) Al-Joboury, M. I., Turner, D. W.: J. Chem. Soc. 1963, 5141; 1964, 4434.

51) Baker, A. D., Baker, C., Brundle, C. R., Turner, D. W.: Intern. J. Mass. Spec. Ion Phys. 1, 285 (1968).

52) Worley, S. D.: Chem. Rev. 71, 295 (1971).

53) Brand, J. C. D.: J. Chem. Soc. 1956, 858.

54) Evans, D. F.: J. Chem. Soc. 1960, 1735.

55) Mierer, A. J., Mulliken, R. S.: Chem. Rev. 69, 639 (1969).

56) McGlynn, S. P., Azumi, T., Kinoshita, M.: The triplet state, pp. 84—85. Englewood Cliffs, N. J.: Prentice-Hall, Inc. 1969.

57) Woodward, R. B., Hoffmann, R.: The conservation of orbital symmetry. New York: Academic Press, Inc. 1970.

58) Paterno, E., Chieffi, G.: Gazz. Chim. Ital. *39*, 341 (1909). — Paterno, E.: Gazz. Chim. Ital. *44*, 463 (1914).

59) Buchi, G., Inman, C. G., Lipinsky, E. S.: J. Am. Chem. Soc. *76*, 4327 (1954).

60) Turro, N. J., Dalton, J. C., Weiss, D. S., in: Organic photochemistry (O. L. Chapman, ed.), Vol. 2, pp. 1—62. New York: Marcel Dekker, Inc. 1969.

61) Dalton, J. C., Turro, N. J.: Ann. Rev. Phys. Chem. *21*, 499 (1970).

62) Turro, N. J., Dalton, J. C., Dawes, K., Farrington, G., Hautala, R., Morton, D., Niemczyk, M., Schore, N.: Accounts Chem. Res. *5*, 92 (1972).

63) Turro, N. J., Dalton, J. C., Weiss, D. S., in: Organic photochemistry (O. L. Chapman, ed.), Vol. 2, pp. 1—62. New York: M. Dekker, Inc. 1969.

64) Lamola, A. A.: In: Energy transfer and organic photochemistry (A. A. Lamola and N. J. Turro, eds.), New York: Interscience Publishers 1969. Chapt. 1 and 2.

65) Engel, P. S., Monroe, B. M.: Advan. Photochem. *8*, 245 (1971).

66) Yang, N. C.: Pure Appl. Chem. *9*, 591 (1964). — Yang, N. C., Nussin, M., Jorgenson, M. J., Murov, S.: Tetrahedron Letters *1964*, 3657.

67) Dalton, J. C., Wriede, P. A., Turro, N. J.: J. Am. Chem. Soc. *92*, 1318 (1970).

68) Turro, N. J., Wriede, P. A., Dalton, J. C.: J. Am. Chem. Soc. *90*, 3274 (1964).

69) Barltrop, J. A., Carless, H. A. J.: Tetrahedron Letters 1968, 3901.

70) Barltrop, J. A., Carless, H. A. J.: J. Am. Chem. Soc. *94*, 1951 (1972).

71) Turro, N. J., Wriede, P. A.: J. Am. Chem. Soc. *90*, 6863 (1968).

72) Turro, N. J., Wriede, P. A.: J. Am. Chem. Soc. *92*, 320 (1970).

73) Walling, C., Huyser, E. S.: Org. Reactions *13*, 91 (1963).

74) Shima, K., Sakurai, H.: Bull. Chem. Soc. Japan *42*, 849 (1969).

75) Schroeter, S. H., Orlando, C. M., Jr.: J. Org. Chem. *34*, 1181 (1969).

76) Yang, N. C., Loeschen, R., Michel, D.: J. Am. Chem. Soc. *89*, 5465 (1967). — Yang, N. C., Loeschen, R.: Tetrahedron Letters (1968), 2571.

77) Kochevar, I. E., Wagner, P. J.: J. Am. Chem. Soc. *92*, 5742 (1970).

78) Kochevar, I. E., Wagner, P. J.: J. Am. Chem. Soc. *94*, 3859 (1972).

79) Epiotis, N. D.: J. Am. Chem. Soc. *94*, 1924, 1935, 1941, 1946 (1972).

80) Liu, R. S. H., Turro, N. J., Hammond, G. S.: J. Am. Chem. Soc. *87*, 3406 (1965).

81) Srinivansan, R., Hill, K. A.: J. Am. Chem. Soc. *88*, 3765 (1966).

82) Kuwata, S., Shigemitsu, Y., Odaira, Y.: J. C. S. Chem. Commun 2 (1972).

83) Schenck, G. O., Hartmann, W., Mannsfeld, S. P., Metzner, W., Krauch, C. H.: Ber., *95*, 1642 (1962).

84) Sartori, G., Turba, V., Valvassori, A., Riva, M.: Tetrahedron Letters *1966*, 4777.

85) Dilling, W. L., Little, J. C.: J. Am. Chem. Soc. *89*, 2471 (1967).

86) Dilling, W. L.: J. Am. Chem. Soc. *89*, 2742 (1969).

87) Turro, N. J., Bartlett, P. D.: J. Org. Chem. *30*, 1849 (1965).

88) Mc-Cullough, J. J., Huang, C. W.: Chem. Commun. 815 (1967); Can. J. Chem. *47*, 758 (1969).

89) Bowman, R. M., McCullough, J. J., Swenton, J. S.: Can. J. Chem. *47*, 4503 (1969).

90) de Mayo, C., Yip, R. W., Reid, J. T.: Proc. Chem. Soc. (London) 1963, 54.

91) Barltrop, J. A., Robson, R.: Tetrahedron Letters *1963*, 597.

92) Dilling, W. L., Tabor, T. E., Boer, F. P., North, P. N.: J. Am. Chem. Soc. *92*, 1399 (1970); Dilling, W. L.: private communication.

93) Hardham, W. M., Hammond, G. S.: J. Am. Chem. Soc. *89*, 3200 (1967).

94) de Mayo, P.: Accounts Chem. Res. *4*, 41 (1971).

W. C. Herndon

95) Eaton, P. E.: J. Am. Chem. Soc. *84*, 2344 (1962). — Eaton, P. E., Hurt, W. S.: J. Am. Chem. Soc. *88*, 5038 (1966).
96) Ruhlen, J. L., Leermakers, P. A.: J. Am. Chem. Soc. *89*, 4944 (1967).
97) Corey, E. J., Bass, J. D., LeMahieu, R., Mitra, R. B.: J. Am. Chem. Soc. *86*, 5770 (1964).
98) Chapman, O. L., Koch, T. H., Klein, F., Nelson, P. J., Brown, E. L.: J. Am. Chem. Soc. *90*, 1657 (1968).
99) de Mayo, P., Pete, J.-P., Tchir, M. F.: Can. J. Chem. *46*, 2535 (1968).
100) Wagner, P. J., Buckeck, D. J.: J. Am. Chem. Soc. *91*, 5090 (1969).
101) Yamazaki, H., Cvetanovic, R. J.: J. Am. Chem. Soc. *91*, 520 (1969).
102) Tada, M., Shinozaki, H., Sato, T.: Tetrahedron Letters 1970, 3897
103) Lenz, G. R.: Tetrahedron *28*, 2211 (1972).
104) v. d. Lugt, W. Th. A. M., Oosterhoff, L. J.: J. Am. Chem. Soc. *91*, 6042 (1969).
105) v. d. Lugt, W. Th. A. M., Oosterhoff, L. J.: Chem. Commun. 1968, 1235.
106) Daudel, R.: Advan. Quantum. Chem. *5*, 1 (1970).
107) Dougherty, R. C.: J. Am. Chem. Soc. *90*, 5780, 5788 (1968).
108) Dougherty, R. C.: J. Am. Chem. Soc. *93*, 7187 (1971).
109) Turner, D. W., Baker, C., Baker, A. D., Brundle, C. R.: Molecular photoelectron spectroscopy. London: Wiley-Interscience 1970.
110) a) Lang, N. C., Eisenhardt, W.: J. Am. Chem. Soc. *93*, 1277 (1971);
     b) J. Am. Chem. Soc., footnote 5.
111) Turro, N. J.: In: Energy Transfer and Organic Photochemistry, pp. 133—296. New York: Interscience Publishers 1969.
112) Sakurai, H., Shima, K., Aono, I.: Bull. Chem. Soc. Japan *38*, 1227 (1965).
113) Barltrop, J. A., Carless, H. A. J.: Chem. Commun. *1970*, 1637; J. Am. Chem. Soc. *93*, 4794 (1971).
114) Dowd, P., Gold, A., Sachdev, K.: J. Am. Chem. Soc. *92*, 5275 (1970).
115) Brown, W. G.: J. Am. Chem. Soc. *90*, 1916 (1968).
116) Williams, J. L. R., Carlson, J. M., Reynolds, G. A., Abel, R. E.: J. Org. Chem. *28*, 1317 (1963).
117) Anet, F. A. L.: Can. J. Chem. *40*, 1249 (1962).
118) Hammond, G. S., Staut, C. A., Lamola, A. A.: J. Am. Chem. Soc. *86*, 3103 (1964).
119) Morrison, H., Curtis, H., McDowell, J.: J. Am. Chem. Soc. *88*, 5415 (1966).
120) Slifkin, M. A.: Nature *200*, 766 (1963).
121) Murrell, J., Tanaka, J.: Mol. Phys. *7*, 363 (1964).
122) Smith, F., Armstrong, A., McGlynn, S.: J. Chem. Phys. *44*, 442 (1966).
123) Schenck, G. O., v. Wilucki, I., Krauch, C. H.: Ber. *95*, 1409 (1962).
124) Lam, E. Y. Y., Valentine, D., Hammond, G. S.: J. Am. Chem. Soc. *89*, 3482 (1967).
125) Burr, J. G.: Advan. Photochem. *6*, 193 (1969).
126) Herndon, W. C.: Mol. Photochem. *5*, 253 (1973).
127) Kroto, H. W., Santry, D. P.: J. Chem. Phys. *47*, 2736 (1967).
128) Freeman, D. E., Klemporer, W.: J. Chem. Phys. *40*, 604 (1964).
129) Chapman, O. L., Nelson, P. J., King, R. W., Trecker, D. J., Griswald, A. A.: Record Chem. Progr. *28*, 167 (1967).
130) Chapman, O. L., Koch, T. H., Klein, F., Nelson, P. J., Brown, E. L.: J. Am. Chem. Soc. *90*, 1657 (1968).
131) Ahlgren, G., Akermark, B.: Tetrahedron Letters (1970, 1885.
132) Porter, G., Suppman, P.: Trans. Faraday Soc. *61*, 1664 (1965).
133) Taft, R. W., Glick, R. E., Lewis, I. C., Fox, I., Ehrenson, S.: J. Am. Chem. Soc. *82*, 756 (1960).

[134] Kaufmann, H. P., SenGupta, A. K.: Ann. *681*, 39 (1965).
[135] Schmidt, G. M. J., in: Reactivity of the photoexcited organic molecule, pp. 227—284. London: Interscience Publishers 1967.
[136] Chadwick, D., Frost, D. C., Weiler, L.: J. Am. Chem. Soc. *93*, 4320 (1971).
[137] Herndon, W. C.: unpublished work.
[138] McGlynn, S. P., Azumi, T., Kinoshita, M.: Molecular spectroscopy of the triplet state, Chapt. 2. Englwood Cliffs, N. J.: Prentice-Hill, Inc.
[139] Kearns, D., Marsh, G.: J. Chem. Phys. *49*, 3316 (1968).
[140] Cox, A., de Mayo, P., Yip, R. W.: J. Am. Chem. Soc. *88*, 1043 (1966).
[141] de Mayo, P., Nickolson, A. A., Tchir, M. F.: Can. J. Chem. *47*, 711 (1969).
[142] Wagner, P. J., Bucheck, D. J.: Can. J. Chem. *47*, 713 (1969).
[143] Herz, W., Nair, M. G.: J. Am. Chem. Soc. *89*, 5474 (1967).
[144] Marsh, G., Kearns, D. R., Schaffner, K.: Helv. Chim. Acata *51*, 1890 (1968).
[145] Lenz, G. R.: Tetrahedron, *28*, 2195 (1972).
[146] Marsh, G., Kearns, D. R., Schaffner, K.: J. Am. Chem. Soc. *93*, 3129 (1971).

Das sonderbare Verhalten elektronen-angeregter 4-Ring-Ketone*

The Peculiar Behavior of Electronically Excited 4-Membered Ring Ketones*

Dozent Dr. Wolf-Dieter Stohrer, Dr. Peter Jacobs, Dr. Klaus H. Kaiser, Dipl.-Chem. Gerhard Wiech und Prof. Dr. Gerhard Quinkert**

Institut für Organische Chemie der Universität Frankfurt (Main), Laboratorium Niederrad

Inhaltsverzeichnis

---

* 7. Essay über lichtinduzierte Reaktionen; 6. Essay[1]
** Herrn Professor Karl Winnacker zum 70. Geburtstag gewidmet

W.-D. Stohrer, P. Jacobs, K. H. Kaiser, G. Wiech und G. Quinkert

0. Überblick

Die aus π^*,n-angeregten Cyclanonen gewöhnlicher Ringgröße resultierenden Photo-produkte (isomere Ketone, Ketene, Enale; niedermolekulare Kohlenwasserstoffe) entstehen über intermediäre 1,n-Alkyl/Acyl-Biradikale als photochemische Primär-verbindungen. Nimmt man dementsprechend 1,4-Alkyl/Acyl-Biradikale auch als photochemische Primärverbindungen aus π^*,n-angeregten Cyclobutanonen an, gewinnt man für die dabei auftretende Fragmentierungsprodukte (niedermolekulare Kohlenwasserstoffe und Ketene) ein klares Bild; wie man von den Biradikalen aus zu den ebenfalls entstehenden ringerweiterten Oxacarbenen (Tetrahydrofuryli-denen) kommt, die mit den jeweiligen ketonischen Edukten isomer sind, ist dagegen unklar. Die Hypothese, daß den Intermediärverbindungen aus Mangel an Alter-native kein anderer Weg offen steht, überzeugt nicht. An konfigurations-isomeren D-Nor-16-keto-Steroiden gemachte Beobachtungen schließen 1,4-Alkyl/Acyl-Biradikale als photochemische Primärverbindungen entweder aus oder schreiben ihnen Eigenschaften zu, wodurch sie in der homologen Reihe der 1,n-Alkyl/Acyl-Biradikale diskriminiert sein würden (Einleitung).

Der Geltungsbereich der lichtinduzierten Cyclobutanon/Tetrahydrofuryliden-Isomerisierung wird durch detaillierte Schilderung der bisher bekannt gewordenen 36 positiven und 5 negativen Fälle strukturell umrissen. Als Testkriterium dient die Einschiebung des carbenischen C-Atoms des entsprechenden Tetrahydrofurylidens in die H—O-Bindung eines hydroxylgruppenhaltigen Reaktionspartners unter Aus-bildung von Acetalen; weniger zuverlässig sind Cycloaddition an Olefine zu Cyclo-propylverbindungen sowie Umsetzung mit Sauerstoff zu Lactonen (Abschnitt 2.1).

Die Photo-Ringerweiterung von Cyclobutanonen in alkoholischer Lösung zu Acetalen verläuft stereospezifisch unter Retention der Konfiguration am Zentrum, das vom C-Atom zum O-Atom der Carbonylgruppe wandert. Diese Information ist für die aufgeworfene Frage nach der möglichen Intermediärrolle von 1,4-Alkyl/Acyl-Biradikalen deshalb nicht eindeutig, weil die Hoffmann'sche Konzeption der „through bond-Kopplung" 1,4-Biradikalen eine erhöhte Konfigurationsstabilität am Zentrum C-4 garantiert. Die Beobachtungen, daß konfigurations-isomere 3-Methylencyclobutanone ebenfalls stereospezifisch ringerweitern und daß im resul-tierenden Photoprodukt weder Konstitutions-Isomere des Edukts noch dessen Ring-erweiterungsprodukte auftreten, sprechen allerdings deutlich gegen eine Zwischen-rolle der Biradikale (Abschnitt 2.2).

Im Regelfall wird bei der Photo-Ringerweiterung von Vierring-Ketonen die-jenige CC-Bindung zum höher-substituierten C_α-Atom regio-selektiv gelöst. Ist dieses Zentrum durch Elektronen-Akzeptoren substituiert, kommt es zugunsten von Fragmentierungen nicht zur Oxacarbenbildung. Faßt man die lichtinduzierte Cyclobutanon/Tetrahydrofuryliden-Isomerisierung als photochemische Baeyer/Villinger-Umlagerung auf, so stellt man allenfalls eine formale Analogie fest, ohne jedoch das sonderbare Verhalten der Vierring-Ketone in der homologen Reihe elek-tronen-angeregter Cyclanone aufzuhellen (Abschnitt 2.2).

Cyclobutanone, die nach Bestrahlung in alkoholischer Lösung bei Raumtempe-ratur zu ringerweiterten Acetalen reagieren, liefern nach Einwirkung von 302 oder 313 nm-Licht bei −186 °C Photoprodukte, die im gleichen Wellenlängenbereich wie die mit den Singulett-Oxacarbenen isoelektronischen Azoverbindungen absor-bieren. In einer Reihe von Beispielen geht der tieftemperatur-elektronenspektros-kopisch erkennbare primäre Photo-Transient in einen längerwellig absorbierenden, kinetisch ebenfalls instabilen, sekundären Thermo-Transienten über. Die Vermu-tung, daß es sich beim Photo-Transienten um das betreffende Oxacarben, beim Thermo-Transienten um das daraus zugängliche 1,4-Alkyl/Acyl-Biradikal handelt, entspricht der Annahme, daß die genannten Biradikale in einer sekundären Dunkelreaktion durch Ringöffnung der im photochemischen Primärprozeß auftretenden Oxacarbene entstehen (Abschnitt 2.3).

Um die Außenseiterrolle zu begründen, die elektronen-angeregte Vierring-Ketone in der Familie der Cyclanone spielen, werden photochemische α-Spaltung „normaler" Cyclanone und Photo-Ringerweiterung der Cyclobutanone im Licht einfacher störungstheoretischer Argumente einander gegenüber gestellt. In beiden Fällen setzt die betreffende Reaktion mit der Dehnung der CC_α-Bindung ein. Während das MO-Korrelationsdiagramm für die α-Spaltung „normaler" Ketone erkennen läßt, daß die nodalen Eigenschaften der relevanten σ-Orbitale bis zur vollständigen Entkopplung zwischen C und C_α erhalten bleiben (Abschnitt 3.1), geht aus dem MO-Korrelationsdiagramm für die Dehnung der CC_α-Bindung beim Cyclobutanon hervor, daß C und C_α hier nicht entkoppeln; ab einem bestimmten Dehnungsabstand wird die zunächst vorherrschende through space-Wechselwirkung von der through bond-Kopplung abgelöst, d.h. man begegnet einem typischen Beispiel der bond/stretch-Isomerie. Während das höchste, im π*,n-angeregten Zustand mit einem Elektron besetzte σ-Orbital bei der α-Spaltung „normaler" Cyclanone sowie im Strukturbereich des bond-Isomers beim Cyclobutanon zwischen den Zentren C_α und O antibindend ist, weist dieses Orbital im Strukturbereich des stretch-Isomers beim Cyclobutanon zwischen den Zentren C_α und O bindenden Charakter auf und fordert dadurch geradezu den Ringschluß zum Oxacarben (Abschnitt 3.2).

Diese qualitative MO-Analyse fordert weder ein 1,4-Alkyl/Acyl-Biradikal als Zwischenverbindung auf dem Reaktionspfad zum Oxacarben, noch schließt sie eine solche Spezies aus; alles wird davon abhängen, ob der Übergang von einer höheren zur untersten Energiehyperfläche im Strukturbereich des Oxacarbens oder des 1,4-Biradikals stattfindet. Die mit der bond/stretch-Isomerisierung verknüpften krassen Änderungen der Bindungsverhältnisse im Molekül wirken sich auf Art und Richtung der Relaxation der Liganden an den Zentren C und C_α aus. Mit Hilfe der Erweiterten-Hückel-Methode wird die molekulare Dynamik untersucht, die sich der Dehnung der Bindung C—C_α überlagert und somit ein Modell für die licht-induzierte Cyclanon/Oxacarben-Isomerisierung hergeleitet. Die vorgebrachten EH-Ergebnisse, die natürlich nur grob-qualitativen Charakter haben, sprechen sich für eine diabatische Photo-Isomerisierung von Cyclobutanonen zu Tetrahydrofuryl-idenen aus, die sich, je nach Substitution und vorhandenem Reaktionspartner, mit unterschiedlicher Neigung auf der Grundzustands-Hyperfläche zum entsprechenden 1,4-Alkyl/Acyl-Biradikal öffnen (Abschnitt 3.3).

1. Einleitung

1.1. Photochemie der Cyclanone gewöhnlicher Ringgröße in Lösung

Die bis auf Ciamician und Silber [2] zurückgehenden, inzwischen sehr zahlreich gewordenen photochemischen Untersuchungen an Cyclanonen gewöhnlicher Ringgröße trugen nicht unwesentlich zur Etablierung der Chemie elektronen-angeregter Moleküle bei. Die an einem willkürlich herausgehobenen Beispiel (s. Abb. 1) aufgezeigte Klassifizierung in verschiedene Reaktionstypen ist längst durch komparative und quantitative Überlegungen [3] erweitert worden [4,5].

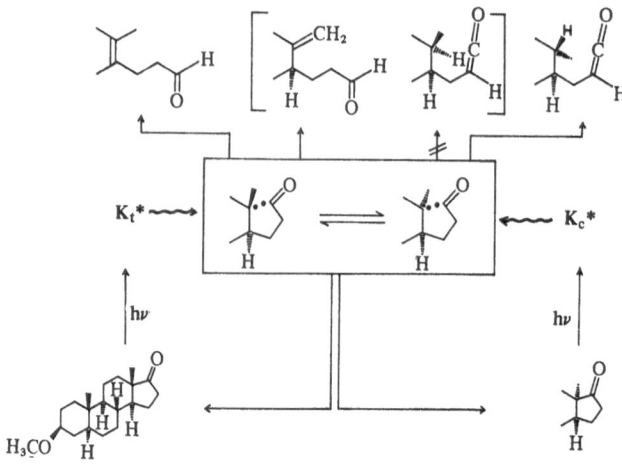

Abb. 1. Die typischen Photo-Isomerisierungen eines Fünfring-Ketons, aufgezeigt am Beispiel eines 17-Keto-Steroids [6-9]; der eingeklammerte Aldehyd tritt nur untergeordnet auf [7], das eingeklammerte Keten wurde nicht aufgefunden [10]

Das zur Diskussion stehende π^*,n-angeregte Keton geht eine α-*Spaltung* (Norrish Type I Process [11]) zum entsprechenden 1,5-, 1,6- oder 1,7-Alkyl/Acyl-Biradikal ein, das

— zum ursprünglichen Edukt (und falls möglich zu seinem Konfigurations-Isomer) *rekombiniert*,
— zu einem ungesättigten Aldehyd oder mehreren konstitutions-isomeren ungesättigten Aldehyden,
— zu einem Keten oder mehreren konstitutions- und/oder konfigurations-isomeren Ketenen *disproportioniert*,
— darüber hinaus evtl. zu einem weiter reagierenden 1,4-, 1,5- oder 1,6-Alkyl/Alkyl-Biradikal *decarbonyliert*.

Die Photo-Konfigurationsisomerisierung wurde erstmals von Butenandt u. Mitarb.[12], die Aldehydbildung von Ciamician und Silber [2] und die Ketenbildung aus unserem Laboratorium [8,13] berichtet.

Für die α-Spaltung gilt als — definitionsgemäß mit Ausnahmen versehene — Regel [14], daß die Bindung zwischen dem C-Atom der lichtabsorbierenden Carbonylgruppe und dem höher alkyl-substituierten Nachbar-C-Atom gelöst wird. Während ein zeitweilig angezweifelter biradikalischer Transient [15] sich zunächst als konzeptionell bequemes Bindeglied zwischen dem elektronenangeregten Edukt und den diversen Produkt-Komponenten durchgesetzt hatte, wird ihm inzwischen eine experimentell gesicherte Existenz zugeschrieben [16-18].

Durch eine Analyse konkurrierender Energieübertragungen [19] vom elektronen-angeregten Cyclanon auf geeignete Löscher (Quencher) ist in vielen Fällen die Frage beantwortbar, in welchem Spin-Zustand sich das photoreaktive Molekül befindet. Die bislang vorliegenden Ergebnisse [4,20] erlauben die Verallgemeinerungen, daß

— die α-Spaltung auf Energiehyperflächen initiiert wird, die untergeordnet vom ersten angeregten Singulett-Zustand und vornehmlich vom zugehörigen Triplett-Zustand des Ketons ausgehen;

— die Lebensdauer des elektronen-angeregten Edukts im Triplett-Zustand beträchtlich geringer ist als im entsprechenden Singulett-Zustand.

1.2. Photochemie der Cyclobutanone in Lösung

Einem elektronen-angeregten Cyclobutanon-Derivat stehen erfahrungsgemäß generell drei verschiedene Produkt-Typen gegenüber. Ein geeignetes Beispiel (s. Abb. 2) führt vor, daß

— unter Extrusion von CO ein Cyclopropan-Derivat,

— durch Cycloeliminierung Keten und ein Äthylen-Derivat,

— in einer Ringerweiterung ein isomeres Oxacarben

entsteht.

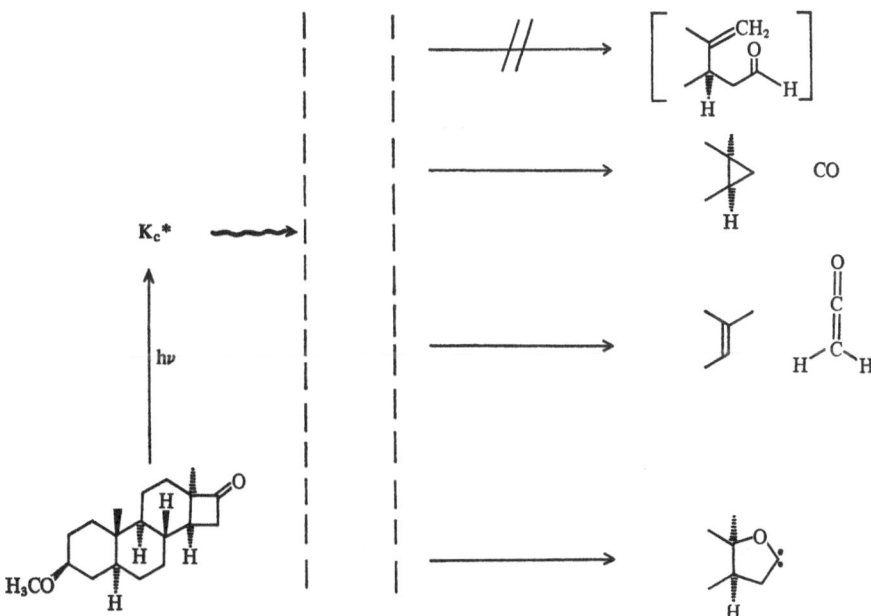

Abb. 2. Die typischen Photo-Reaktionen eines Vierring-Ketons, aufgezeigt am Beispiel eines D-Nor-16-keto-Steroids [21]; der eingeklammerte γ,δ-ungesättigte Aldehyd wurde nicht aufgefunden

Photo-Decarbonylierung und Photo-Cycloeliminierung eines Vier-ring-Ketons in kondensierter Phase wurden unseres Wissens erstmals aus unserem Laboratorium berichtet [21]; die erste Photo-Ringerweite-rung eines Cyclobutanon-Derivats geht auf Hostettler [22] zurück, der an die Beobachtungen von Yates und Kilmurry [23,24] anknüpfen konnte, nach denen Cyclocamphanon bei der Lichteinwirkung in alkoholischer Lösung in ein konstitutionell unsymmetrisches, semicyclisches Acetal, in wäßriger Lösung in ein konstitutionell symmetrisches Bisacetal übergeht (s. Abb. 3).

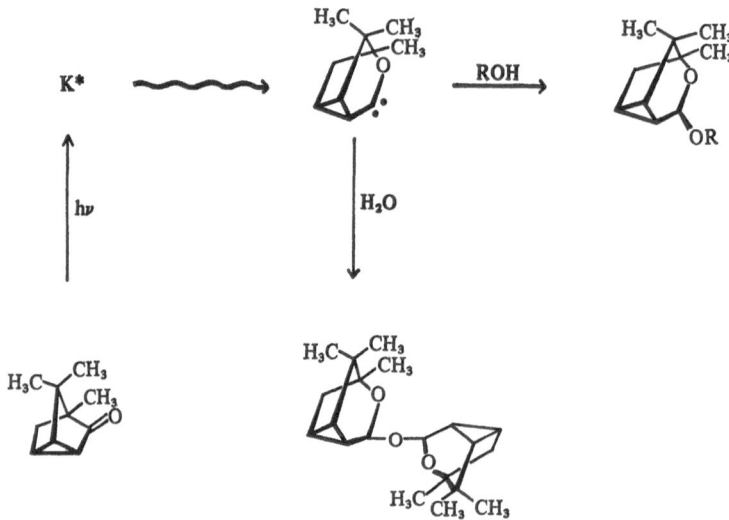

Abb. 3. Das erste bekannt gewordene Beispiel einer lichtinduzierten Ringerweiterung eines Cyclanons zum entsprechenden Oxacarben, das in Gegenwart hydroxyl-gruppen-haltiger Lösungsmittel zu Acetalen reagiert [25]

Von Anfang an hat man verständlicherweise versucht, die Photo-chemie der Vierring-Ketone in diejenige der Cyclanone mit gewöhn-licher Ringgröße zu integrieren: Auf eine einleitende α-Spaltung zu einem 1,4-Alkyl/Acyl-Biradikal sollen weitere Umsetzungen dieses kine-tisch instabilen Transienten folgen. Während Decarbonylierung und Grob-Fragmentierung [25a,b] sich hierfür nahezu aufdrängen, ist die Cycli-sierung zu einem Oxacarben als eine Verlegenheitslösung gespannter Ringsysteme angesehen worden [25], indem die sonst üblichen intra-molekularen Reaktionen — Rekombination zum carbocyclischen System und Disproportionierungen zu Aldehyden und/oder Ketenen — aus ener-getischen und/oder entropischen Gründen als benachteiligt galten.

Am Beispiel des D-Nor-16-keto-Steroids (s. Abb. 2) fällt auf, daß der
γ,δ-ungesättigte Aldehyd mit exocyclischer CC-Doppelbindung nicht im
Produkt zu finden ist, obwohl er als einzige der formal möglichen Dispro-
portionierungs-Komponenten aus einem 1,4-Alkyl/Acyl-Biradikal favo-
risiert sein sollte. Diese Beobachtung, zusammen mit der noch im Detail
zu erörternden stereochemischen Besonderheit der lichtinduzierten
Cyclanon/Oxacarben-Isomerisierung (s. Abschnitt 2.2), veranlaßte
uns [21], ein „freies" Alkyl/Acyl-Biradikal als Intermediärverbindung
auszuschließen: Entweder tritt ein solcher Transient überhaupt nicht
auf oder aber er besitzt Eigenschaften, die ihn als Mitglied der homologen Reihe der 1,n-Alkyl/Acyl-Biradikale diskriminieren.

Ketone sind jüngst als die besonderen Lieblingskinder organischer
Photochemiker apostrophiert worden [26]. A la bonne heure! Solange
man das sonderbare Verhalten elektronen-angeregter Vierring-Ketone
nicht deuten kann, ist es mit dem Verständnis der Photochemie von Cyclanonen nicht allzu weit her. Da sich die lichtinduzierte Cyclobutanon/
Tetrahydrofuryliden-Isomerisierung inzwischen zur *crux interpretum*
entwickelt hat, konzentrieren wir uns im vorliegenden Übersichtsartikel
bewußt ausschließlich auf diese Umlagerung in kondensierter Phase.

2. Die Photo-Ringerweiterung des Cyclobutanons und seiner Derivate

2.1. Gültigkeitsbereich der Reaktion und Natur ihrer Produkte

Die uns bislang bekannt gewordenen Fälle der Photo-Ringerweiterung
bei Vierring-Ketonen sind in der nachfolgenden Tabelle 1 zusammen-
gestellt. Die Edukte lassen sich in fünf Gruppen gliedern: Die erste umfaßt den unsubstituierten Grundkörper und seine Derivate, die in den
2,2-, 2,3-, 2,2,3-, 2,2,4,4-, 2,2,3,4- sowie in den 2,2,3,4,4-Positionen durch
Alkyl-, Alkoxy-, Aryl-, Hydroxy-, substituierte Amino-Reste oder
Halogen-Atome substituiert sind und schließt Fälle mit ein, in denen der
Cyclobutanon-Ring in 2,3-Stellung einem gesättigten 6- oder 8-gliedri-
gen bzw. einem ungesättigten 5-gliedrigen Ring ankondensiert ist. In
der zweiten Gruppe ist C-3 des Cyclobutanon-Ringes entweder an einer
CC- oder CN-Doppelbindung beteiligt oder es ist das Spiro-Zentrum eines
carbocyclischen oder heterocyclischen Spiro|2,3|hexan-Systems. In der
dritten Gruppe nimmt C-2 des Cyclobutanon-Ringes entweder an einer
CC-Doppelbindung teil oder es bildet das Spiro-Zentrum eines carbocyc-
lischen Spiro|2,3|hexan-Systems. Die vierte Gruppe schließt Benzo-
cyclobuten-3,4-dion sowie dasjenige α-Diketon mit ein, bei dem die beiden
verbleibenden Ring-C-Atome noch jeweils einem Cyclopropan-System

angehören. Die fünfte, in der Tabelle nicht aufgeführte Gruppe enthält schließlich die Vierring-Ketone, die im Gegensatz zu den Mitgliedern der vier erstgenannten Gruppierungen bei der Bestrahlung in alkoholischer Lösung nicht zu Acetalen führen (s. Abb. 4).

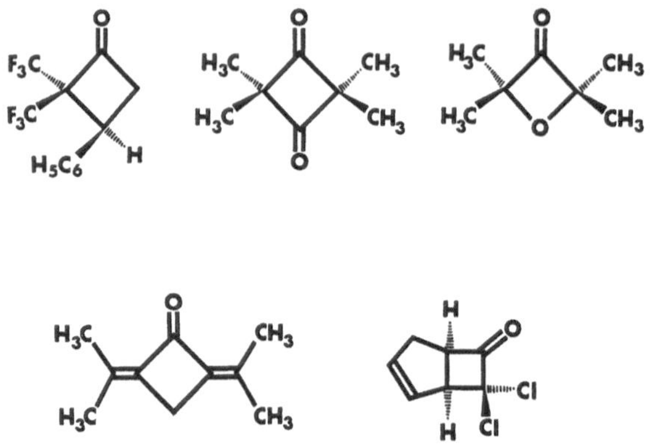

Abb. 4. Vierring-Ketone, die bei der Bestrahlung in alkoholischen Solventien nicht zu Acetalen reagieren: Das bis-trifluormethylierte Keton [27] geht ausschließlich Photo-Cycloeliminierung ein, das tetramethyl-substituierte 1,3-Dion [28] wie auch das Oxetanon-Derivat [29] folgen daneben z. T. der Photo-Decarbonylierungsroute; das gekreuzt-konjugierte Dienon [30] verhält sich photochemisch inert, und das bicyclische Dichlor-keton [31] unterliegt der Photo-Cycloeliminierung

Die im Beisein von Alkoholen oder Wasser gewonnenen Acetale und/oder Halbacetale werden als reaktions-erhellende Abfangprodukte kinetisch instabiler Oxacarbene durch die hydroxylgruppen-haltigen Solventien angesehen. Durch gelegentliche Verwendung von RO-D ist jedesmal sichergestellt worden, daß dem bei der Ringerweiterung neu erzeugten Chiralitätszentrum die Liganden OR und D zuzuordnen sind, daß mit anderen Worten eine Einschiebung des carbenischen C-Atoms eines Tetrahydrofurylidens in die O—D-Bindung des Reaktionspartners stattgefunden haben kann (s. Tabelle 1). Die hin und wieder in Anwesenheit von Olefinen oder Sauerstoff isolierten Verbindungen — Cycloadditionsprodukte oder Lactone (s. Abb. 5) — sprechen ebenfalls für intermediär auftretende Oxacarbene. Zu Bemühungen, diese Transienten direkt nachzuweisen sowie zur Frage, ob sie unmittelbar im photochemischen Primärprozeß oder in einer anschließenden Dunkelreaktion entstehen, siehe die Abschnitte 2.3 bzw. 2.2 und 3.3.

Tabelle 1. Übersicht zur lichtinduzierten Ringerweiterung von Cyclobutanon-Derivaten

Edukt	Bestrahlungsbedingungen	Produkt	Lit.
(Cyclobutanon)	 Hg-Hochdrucklampe Quarz Raumtemperatur CH₃OH (CH₃OD)	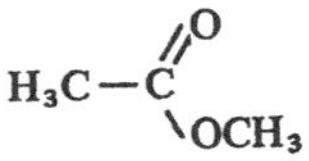 $\varnothing_{313} = 0,02$; (Struktur) 8%[1]) ; $H_3C-\overset{O}{C}-OCH_3$ 48%[1])	30,34, 25)
(2,2-Dimethylcyclobutanon)		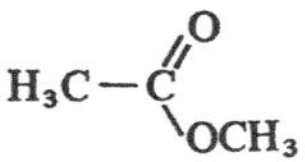 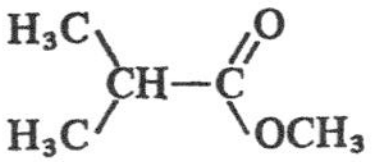 (Struktur) 41%[2]) ; (Cyclopropan mit CH₃, CH₃) 8%[3]) ; $H_3C-\overset{O}{C}-OCH_3$ 32%[2]) ; $H_3C-\underset{H_3C}{CH}-\overset{O}{C}-OCH_3$ 3%[2])	30,34)

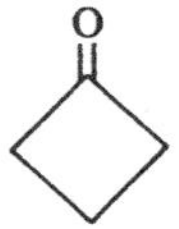

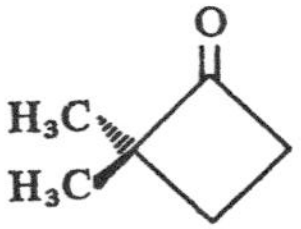

Tabelle 1 (Fortsetzung)

Edukt	Bestrahlungs-bedingungen	Produkt	Lit.
			35)
	313nm-Licht CH₃OH		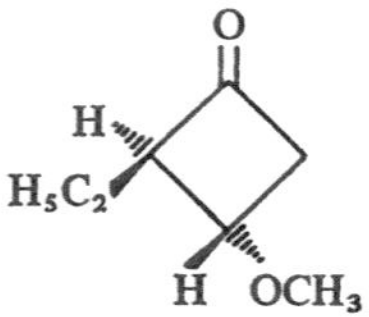 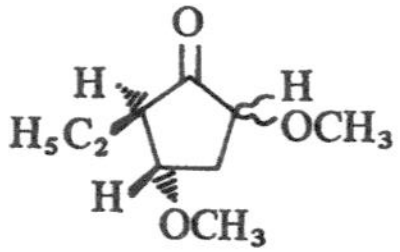 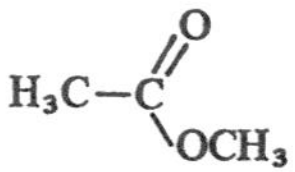 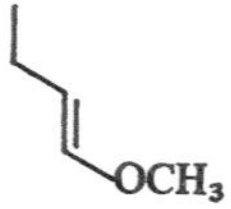  35)
	313nm-Licht 27 °C CH₃OH		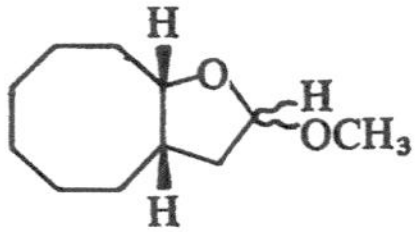 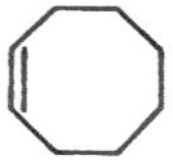 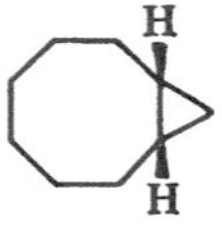 36)

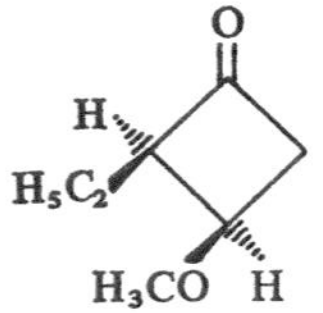

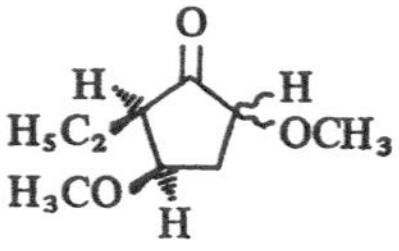

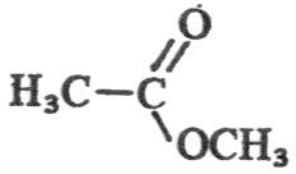

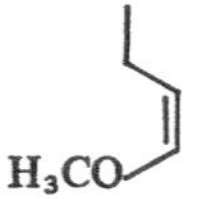

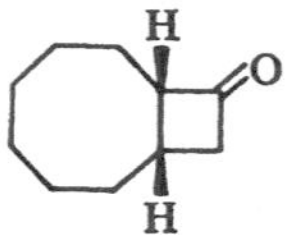

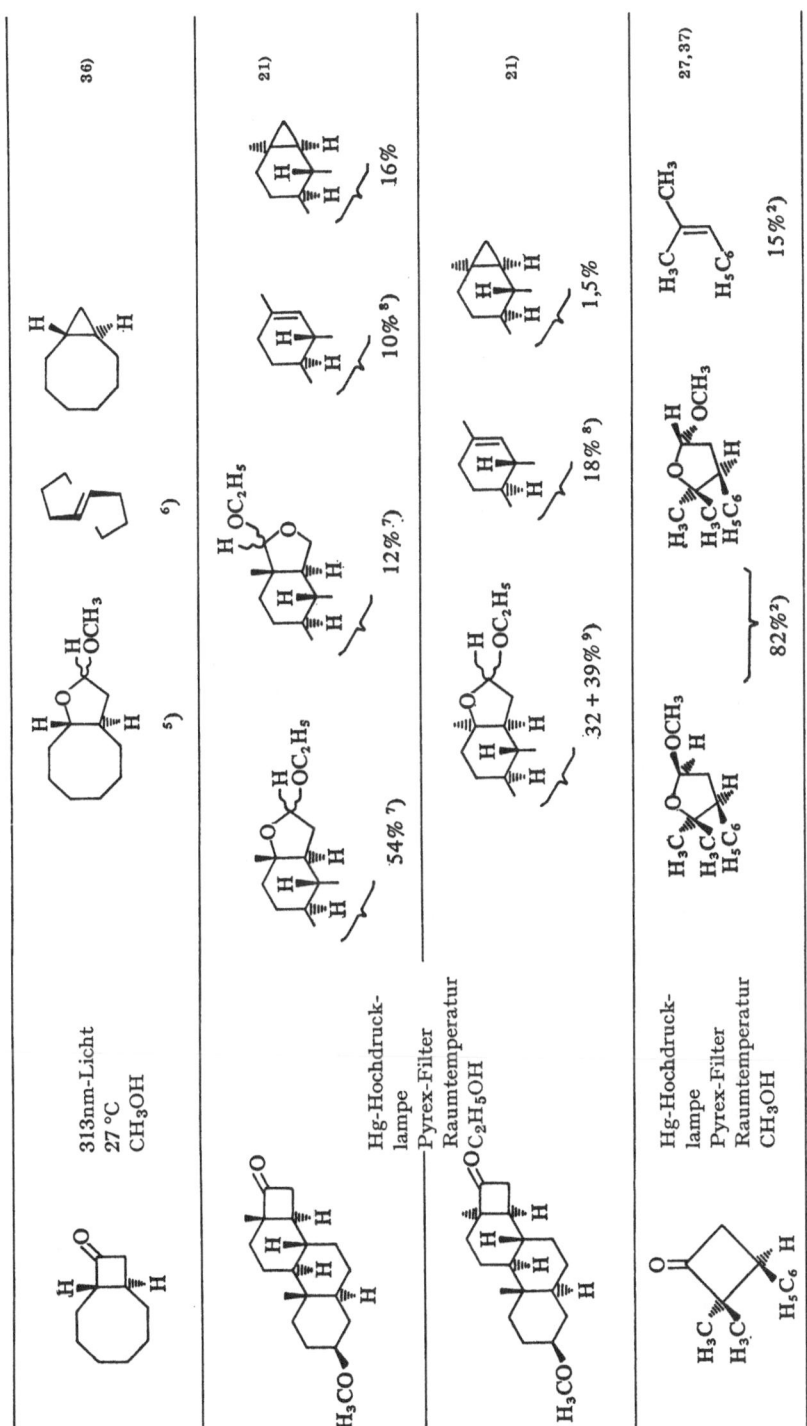

191

Tabelle 1 (Fortsetzung)

Edukt	Bestrahlungs-bedingungen	Produkt	Lit.
	Hg-Hochdruck-lampe Quarz Raumtemperatur CH₃OH	$\varnothing_{313} = 0{,}14$ 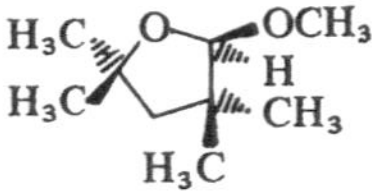 68% [2] 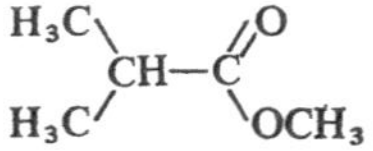 13% [2]  11% [3]	30,34)
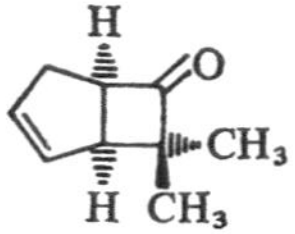	Hg-Hochdruck-lampe Raumtemperatur CH₃OH	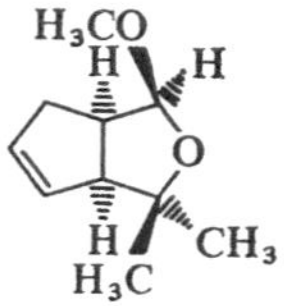 10) 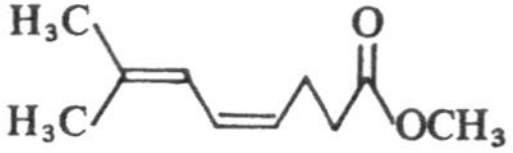 10)	22)
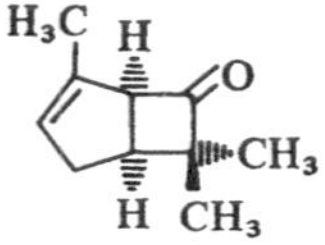	Hg-Hochdruck-lampe Vycor-Filter CH₃OH	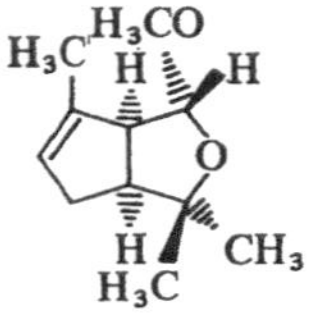 11% 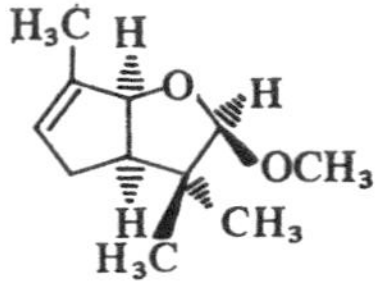 3% 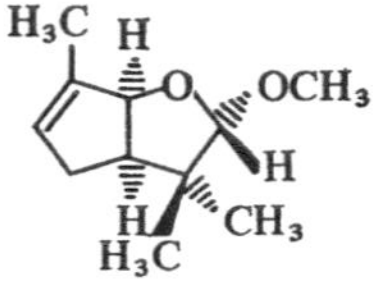 2% [11]	38)

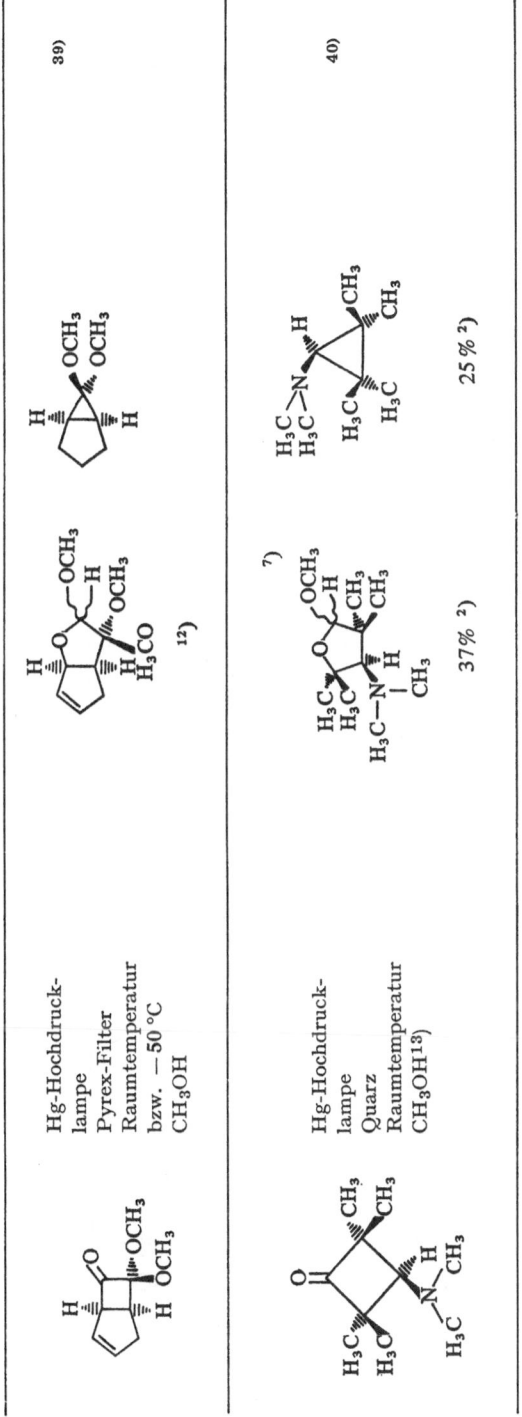

39)

Hg-Hochdruck-
lampe
Pyrex-Filter
Raumtemperatur
bzw. – 50 °C
CH₃OH

12)

40)

Hg-Hochdruck-
lampe
Quarz
Raumtemperatur
CH₃OH[13]

7)

37 % [2]

25 % [2]

Tabelle 1 (Fortsetzung)

Edukt	Bestrahlungs-bedingungen	Produkt	Lit.

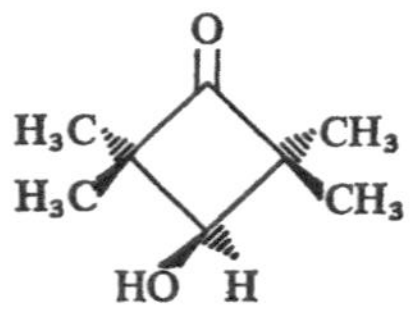

Row 1 — Bestrahlungsbedingungen: Hg-Hochdruck-lampe Quarz Raumtemperatur CH_3OH

$\emptyset_{313} = 0,16$

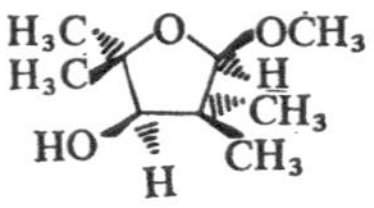

Produkte: 36 %[1]   31 %[1]   17 %[1]   14)

Lit.: 30, 34, 40)

Row 2 — Bestrahlungsbedingungen: Raumtemperatur CH_3OH

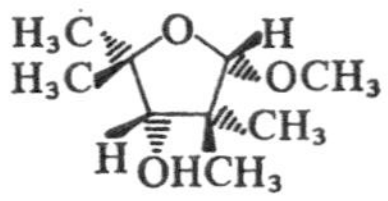

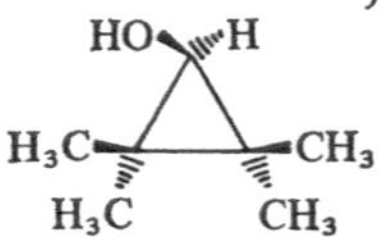

Produkte: 30 %[2]   35 %[2]

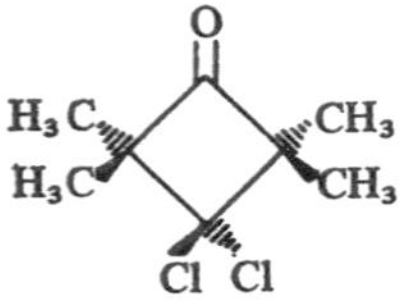

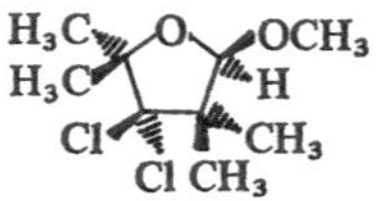

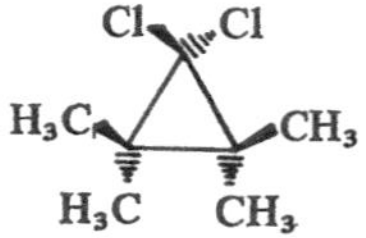

Lit.: 40)

Row 3 — Bestrahlungsbedingungen: Hg-Hochdruck-lampe Pyrex-Filter Raumtemperatur CH_3OH

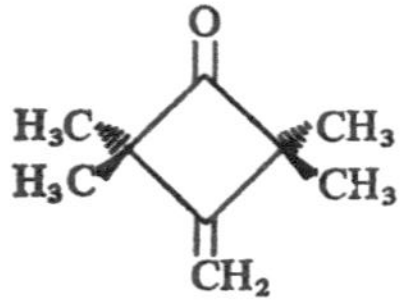

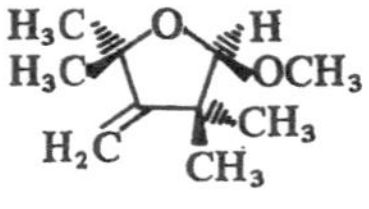

Produkte: 89 %[2]   15)   15)

Lit.: 37)

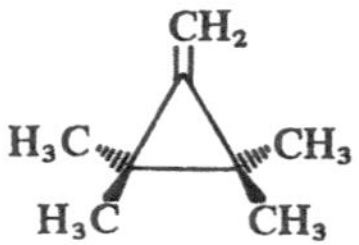

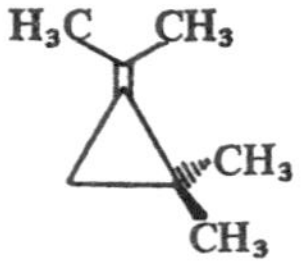

194

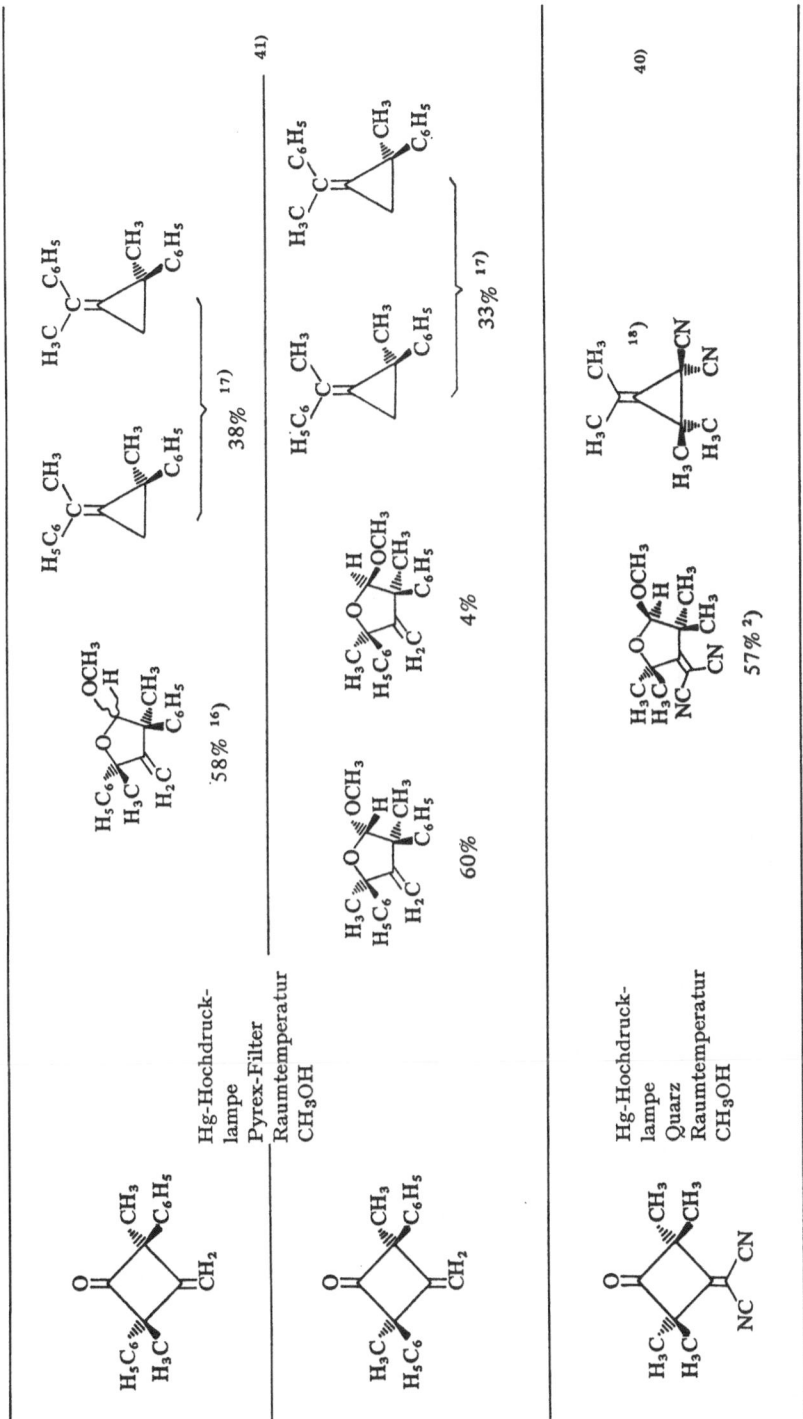

Tabelle 1 (Fortsetzung)

Edukt	Bestrahlungs-bedingungen	Produkt	Lit.
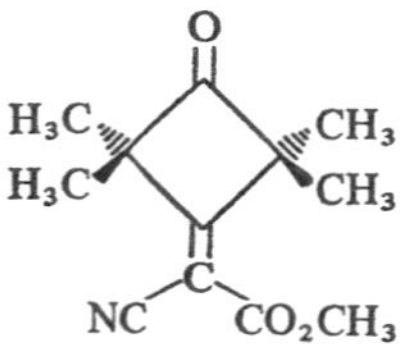	Hg-Hochdruck-lampe Quarz Raumtemperatur CH_3OH	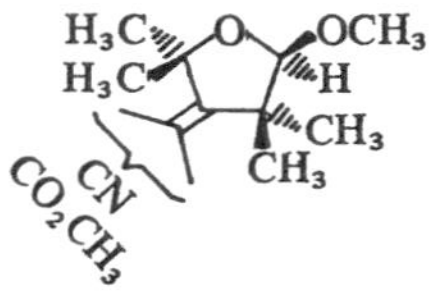 58% [19] 73% [2,21] 23% [2]	[40]
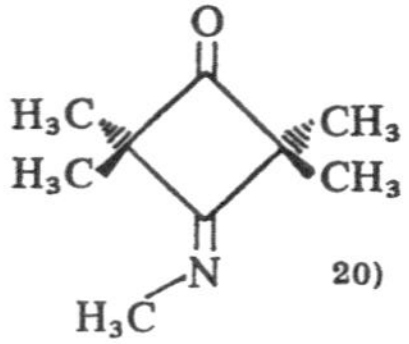 20)		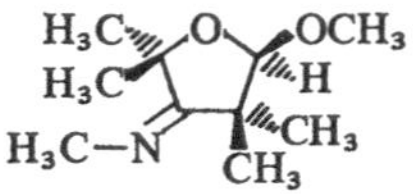 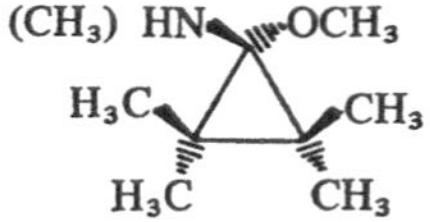	
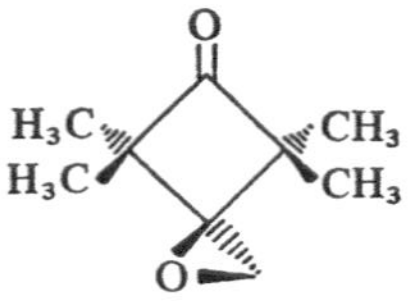	Pyrex-Filter CH_3OH	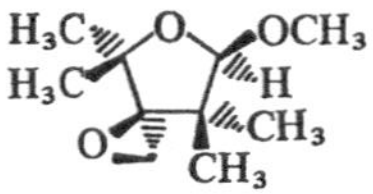 55% [1] 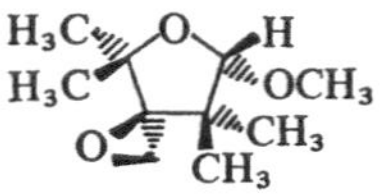 31% [1] 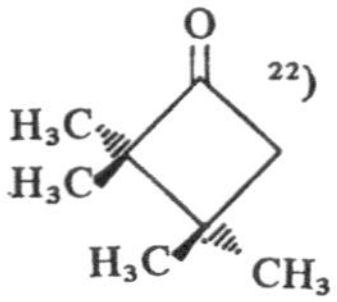 12–14% [22]	[42]

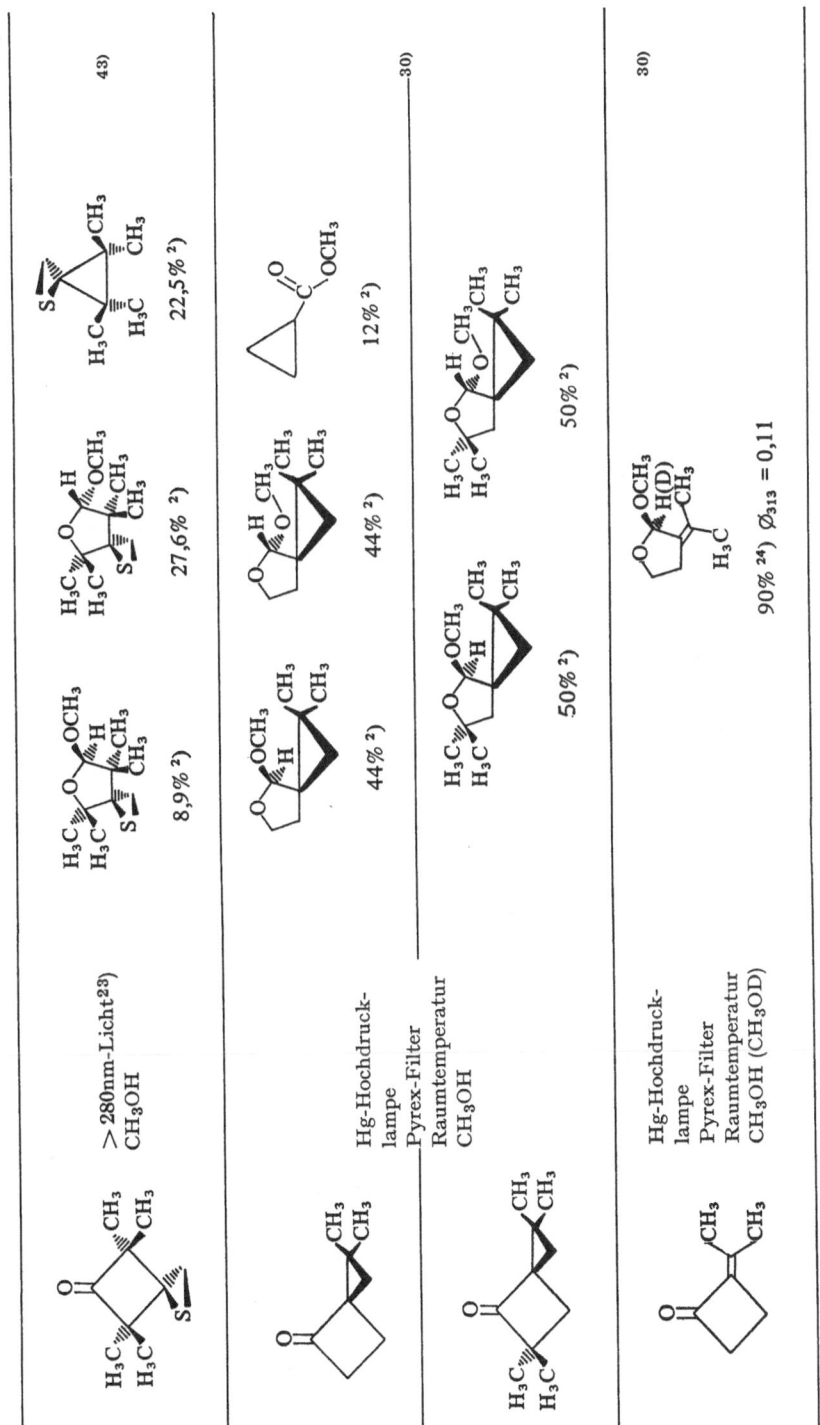

43)

> 280nm-Licht[23)]
CH₃OH

8,9%²⁾   27,6%²⁾   22,5%²⁾

.30)

Hg-Hochdruck-
lampe
Pyrex-Filter
Raumtemperatur
CH₃OH

44%²⁾   44%²⁾   12%²⁾

50%²⁾   50%²⁾

30)

Hg-Hochdruck-
lampe
Pyrex-Filter
Raumtemperatur
CH₃OH (CH₃OD)

90%²⁴⁾ $\emptyset_{313} = 0{,}11$

Tabelle 1 (Fortsetzung)

Edukt	Bestrahlungs-bedingungen	Produkt	Lit.
E oder Z	Hg-Hochdruck-lampe Quarz Raumtemperatur CH₃OH	~ 90 %[25] 100 %[25]	30) 30, 44)
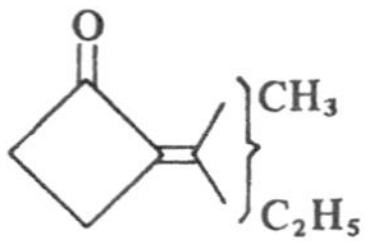 E oder Z	Hg-Hochdruck-lampe Raumtemperatur Pyrex-Filter CH₃OH	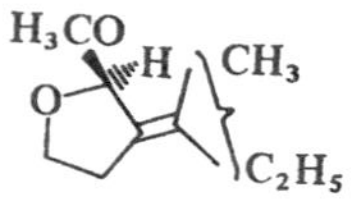 E und Z	45)

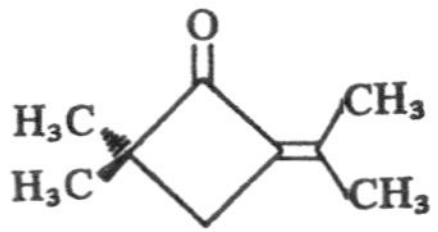

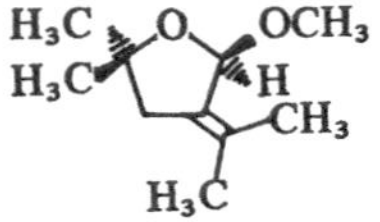

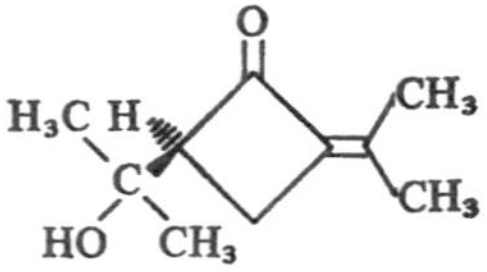

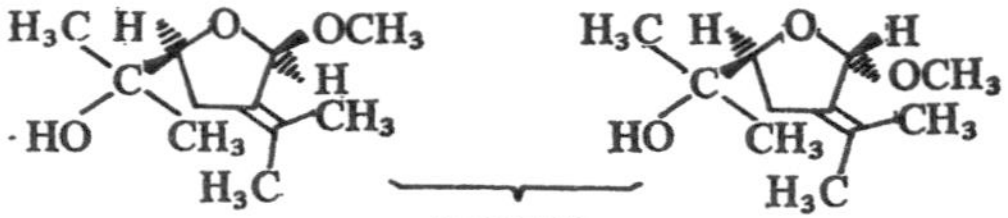

198

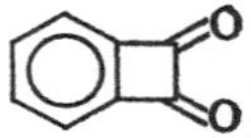

32, 33)

Hg-Hochdruck-
lampe
Raumtemperatur
C_2H_5OH

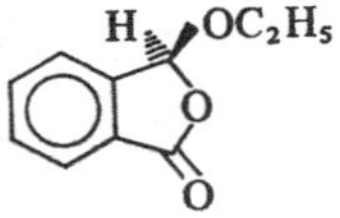

39% [26])

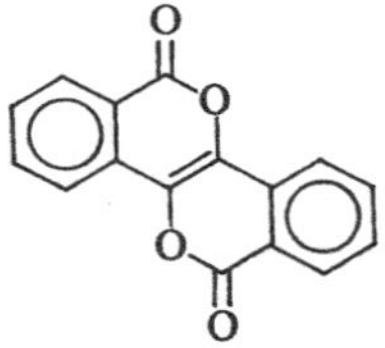

4%

46)

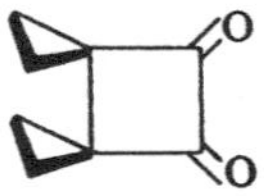

Sonnenlicht von
Paris
CH_3OH

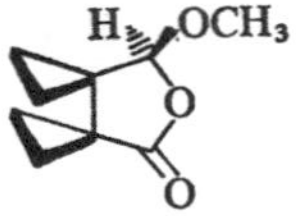

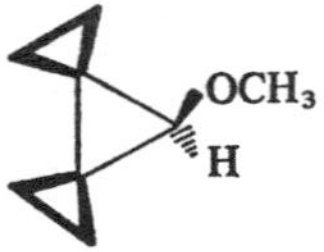

27)

OC_2H_5 H

OCH_3 H

OCH_3 H

W.-D. Stohrer, P. Jacobs, K. H. Kaiser, G. Wiech und G. Quinkert

Fußnoten zu Tabelle 1

[1]) Die gaschromatographisch bestimmten chemischen Ausbeuten erhielt man durch Extrapolation der für 88,5 (93,1)% bzw. einen nicht näher spezifizierten Umsatz ermittelten Wert auf 100%.

[2]) Gaschromatographisch bestimmte Ausbeuten für praktisch vollständigen Umsatz.

[3]) Die ungefähre Ausbeute wurde NMR-spektroskopisch ermittelt.

[4]) Die konfigurations-isomeren Edukte wurden unter den Belichtungsbedingungen nicht ineinander umgewandelt; keine Komponente des Reaktionsprodukts ließ sich durch Piperylen (1m) quenchen.

[5]) Die Originalzuschrift nimmt zur Stereochemie der Acetalbildung in bezug auf das neu entstehende Chiralitätszentrum keine Stellung; sie enthält relative Quantenausbeuten, aber keine chemischen Ausbeuten für die einzelnen Produkt-Komponenten.

[6]) Da neben E-Cycloocten das mit zunehmender Belichtungsdauer anwachsende Z-Stereo-Isomer auftritt, ist mit einer Isomerisierung des ketonischen Edukts und/oder des olefinischen Produkts zu rechnen.

[7]) Die Konfiguration am neu erzeugten Chiralitätszentrum ist unbekannt;

[8]) Bei der Cycloeliminierung mit entstandenes Keten wurde in einem separaten Experiment durch Cyclohexylamin abgefangen.

[9]) Es sind beide Stereo-Isomere in bezug auf C-16 entstanden; ihre jeweilige Konfiguration ist unbekannt.

[10]) Die beiden Produkt-Komponenten entstehen in etwa äquivalenter Menge.

[11]) Weitere Produkt-Komponenten sind: (3Z)-3,7-Dimethyl-octa-3,6-diensäure (5%) und ein nicht näher bestimmter Aldehyd.

[12]) Die in ihrer Konfiguration aufgeklärten endo- und exo-Isomere treten von der Temperatur weitgehend unabhängig im Verhältnis 3:2 auf.

[13]) Statt Methanol sind auch t-Butanol, 2n HCl, Äther/Wasser und Äther/CH$_3$COOH verwendet worden; es wurden jeweils die zu erwartenden Produkte isoliert.

[14]) Isobutyraldehyd (14%) und Isobuttersäuremethylester (16%) wurden als Produkt-Komponenten unter anderen Analysebedingungen nachgewiesen.

[15]) Die Decarbonylierungsprodukte wurden gaschromatographisch nicht voneinander getrennt; ihre Identifizierung geschah NMR-spektroskopisch.

[16]) Ausbeute der beiden Konfigurations-Isomeren [7]) (im Verhältnis 3:1) in bezug auf das neu gebildete Chiralitätszentrum nach Abtrennung von Decarbonylierungsprodukten.

[17]) Die beiden stereo-isomeren Methylencyclopropan-Derivate (photochemisch instabil) sind NMR-spektroskopisch nachgewiesen worden.

[18]) Das umgelagerte Decarbonylierungsprodukt wurde neben 1,1-Dicyano-2-isopropenyl-3-methyl-buten-1 isoliert; die letztere Verbindung entsteht während der Gaschromatographie.

[19]) NMR-spektroskopisch ließen sich zwei Isomere (1:1) feststellen, deren Trennung nicht gelang; obwohl bei einem 90 proz. Umsatz des Edukts 40% CO ermittelt wurden, ist ein weiteres Decarbonylierungsprodukt nicht erwähnt worden.

[20]) Weiterhin sind Verbindungen untersucht worden, die anstelle der N—CH$_3$-Gruppe die n-Butyl-, n-Hexyl-, n-Cyclohexyl- oder n-Phenyl-Gruppierung enthielten.

[21]) Hier wie bei den analogen Ringerweiterungsprodukten ist von einem nicht näher spezifizierten Isomeren-Gemisch (2:3) die Rede, das nicht aufgetrennt werden konnte.

[22]) Das Cyclobutanon-Derivat entsteht durch Umlagerung aus der Photo-Decarbonylierungskomponente.

23) Durch 254 nm-Licht entsteht in Methanol hauptsächlich 2,2,4,4-Tetramethyl-3-methylen-cyclobutanon.

24) In einem separaten Ansatz ließ sich nach 95 proz. Umsatz des Edukts NMR-spektroskopisch nur eine Produkt-Komponente nachweisen.

25) Nach der Entfernung des Lösungsmittels anfallendes Rohprodukt.

26) Rohprodukt.

27) Die nur untergeordnet auftretende Produkt-Komponente war schwierig zu reinigen; die angegebene Konstitution wurde NMR-spektroskopisch geschlossen.

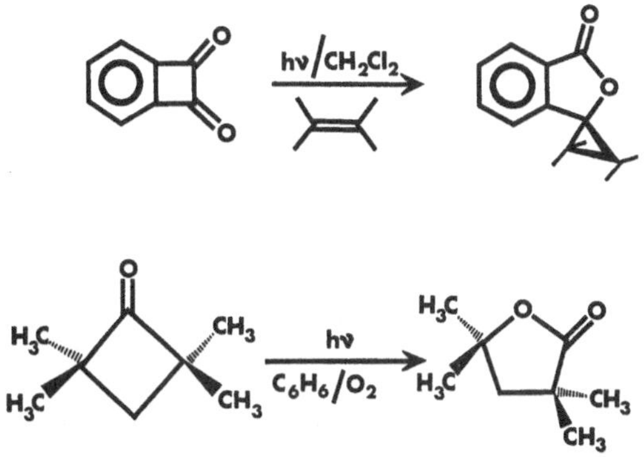

Abb. 5. Beispiele von Vierring-Ketonen, die bei der Bestrahlung in Gegenwart eines Olefins zum entsprechenden Cycloaddukt [32,33)] und in Gegenwart von Sauerstoff zum entsprechenden Lacton [34)] reagieren

2.2. Stereochemische Ergebnisse und regioselektive Befunde

Vierring-Ketone, die derart konstituiert sind, daß sie im Verlauf der Photo-Ringerweiterung die Lösung einer CC-Bindung zwischen der Carbonylgruppe und einem Chiralitätszentrum erfahren und von denen jeweils die beiden möglichen Stereo-Isomeren vorliegen, sind willkommene stereochemische Informanten über den Reaktionsablauf. Abb. 6 enthält die bislang bekannt gewordenen Paare konfigurations-isomerer Cyclobutanon-Derivate, die in alkoholischer Lösung bestrahlt worden sind: Ausnahmslos tritt dasjenige acetalische Reaktionsprodukt auf, bei dem die Konfiguration am wandernden C_α-Atom erhalten geblieben ist.

Der Nachweis der Stereospezifität und ihrer Richtung gelingt bequem und eindeutig auf der Stufe des zugehörigen Lactons. Letzteres entsteht unter Erhalt der Konfiguration durch Baeyer/Villiger-Reaktion [47)] aus dem zugrunde liegenden Keton wie auch durch Chromsäure-Oxydation aus dem acetalischen Photoprodukt. Die Festlegung der Konfiguration auf der Lacton-Stufe hat den Vorteil, daß hier das zu-

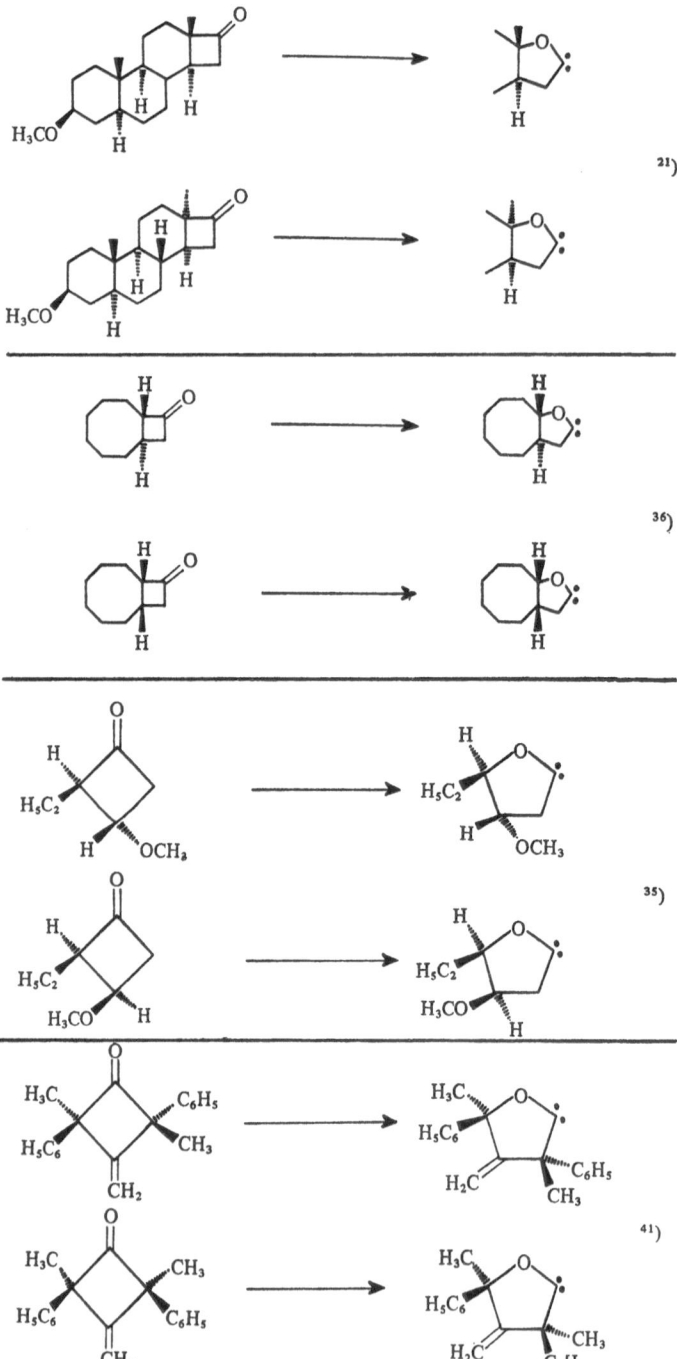

Abb. 6. Die bisher untersuchten Paare konfigurations-isomerer Vierring-Ketone, die nach UV-Bestrahlung stereospezifisch (mit Retention der Konfiguration am wandernden Zentrum) ringerweitern

sätzliche Chiralitätszentrum wieder beseitigt worden ist, das bei der Einschiebung des Oxacarbens in das hydroxylgruppen-haltige Lösungsmittel vorübergehend geschaffen worden war.

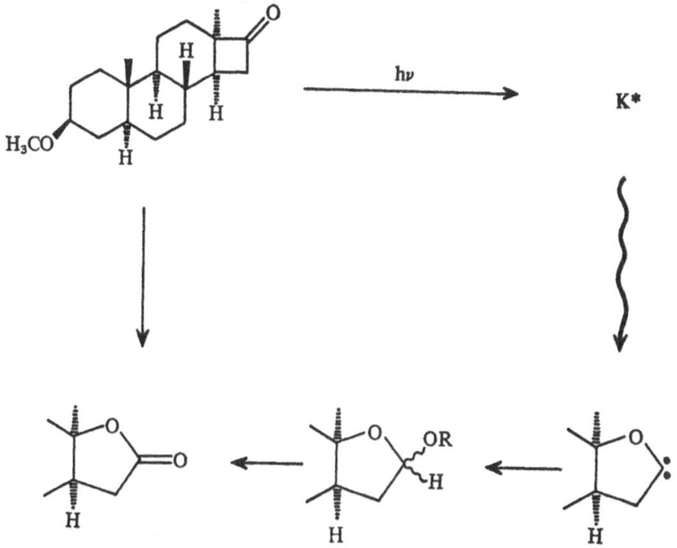

Abb. 7. Beispielhafte Illustration zur Ermittlung der Stereochemie der lichtinduzierten Cyclobutanon/Tetrahydrofuryliden-Isomerisierung [21)]

Abb. 7 zeigt deutlich, daß eine solche Konfigurations-Zuordnung schlüssig ist und bedient sich desjenigen Beispiels [21)], mit dem die reaktionsmechanistische Diskussion der Photo-Ringerweiterung von Cyclobutanon-Derivaten in die entscheidende Phase eintrat und sich erneut die Frage stellte: Wird das zum ketonischen Edukt isomere Oxacarben unmittelbar im photochemischen Primärprozeß gebildet oder geht ihm ein 1,4-Alkyl/Acyl-Biradikal voraus? Die Antwort lautet ähnlich wie zuvor (s. Abschnitt 1.2): Mit der geschilderten stereochemischen Erfahrung ist ein konformationell leicht beweglicher, acyclischer Transient nicht vereinbar. Nun zeigt das Hoffmann'sche Konzept der „through bond-Kopplung" [48,49)], daß die Rotationen um die drei Bindungen eines offenkettigen 1,4-Biradikals differenziert betrachtet werden müssen; die Wechselwirkung zwischen der 2,3σ-Bindung und den terminalen Orbitalen
— ist gegen eine Drehung um die mittlere Bindung unempfindlich;
— erreicht den jeweiligen Maximalwert der Stabilisierung in denjenigen Konformationen, in welchen die endständigen Orbitale möglichst parallel zur 2,3-Bindung angeordnet sind.

W.-D. Stohrer, P. Jacobs, K. H. Kaiser, G. Wiech und G. Quinkert

Die hierdurch gesteigerte Rotationsbarriere der 3,4-Bindung in einem
1,4-Alkyl/Acyl-Biradikal erhöht die Konfigurations-Stabilität dieser
Spezies. Die Mehrzahl der in Abb. 6 aufgeführten Paare von Konfigurations-Isomeren verliert daher an Überzeugungskraft, wenn der Befund
der stereo-spezifischen Ringerweiterung gegen einen Reaktionsweg zum
Oxacarben über das Biradikal verwendet werden soll. Die stereo-isomeren 2,4-Diphenyl-2,4-dimethyl-3-methylen-cyclobutanon-Derivate (s.
Abb. 6) machen hier allerdings eine Ausnahme.

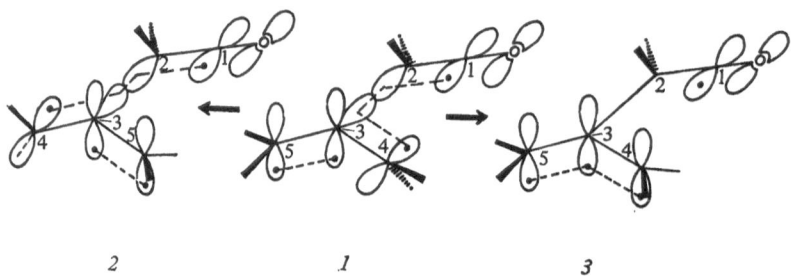

2   *1*   *3*

Abb. 8. Verschiedene Konformationen der biradikalischen Spezies, die durch α-
Spaltung aus einem 3-Methylen-cyclobutanon entsteht. 1: unmittelbar nach der
α-Spaltung, through bond-Stabilisierung; 2: aus 1 durch Drehung um die 2,3-Bin-
dung, through bond-Stabilisierung; 3: aus 1 durch Drehung um die 3,4-Bindung,
Stabilisierung durch Konjugation mit der 3,5-Doppelbindung

Zur chemischen Dynamik der ihnen entsprechenden 1,4-Alkyl/Acyl-
Biradikale läßt sich folgendes sagen: Wie bereits erwähnt, wirkt die
through bond-Wechselwirkung auf das alkyl-radikalische Zentrum kon-
figurations-erhaltend; die im vorliegenden Fall sicherlich vorherrschende
Tendenz zur Ausbildung des planaren Allyl-Systems ist dagegen konfi-
gurations-verändernd. Trotzdem verläuft die Photo-Ringerweiterung
in beiden Fällen stereo-spezifisch [41] (s. Abb. 8). Ferner sollte erwähnt
werden, daß die zur Diskussion stehenden symmetrischen Vierring-
Ketone keine lichtinduzierte Gerüstumlagerung zu einem unsymmetri-
schen Cyclobutanon-Derivat eingehen [50], obwohl die through bond-
Kopplung durch eine Drehung des Acyl/Allyl-Biradikals um die 2,3-
Bindung nicht beeinträchtigt würde. *Beide Beobachtungen sind mit
einem Biradikal, das dem Oxacarben vorgelagert ist, schwerlich zu verein-
baren!*
Wie bei der α-Spaltung eines elektronen-angeregten Cyclanons
gewöhnlicher Ringgröße (s. Abschnitt 1.1) wird bei der Photo-Ringerweiterung eines Cyclobutanon-Derivats im Regelfall die CC-Bindung

zwischen der Carbonylgruppe und dem höher alkylsubstituierten C_α-Atom gelöst (s. die überwiegende Anzahl der Beispiele in Tabelle 1). Diese Tatsache — wie auch die Beobachtung, daß im Gegensatz zum 2,2-Dimethyl-3-phenyl-cyclobutanon (s. Tabelle 1) 2,2-Bistrifluormethyl-3-phenyl-cyclobutanon bei der Bestrahlung in Methanol nicht zu einem Acetal führt [27] (s. Abb. 4) — läßt sich als Argument für eine nucleophile Wanderung des C_α-Atoms vom C-Atom der ursprünglichen Carbonyl-gruppe zu ihrem O-Atom verwenden. Ein solcher Gedankengang weckt leicht Assoziationen zur bereits erwähnten Baeyer/Villiger-Reaktion [51]. Die Analogie wächst gar bei einem Vergleich z.B. der Ringerweiterungs-Produkte aus den konfigurations-isomeren D-Nor-16-keto-Steroiden: Das 13α-Isomer löst bei der Photo-Ringerweiterung als auch bei der Baeyer/Villiger-Oxydation die Bindung zwischen C—13 und C—16 und zwar nur diese (s. Tabelle 1); beim zweifellos gespannteren 13β-Isomer sind in beiden Reaktionen beide von C—16 ausgehende CC-Bindungen betrof-fen [21] (s. Tabelle 1 sowie Abb. 9). Die erst kürzlich geäußerte Fest-stellung [52], die lichtinduzierte Cyclobutanon/Tetrahydrofuryliden-Um-

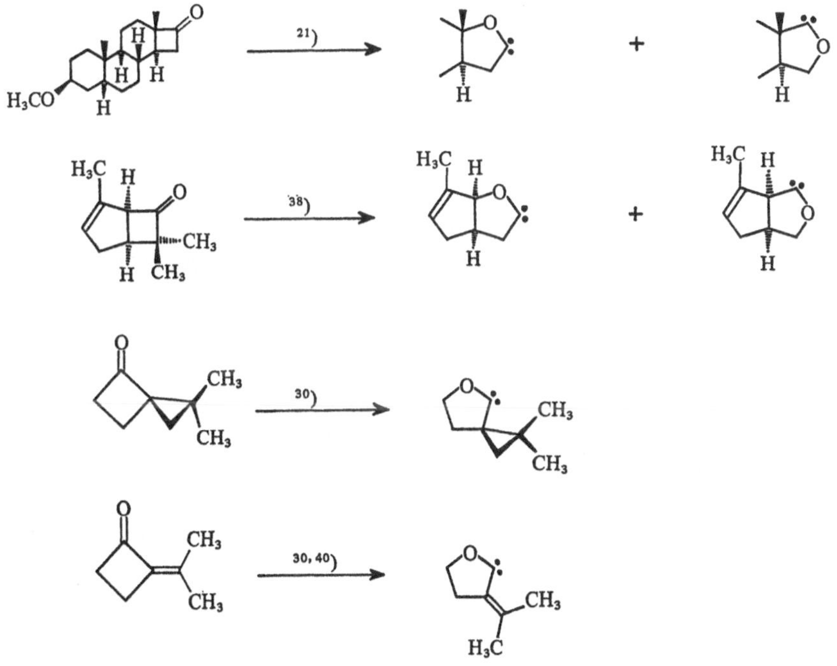

Abb. 9. Beispiele für Photo-Ringerweiterungen von Cyclobutanon-Derivaten, bei denen die CC-Bindung zwischen der Carbonylgruppe und dem höher substituierten Nachbar-C-Atom nur teilweise oder gar nicht gelöst wird.

205

lagerung sei als photochemische Baeyer/Villiger-Reaktion anzusehen, wird dieser Übereinstimmung gerecht; sie vermag jedoch die geheimnisvolle Sonderstellung des Vierringes bei der Photoreaktion keineswegs zu deuten.

2.3. Spektroskopische Beobachtungen

Die Isolierung von Acetalen, Cycloaddukten oder Lactonen nach der UV-Bestrahlung von Vierring-Ketonen im Beisein von hydroxylgruppenhaltigen Lösungsmitteln, Olefinen oder Sauerstoff (s. Abschnitt 2.1). wird allgemein als überzeugender Indizienbeweis für intermediär auftretende Oxacarbene angesehen [25]. Bemühungen um einen direkten Nachweis dieser kinetisch instabilen Transienten sind daher nicht durch einen Rest von Mißtrauen gegen ihre Existenz motiviert, sondern suchen die genaue Stellung dieser Verbindungen im reaktions-historischen Ablauf zu erfahren.

Da eine Isolierung der Oxacarbene oder wenigstens ihr spektroskopischer Nachweis bei Raumtemperatur ihrer Kurzlebigkeit wegen entfällt, kommt neben der Blitzlicht-Spektroskopie die Anwendung spektroskopischer Methoden bei herabgesetzten Temperaturen in Betracht. Wir haben in den vergangenen Jahren mit der Tieftemperatur-Technik gute Erfahrungen im Umfang mit kinetisch instabilen Transienten gemacht [1] und sollten daher in der Lage sein, Tetrahydrofuryliden-Derivate nachzuweisen, sofern sie folgende Voraussetzungen erfüllen:

Sie müssen

— auch noch bei herabgesetzten Temperaturen photochemisch entstehen;

— sich im experimentell zugänglichen Temperatur- und/oder Viskositätsbereich „einfrieren" lassen;

— durch eine geeignete spektroskopische Methode erkennbar sein.

Welche der spektroskopischen Methoden hierfür in Frage kommt, geht aus der Erörterung der Elektronenstruktur von cyclischen Oxacarbenen gewöhnlicher Ringgröße hervor. Durch Wechselwirkung zwischen dem p_o-Donator- und dem p_c-Akzeptor-Orbital zu einer π_{oc}-Bindung werden die beiden von Liganden nicht mitbeanspruchten Elektronen am carbenischen C-Atom in das dort vorhandene n_σ-Orbital hineingedrängt. Hierdurch

— wird das Methylen-C-Atom nucleophiler als in Abwesenheit einer solchen Wechselwirkung [53];

— gewinnt der elektronen-energetische Grundzustand des Oxacarbens Singulett-Charakter [53,54];

— kommt es zu einer alles in allem repulsiven Wechselwirkung zwischen den mit Elektronen besetzten n_σ-Orbitalen an den beiden benachbarten Zentren;

— resultiert eine relative langwellige π^*,n-Elektronenanregung wie bei den isoelektronischen Azoverbindungen.

Da die Chromophore beider isoelektronischer Verbindungen in der Summe der Kernladungszahlen sowie in ihren Bindungsverhältnissen übereinstimmen, ist für die energie-ärmste π^*,n-Anregung eine ähnliche Lage der betreffenden Bande im jeweiligen Absorptionsspektrum zu erwarten. Aliphatische Z-Azoverbindungen absorbieren in der Gegend von 360 nm mit einer molaren Extinktion von ca. 300 [55]. Die in Tabelle 2 aufgeführten Vierring-Ketone, die nach Bestrahlung in alkoholischer Lösung bei Raumtemperatur zu ringerweiterten Acetalen reagieren (s. Tabelle 1), liefern nach Einwirkung von 302 oder 313 nm-Licht bei −186 °C Photoprodukte, die im gleichen Wellenlängenbereich wie Azoverbindungen absorbieren [56]. Andererseits läßt sich im Fall des 2,2-Bistrifluormethyl-3-phenyl-cyclobutanons (s. Abb. 4) weder bei Raumtemperatur das zugehörige Acetal noch bei −186 °C ein Photoprodukt mit einem Absorptionsgebiet an der kurzwelligen Grenze des sichtbaren Teils des elektro-magnetischen Spektrums beobachten. *In den spektroskopisch erkennbaren Photoprodukten mit Alkoholen zu den entsprechenden Acetalen reagierende Oxacarbene zu sehen, ist sicherlich nicht an den Haaren herbeigezogen*; die aus manchen elektronen-angeregten Cyclobutanon-Derivaten entstandenen ketenischen Cycloeliminierungskomponenten, die sich bei herabgesetzter Temperatur ebenfalls anreichern lassen, weisen jedenfalls andersartige Absorptions-Eigenschaften auf.

Mit einem tieftemperatur-spektroskopischen Nachweis oxacarbenischer Transienten wäre natürlich die schon mehrfach angerührte Frage (s. Abschnitte 1.2 und 2.2) nicht beantwortet, ob diese Verbindungen durch Cyclisierung eines zunächst entstandenen 1,4-Alkyl/Acyl-Biradikals gebildet werden. Vielleicht führt die in einigen Fällen elektronen-spektroskopisch wahrnehmbare Veränderung der bei tiefen Temperaturen faßbaren Spezies zu einer weiteren Klärung.

Die kinetische Stabilität der photochemisch gewonnenen Transienten wird von der Viskosität der verwendeten Matrix beeinflußt [56]; 2,2,4,4-Tetramethyl-cyclobutanon z. B. legt hiervon Zeugnis ab (s. Tabelle 2): In 2-Methyl-tetrahydrofuran sowie in einem Gemisch aus Methylcyclohexan/3-Methyl-pentan (4:1) kann die um 360 nm maximal absorbierende Spezies bei −186 °C noch genügend stabilisiert werden, während sie sich im erheblich weniger viskosen Lösungsmittel Methylcyclohexan/Isopentan (1:4) nicht mehr zu einer für den spektroskopischen Nachweis erforderlichen Konzentration anreichern läßt. In dem allgemein verwendeten Gemisch aus Äther/Isopentan/Äthanol (5:2:5), das in seiner

Viskosität zwischen denjenigen der zuvor genannten Lösungsmittel liegt, läßt sich bei − 186 °C ein langsamer Abbau des Absorptionsgebiets mit einem Maximum bei 357 nm zugunsten eines neu auftretenden Absorptionsgebiets mit einem Maximum bei 437 nm feststellen. Die thermische Umwandlung beginnt bereits während der Bestrahlung und ist durch einen isosbestischen Punkt bei 400 nm bzw. durch ein lineares Mauser-Diagramm der Extinktionsdifferenzen [57] gekennzeichnet.

Im Fall des Benzocyclobuten-3,4-dions erhält man ein besonders instruktives Bild: Bei der Bestrahlung dieser Vierring-Verbindung bei − 186 °C in 2-Methyl-tetrahydrofuran mit 302 nm-Licht beobachtet man während der spektroskopischen Reaktionskontrolle eine Kurvenschar, die sich in einem isosbestischen Punkt (307,5 nm) schneidet; das zugehörige Diagramm der Extinktionsdifferenzen ist linear. Das hypothetische Oxacarben (T_1) mit einem Absorptionsmaximum bei 393 nm läßt sich bei der angegebenen Temperatur durch 405 nm-Licht zu 68% wieder in das Edukt zurückführen. Thermisch wandelt sich der Transient T_1 in eine Verbindung T_2 um, die bei 419 nm ein Absorptionsmaximum aufweist und bei Erhöhung der Temperatur bzw. Erniedrigung der Viskosität z.T. das ursprüngliche Dion (E) zurückbildet (s. Abb. 10). Wenn der Platz von T_1 vom entsprechenden Oxacarben ausgefüllt wird, welche Struktur steht dann für T_2?

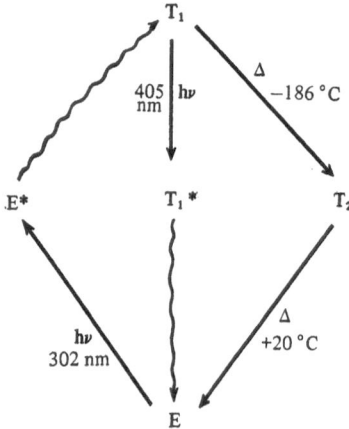

Abb. 10. Beziehungen zwischen dem Edukt Benzocyclobuten-3,4-dion (E) und seinem Photo- (T_1) und Thermo-Transienten (T_2)

Abb. 11 schlägt hierfür das konjugierte Dien-diketen vor, eine sicherlich relativ energie-arme Konformation in der Gruppe der „1,4-Diacyl-Biradikale". Diese Anregung kehrt die von Staab und Ipaktschi [32,33] vorgebrachte Reihenfolge — primäre Photo-Isomerisierung des Benzo-cyclobuten-3,4-dions zum o-chinoiden Bisketen und dessen sekundäre Umlagerung zum cyclischen Carben — um und wird der Beobachtung gerecht, daß nach der Bestrahlung des α-Diketons bei der Siedetemperatur des Stickstoffs die in anderen Fällen [58] tieftemperatur-IR-spektroskopisch leicht zu erkennende Keten-Absorption bei ca. 2100 cn⁻¹ nicht auftritt [59,60] (s. jedoch [59a]).

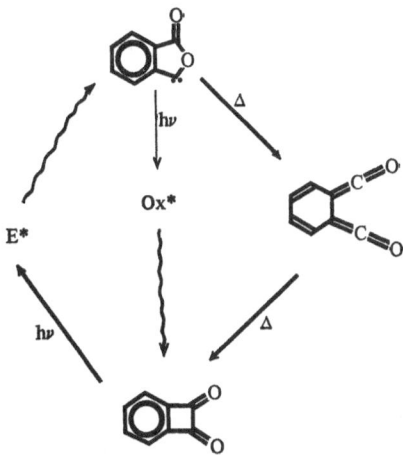

Abb. 11. Strukturelle Interpretation von Abb. 10

Im vorliegenden Beispiel wie auch in allen anderen Fällen, wo auf einen primären Photo-Transienten ein sekundärer Thermo-Transient folgt, steht der experimentelle Beleg für die Struktur, besonders der jeweiligen sekundären Zwischenverbindung, noch aus.

Nachdem in den voraufgegangenen Abschnitten die wesentlichen Fakten zum Reaktionsmechanismus der Photo-Ringerweiterung von Cyclobutanonen zusammengetragen worden sind, läßt sich resümieren, daß

— die von P. Yates [24] vorgeschlagenen *Oxacarbene im photochemischen Primärprozeß*,

— die evtl. auftretenden *1,4-Alkyl/Acyl-Biradikale in einer sekundären Dunkelreaktion durch Ringöffnung der Oxacarbene*

gebildet werden.

E. K. C. Lee u. Mitarb. [61] kommen unserer Auffassung am nächsten und halten es darüber hinaus für möglich, daß Tetrahydrofurylidene und 1,4-Alkyl/Acyl-Biradikale miteinander im Gleichgewicht stehen. Die übrigen Autoren, die sich, nach Bekanntgabe des ersten Beispiels [22]

W.-D. Stohrer, P. Jacobs, K. H. Kaiser, G. Wiech und G. Quinkert

Tabelle 2. Zur Tieftemperatur-Bestrahlung von Cyclobutanonen

Keton	Lösungs-mittel[1]	UV-Daten [nm] −186 °C	Bestrahlungsbedingungen[2]		UV-Daten [nm] −186 °C		Lit.
			[°C]	[nm]	Photo-produkt	therm. Fol-geprodukt	
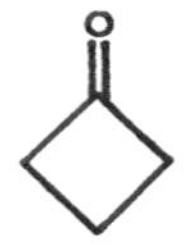	MTHF	275 (11) 284 (12)	−186	302	349	460[3]	93)
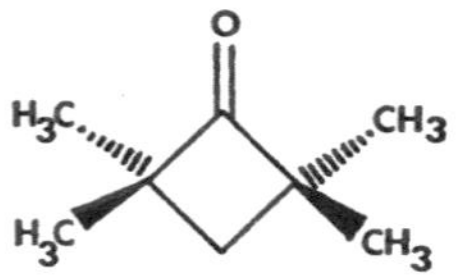	MTHF	298 (14) 310 (18)	−186	302	357 (400)[4]	—	
	EPA	298 (18,5) 310 (18)	−186	302	357	437[3]	56)
	MCP	297 (21) 310 (22)	−186	302	363	—	56)
	MCJ	297 (17) 310 (16)	−186	302	—	—	
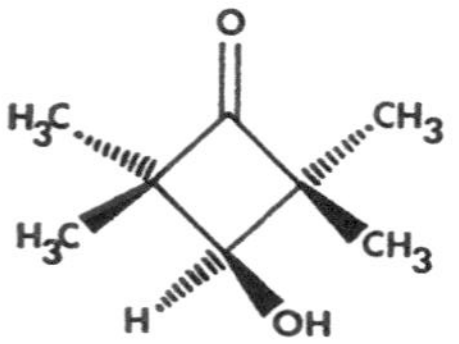	MTHF	300 (17) 313 (17)	−186	313	365	—	93)

Structure	Solvent						Ref.
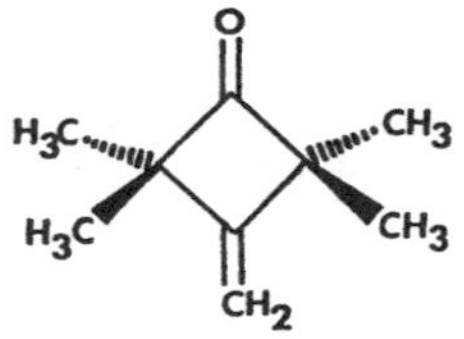 (structure 1)	MTHF	300 (17) 314 (16)	-186	313	360	—	93)
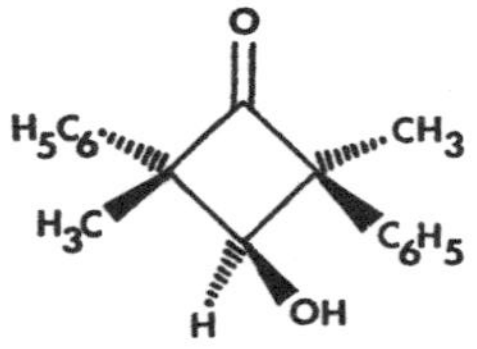 (structure 2)	MTHF	306 (47) 317 (48) 329 (35)	-186	313	365	—	93)
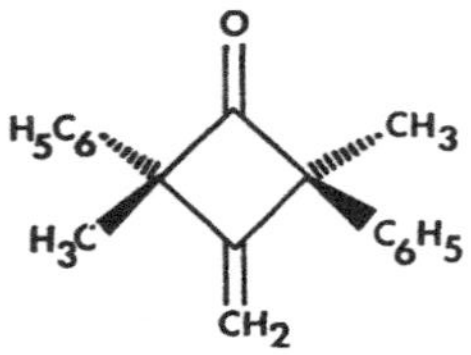 (structure 3)	MTHF	309 (79) 319 (103) 331 (81)	-186	313	365 (400)[4]	—	93)
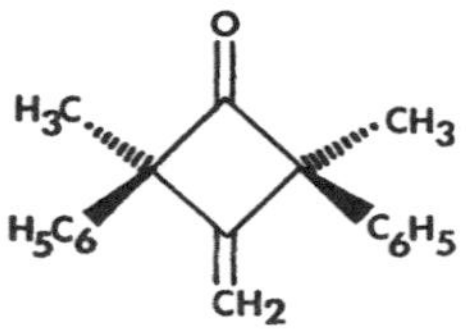 (structure 4)	MTHF	310 (358) 321 (562) 333 (588)	-186	313	363 (400)[4]	—	93)

211

Tabelle 2 (Fortsetzung)

Keton	Lösungsmittel[1]	UV-Daten [nm] −186 °C	Bestrahlungsbedingungen[2]		UV-Daten [nm] −186 °C		Lit.
			[°C]	[nm]	Photoprodukt	therm. Folgeprodukt	
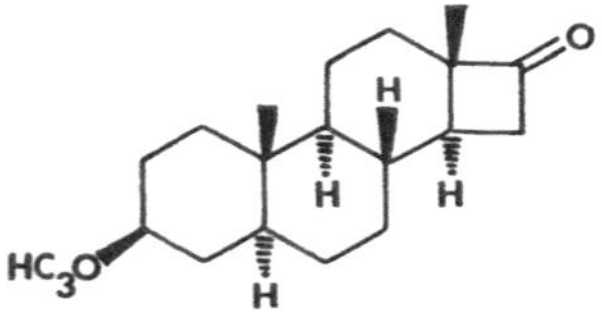	MTHF	292 (39)	−186	302	370	—	93)
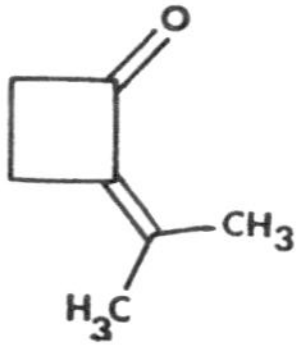	MTHF	324 (44) 338 (65) 354 (62) 372 (25)	−186	313	445 (300)4)	—	93)
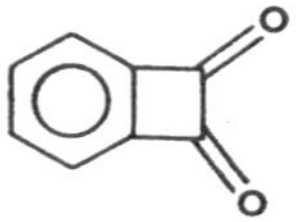	MTHF	285 (4030) 293 (7415) 301 (9420) 402 (36) 424 (55)	−186	302	393 (2600)4)	419	93)

1) MTHF: Methyltetrahydrofuran; EPA: Äther:Isopentan:Äthanol (5:2:5); MCP: Methylcyclohexan: 3-Methylpentan (4:1); MCI: Methylcyclohexan:Isopentan (1:4).
2) Hg-Hochdruckbrenner HBO 200; Bausch & Lomb 500 nm-Gittermonochromator.
3) Bildet sich während der Photolyse.
4) Größenordnung des Extinktionskoeffizienten.

der lichtinduzierten Oxacarbenbildung aus einem Vierring-Keton im Jahre 1965, mit dieser Photo-Isomerisierung beschäftigten, lassen dagegen dem jeweiligen Oxacarben, das dann freilich in einer Dunkelreaktion gebildet würde, das zugehörige 1,4-Alkyl/Acyl-Biradikal vorangehen [25,27,34,35,38–40,45]. N. J. Turro u. Mitarb., die einen substantiellen Beitrag zur Kenntnis der Photochemie des Cyclobutanons und seiner Derivate leisteten, äußern gelegentlich Skrupel an dieser, sonst von ihnen verfolgten Auffassung und sprechen dann von einem biradikalischen Übergangszustand [30,36]. In ihrer jüngsten Mitteilung [45] ist jedoch wieder von einer Zwischenverbindung die Rede, die als „virtuelles Biradikal", als „ein Biradikal mit partieller Bindungsordnung zwischen den Zentren 1 und 4", angesprochen wird.

Dieser Versuch der semantischen Zähmung einer widerspenstigen Struktur erinnert uns an die eigenen Bemühungen, durch Ausschalten eines „freien" Alkyl/Acyl-Biradikals [21] (s. Abschnitt 1.2) die stereochemischen Besonderheiten von reagierenden, elektronen-angeregten Cyclobutanon-Derivaten zu berücksichtigen, ohne zugleich einen voreiligen Bruch mit der Photochemie der höheren Homologen zu riskieren. Solche Umschreibungen weisen gewöhnlich darauf hin, daß dort, wo sie gebraucht werden, mehr als ein Problem zu lösen ist.

Nachdem in zäher Kleinarbeit bekannt geworden war, daß in alkoholischem Medium bestrahlte Cyclobutanone in der Regel ringerweiterte Acetale liefern, und daß intermediär auftretende Tetrahydrofurylidene die eigentlichen Photoprodukte dieser Umsetzung sind, die unmittelbar aus den elektronen-angeregten Edukten hervorgehen, wurde der eigentliche Kern des Problem-Komplexes sichtbar: Warum ist dieses Verhalten für Vierring-Ketone typisch, für Cycloanone gewöhnlicher Ringgröße dagegen unüblich? Auf diese Frage wird anschließend eine Antwort gegeben, die sich der MO-Sprache bedient und die durch Ergebnisse semi-empirischer Berechnungen gestützt wird.

3. MO-theoretische Überlegungen und semi-empirische Rechnungen zur Ringöffnung von Cyclanonen

Die Frage nach der lichtinduzierten α-Spaltung von Cyclanonen zu Alkyl/Acyl-Biradikalen kehrte wie ein Leitmotiv in allen voraufgegangenen Abschnitten dieser Abhandlung wieder. Für die Mehrzahl der Vierring-Ketone wurde sie verneint (s. Abschnitt 2.2 und 2.3; s. ferner Abb. 12), da sich die Deutung der für sie typischen Photo-Isomerisierung zu ringerweiterten Oxacarbenen nur in unnötige Widersprüche verstrickt, falls man zwischendurch 1,4-Alkyl/Acyl-Biradikale als Transienten in einem Minimum auf der Energiehyperfläche annimmt. Elektronen-angeregte Cyclanone gewöhnlicher Ringgröße verhalten sich dagegen anders und reagieren unter α-Spaltung zu den entsprechenden biradikalischen Zwischenverbindungen (s. Abschnitt 1.1; s. ferner Abb. 12).

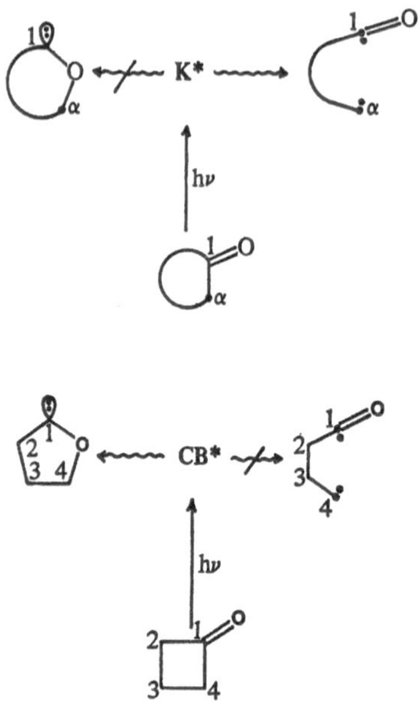

Abb. 12. In der Regel unterliegen elektronen-angeregte Cyclanone gewöhnlicher Ringgröße der α-Spaltung zu den entsprechenden Alkyl/Acyl-Biradikalen, während Cyclobutanone üblicherweise zu den entsprechenden Oxacarbenen photo-isomerisieren

Um die Sonderstellung der Vierring-Ketone innerhalb der Familie der Cyclanone ins rechte Licht zu rücken, ist es angebracht, die MO-theoretische Erörterung ihres Reaktionsverlaufs zunächst mit der „normalen" α-Spaltung zu beginnen.

3.1. Störungstheoretische Überlegungen [62] zur α-Spaltung von Cyclanonen

Das Wechselwirkungsdiagramm der Abb. 13 konstruiert aus den Orbitalen σ und σ^* der C_1—C_α-Bindung sowie dem in der Gerüstebene plazierten Orbital des einsamen Elektronenpaares am O-Atom der Carbonylgruppe die an der α-Spaltung beteiligten Orbitale ϕ_1, ϕ_2 und ϕ_3.

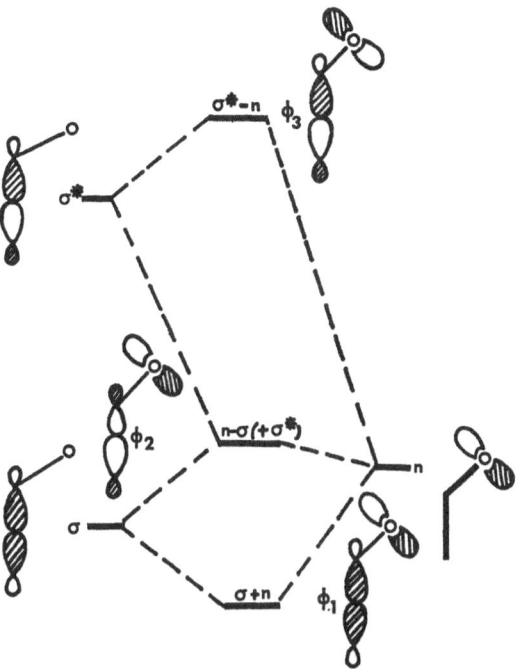

Abb. 13. Konstruktion der an der α-Spaltung eines Ketons beteiligten σ-Orbitale ϕ_1, ϕ_2 und ϕ_3 (Mitte) aus den Orbitalen σ und σ* der zu lösenden CC-Bindung (links) und dem Orbital n des einsamen Elektronenpaares am Sauerstoff (rechts).

Dem Korrelationsdiagramm der Abb. 14 zufolge gehen die Keton-Orbitale ϕ_1 und ϕ_3 bei der α-Spaltung in die Acyl-Orbitale π_y und π_y* über, und das Orbital ϕ_2, das der Spektroskopiker meint, wenn er vom „einsamen" Elektronenpaar am O-Atom in Ketonen spricht, wird in das Orbital Pc_α transformiert. Entlang der Reaktionskoordinate der α-Spaltung wird demnach die Koeffizientendichte an C_α in den Orbitalen ϕ_1 und ϕ_3 immer kleiner, um bei vollständiger Entkopplung zu verschwinden, während im Orbital ϕ_2, die Koeffizientendichte an C_α auf Kosten von derjenigen an den Zentren C_1 und O immer größer wird. Für die weitere Diskussion bleibt festzuhalten, daß sich zwar die Koeffizientendichten verändern, daß aber die nodalen Eigenschaften der Orbitale ϕ_1, ϕ_2 und ϕ_3 (s. Abb. 14) bis hin zur vollständigen Entkopplung erhalten bleiben. Die senkrecht zur Molekülebene stehenden Orbitale π_z und π_z^* der Carbonylgruppe bleiben in diesem Modell — sofern man von möglichen induktiven und hyperkonjugativen Effekten der Gruppe C_α absieht — unverändert.

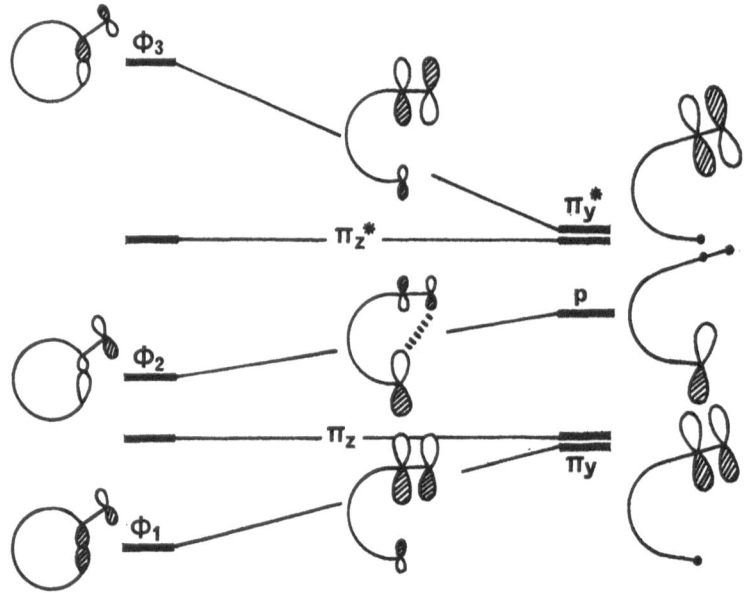

Abb. 14. Korrelationsdiagramm für die α-Spaltung eines Cyclanons; die Orbitale π_z und π_z^* sind die senkrecht zur Papierebene stehenden „eigentlichen" π-Orbitale der Carbonylgruppe

Nach Abb. 14 korreliert die für den Grundzustand des Ketons repräsentative Elektronenkonfiguration $\phi_1^2 \pi_z^2 \phi_2^2$ mit denjenigen des Acylium-Kations und des α-Carbanions, $\pi_y^2 \pi_z^2$ und p_{ca}^2. Weiter korreliert die für den π*,n-Anregungszustand charakteristische Elektronenkonfiguration $\phi_1^2 \pi_z^2 \phi_2 \pi_z^*$ mit denjenigen des Acylradikals und des α-C-Radikals, $\pi_y^2 \pi_z^2 \pi_z^*$ [65)] und p_{ca}. Im Grundzustand der Carbonylverbindung ist das Orbital ϕ_2 entlang der gesamten Reaktionskoordinate von zwei Elektronen besetzt, im π*,n-Zustand nur von einem. Dieses besetzte Orbital ist im Bereich zwischen den Zentren C_a und O, die bei einer zur α-Spaltung konkurrierenden, konzertierten Oxacarbenbildung eine Bindung eingehen, antibindend (s. Abb. 14) und widersetzt sich somit einer 1,2-Wanderung des Zentrums C_a vom C-Atom zum O-Atom der Carbonylgruppe.

3.2. Störungstheoretische Überlegungen und semi-empirische Berechnungen zur Oxacarbenbildung bei Cyclobutanonen

Das Korrelationsdiagramm der Abb. 14 gilt gleichermaßen für offenkettige wie für cyclische Carbonylverbindungen. Bei einem Cyclobutanon-Derivat ist es in der Regel durch das Korrelationsdiagramm der Abb. 15

zu ersetzen, das von einem bestimmten Dehnungsabstand C_1—C_4 ab die Orbitale ψ_2 und ψ_3 in vertauschter Reihenfolge enthält.

Der Grund hierfür liegt in der „through bond-Wechselwirkung"[48,49] der Orbitale ϕ_1, ϕ_2 und ϕ_3 mit den Orbitalen der σ-Bindung C_2—C_3 (s. Abb. 16 und 17).

Bei relativ geringem Dehnungsabstand ändern die Orbitale nur ihre Energielagen und ihre Koeffizientendichten an den jeweiligen Zentren,

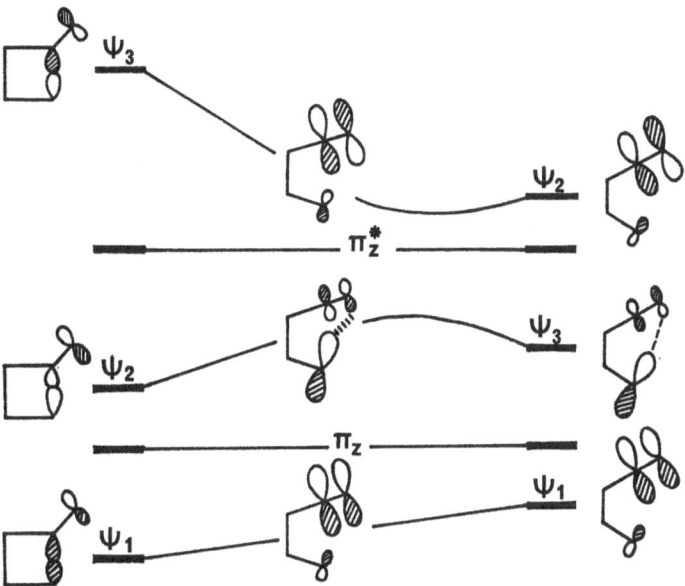

Abb. 15. Korrelationsdiagramm für die Spaltung der Bindung C_1—C_4 im Cyclobutanon. (Zur Interpretation der Diagramme hier und in den folgenden drei Abbildungen sei bemerkt, daß die nodalen und energetischen Eigenschaften der Orbitale bei „Multiplikation" mit −1 unverändert bleiben)

nicht aber ihre nodalen Eigenschaften (s. das Wechselwirkungsdiagramm der Abb. 16); bei relativ großem Abstand der Zentren C_1 und C_4 voneinander ändern sich auch diese [67] (s. das Wechselwirkungsdiagramm der Abb. 17). Insgesamt resultiert ein im Vergleich zu demjenigen der Abb. 15 komplexeres Korrelationsdiagramm (s. Abb. 18).

Für die weitere Betrachtung genügt die Beschränkung auf die jeweiligen Orbitale ψ_2 und ψ_3 der Abb. 19; dieses Schema sagt aus, daß
— sich die formalen *Korrelationslinien*, welche die Orbitalpaare vom Type ψ_2 und der Prägung ψ_3 jeweils miteinander verknüpfen, an einem

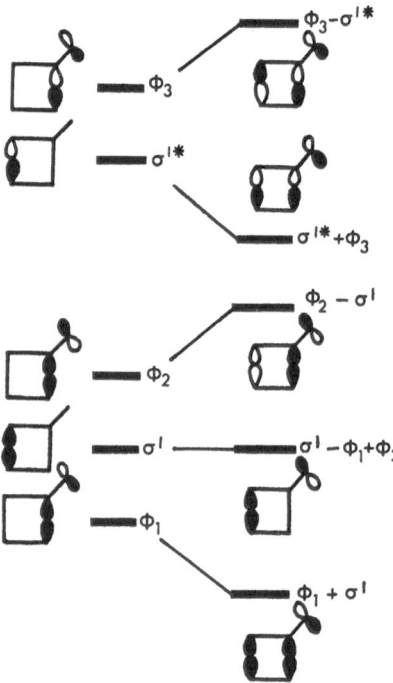

Abb. 16. Wechselwirkung der Orbitale ϕ_1, ϕ_2 und ϕ_3 bei geringem Abstand C_1—C_4 mit den Orbitalen σ' und σ'^* (links) zu σ-Orbitalen des bond-Isomers (rechts) (s. Abb. 19)

ganz bestimmten Punkt entlang der Reaktionskoordinate der Dehnung der 1,4-Bindung *kreuzen*;

— aufgrund vorhandener Pseudosymmetrie ein *Kreuzen der Orbitale* gerade eben *vermieden* wird, so daß beim Fortschreiten entlang der Reaktionskoordinate eine kontinuierliche Änderung des HOMO's von $\psi_2 \rightarrow \psi_3$ und des untersten unbesetzten σ-Orbitals von $\psi_3 \rightarrow \psi_2$ stattfindet;

— im Verlauf der Dehnung — bildlich gesprochen — das Orbital ψ_3 auf Kosten des Orbitals ψ_2 populiert wird, daß mit anderen Worten ein Elektronen-Transfer von ψ_2 nach ψ_3 erfolgt, wie er sonst nur durch eine (photochemische) Anregung erreicht werden kann;

— hier ein Fall der *bond/stretch-Isomerie* [68,69)] vorliegt.

Der strukturelle Bereich des bond-Isomers liegt links vom (verhinderten) Kreuzungspunkt der Orbitale ψ_2 und ψ_3 und schließt alle Kon-

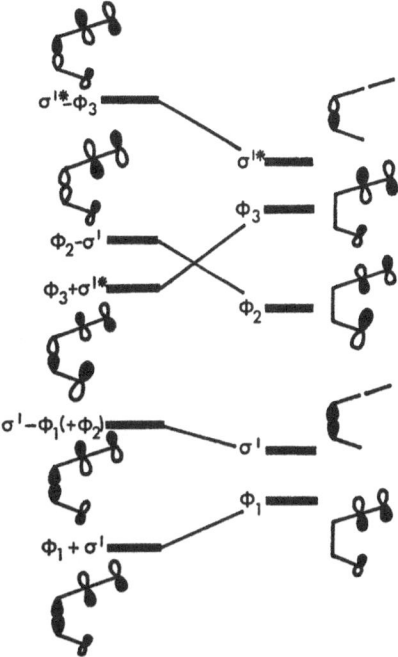

Abb. 17. Wechselwirkung der Orbitale ϕ_1, ϕ_2 und ϕ_3 bei großem Abstand C_1—C_4 mit den Orbitalen σ' und σ'^* (rechts) zu σ-Orbitalen des stretch-Isomers (links) (s. Abb. 19)

formationen mit ein, bei denen ψ_2 energetisch tiefer liegt als ψ_3; das stretch-Isomer erstreckt sich rechts davon und umfaßt alle Konformationen, bei denen ψ_3 unter ψ_2 angeordnet ist.

Links vom (verhinderten) Kreuzungspunkt und damit bei relativ geringem Abstand zwischen den Zentren C_1 und C_4
— dominiert die through space-Wechselwirkung zwischen diesen Zentren über die through bond-Kopplung durch die C_2—C_3-Bindung;
— weist das HOMO ψ_2 im Bereich C_1—C_4 bindenden und zwischen den Zentren C_4 und O antibindenden Charakter auf;
— ist die Oxacarbenbildung benachteiligt.

Rechts vom (verhinderten) Kreuzungspunkt und damit bei relativ großem Abstand zwischen den Zentren C_1 und C_4
— dominiert die through bond-Kopplung durch die 2,3 σ-Bindung über die through space-Wechselwirkung zwischen den Zentren C_1 und C_4;

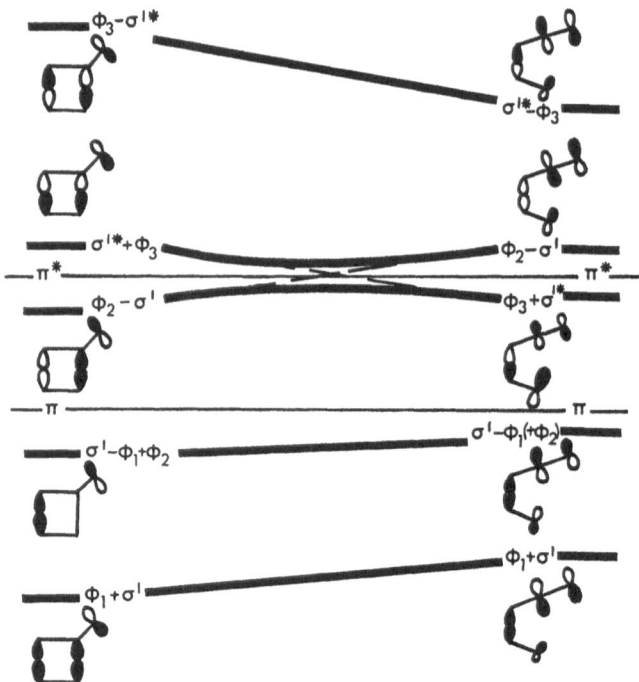

Abb. 18. Korrelation der Orbitale des bond-Isomers (links; s. Abb. 16) mit den Orbitalen des stretch-Isomers (rechts; s. Abb. 17) bei der Dehnung der Bindung C_1—C_4

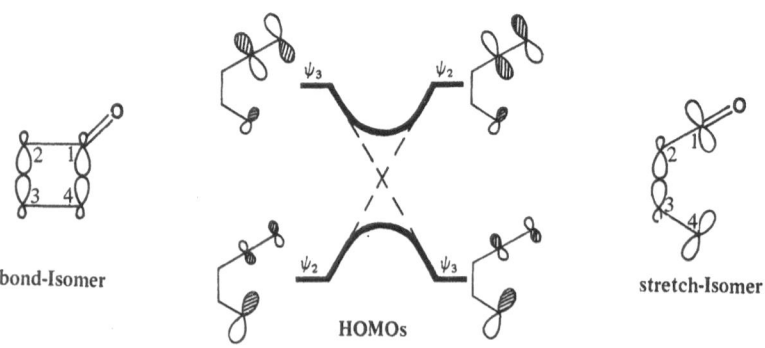

Abb. 19. Das Verhalten der σ-Orbitale ψ_2 und ψ_3 bei der hier vorliegenden bond/stretch-Isomerie

- ist das HOMO ψ_3 im Bereich C_1–C_4 antibindend, zwischen den Zentren C_4 und O dagegen bindend;
- ist die 1,2-Wanderung des Zentrums C_4 vom C-Atom zum O-Atom der Carbonylgruppe unter Ausbildung des entsprechenden Oxacarbens begünstigt.

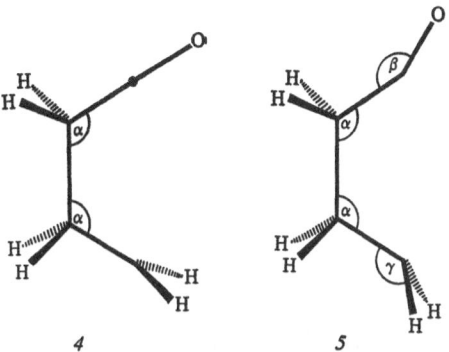

Abb. 20. Den EH-Berechnungen zugrunde gelegte Modell-Strukturen für die bond/stretch-Isomerie: *4*, mit einem Freiheitsgrad (α); *5*, mit drei Freiheitsgraden (α, β, γ)

Diese qualitative Charakteristik [70] läßt sich durch EH-Rechnungen [71] stützen, denen die einfache Modell-Struktur *4* mit nur einer Variablen (α, s. Abb. 20) zugrunde gelegt wurde und deren wesentliche Ergebnisse Abb. 21 ausdrückt.

Im unteren Schema ist die für den Grundzustand von *4* berechnete Gesamtenergie als Funktion von α aufgetragen; außer dem ausgeprägten Minimum für das bond-Isomer tritt bei einem Aufweitungswinkel von 117,5° das weniger deutlich ausgeprägte Minimum des stretch-Isomers auf.

Der obere Teil von Abb. 21 gibt für die hier interessierenden Bindungen die Abhängigkeit der „reduced Mulliken overlap populations" ebenfalls von der Größe von α wieder und erhärtet den mit zunehmendem Abstand zwischen den Zentren C_1 und C_4 einsetzenden Elektronen-Transfer von $\psi_2 \to \psi_3$ (s. Abb. 19): unterhalb von 115° ist der Bereich C_1–C_4 bindend, derjenige von C_4–O dagegen antibindend; oberhalb von 115° gilt umgekehrt, daß der Bereich C_1–C_4 antibindend, der Bereich C_4–O hingegen bindend ist. Bei $\alpha \approx 115°$ durchläuft der Wert der „reduced Mulliken overlap population" für den Bereich C_1–O ein Maximum; dieses Verhalten stimmt ebenfalls mit den zuvor entwickelten Modellvorstellungen überein: Bei Dehnung der Bindung C_1–C_4 „ent-

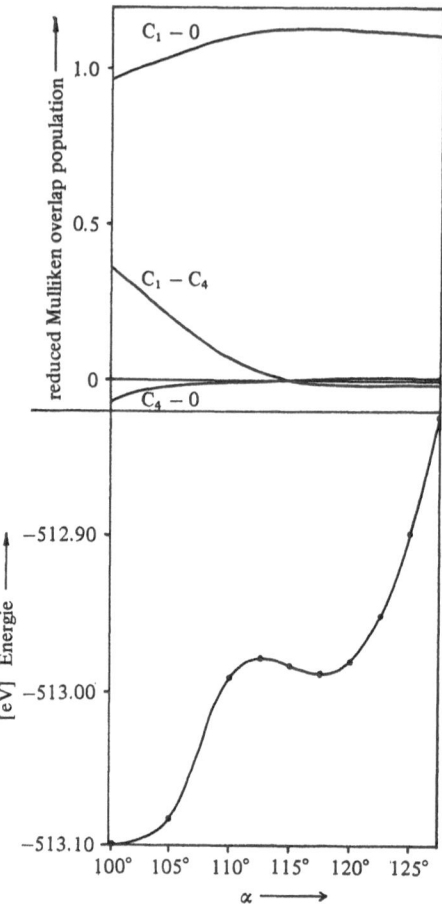

Abb. 21. Gesamtenergie von *4* (unten) und „reduced Mulliken overlap populations"
für C_1—C_4, C_1—O und C_4—O (oben) als Funktion des Winkels α

mischt" sich ψ_2 zusehends und verliert π_y^*-Charakter (s. Abb. 15). Damit
wird die Elektronendichte im π_y^*-Bereich laufend kleiner, die Elektronen
konzentrieren sich auf C_4 (s. Abschnitt 3.1 und insbesondere Abb. 14).
Hierdurch kommt der antibindende Charakter in π_y^* weniger zur Aus-
wirkung, so daß sich die Bindung C_1—O festigt. Vom (vermiedenen)
Kreuzungspunkt von ψ_2 und ψ_3 ab wird die schwach gewordene through
space-Wechselwirkung durch die through bond-Kopplung überkompen-
siert. Die sich mit fortschreitender Dehnung der Bindung C_1—C_4 am
Zentrum C_4 angehäufte Elektronendichte wird nun teilweise durch die
through bond-Wechselwirkung wieder nach π_y^* rücktransformiert [74],
so daß die CO-Bindung wieder schwächer wird.

3.3. Information über die molekulare Dynamik der Cyclobutanon/ Tetrahydrofuryliden-Isomerisierung mit Hilfe der EH-Methode

Die voraufgegangene, auf der Konzeption der through bond-Wechselwirkung basierende qualitative MO-Analyse [70] bringt für die lichtinduzierte Cyclobutanon/Tetrahydrofuryliden-Isomerisierung eine Konformation rechts vom (vermiedenen) Kreuzungspunkt mit ins Spiel; sie fordert allerdings weder, daß es sich hierbei um ein 1,4-Alkyl/Acyl-Biradikal handelt, noch schließt sie — im Gegensatz zur stereochemischen Argumentation (s. Abschnitt 2.2) — eine solche Spezies aus. Alles wird davon abhängen, ob der Übergang von einer höheren zur untersten Energiehyperfläche im Strukturbereich des Oxacarbens oder des 1,4-Biradikals stattfindet. Um ein Gefühl für Art und Ausmaß der molekularen Dynamik und insbesondere für die Relaxation der Liganden an C_1 und C_4, die mit der bond/stretch-Isomerisierung verknüpft ist, zu entwickeln, haben wir — ungeachtet des grob-qualitativen Charakters ihrer Aussage — mit der EH-Methode eine vollständige Energiehyperfläche für den Grundzustand sowie für den untersten Anregungszustand[75] der bond/stretch-Isomerisierung berechnet [76].

Dazu wurde die einfache Modell-Struktur *4* (s. Abb. 20) mit linearer Anordnung der Zentren O—C_1—C_2, coplanarer Orientierung von C_4 mit seinen Liganden und nur einem Freiheitsgrad zugunsten der subtileren Modell-Struktur *5* (s. Abb. 20) mit den Freiheitsgraden α, β und γ aufgegeben [71]. Abb. 22 gibt die Energiehyperfläche für den Grundzustand, Abb. 23 diejenige für den ersten Anregungszustand [75] wieder; Abb. 24 zeigt schließlich Schnitte durch diese Hyperflächen entlang der eingezeichneten Reaktionspfade.

Die Energiehyperfläche der Modell-Struktur *5* (s. Abb. 20) im elektronen-energetischen Grundzustand (s. Abb. 22) enthält einen Reaktionsweg, der im Tal beim Ausgangspunkt A (Cyclobutanon) beginnt, in einer bond/stretch-Isomerisierung über den Gipfel B zum Hochtal C (1,4-Alkyl/Acyl-Biradikal) führt und schließlich über einen weiteren Gipfel D den in einem zweiten Hochtal gelegenen Zielort E (Oxacarben) erreicht. Abb. 24 gibt Aufstieg und Höhenwanderung in einer anderen Perspektive wieder und weist darüber hinaus die schematisch ausgedrückten Strukturen der fünf Extrempunkte unter Mitberücksichtigung der Änderungen des Winkels γ aus.

Der formal symmetrie-verbotene Übergang vom bond-Isomer zum stretch-Isomer erzeugt definitionsgemäß das Maximum B. Der davor liegende Konformationsbereich ist dadurch gekennzeichnet, daß mit wachsendem Wert von β auch derjenige von γ zunimmt (s. Abb. 24); er umfaßt das „normale" Cyclobutanon (A) sowie Cyclobutanon-Konformationen mit gedehnter C_1—C_4-Bindung. Je größer im Bereich des

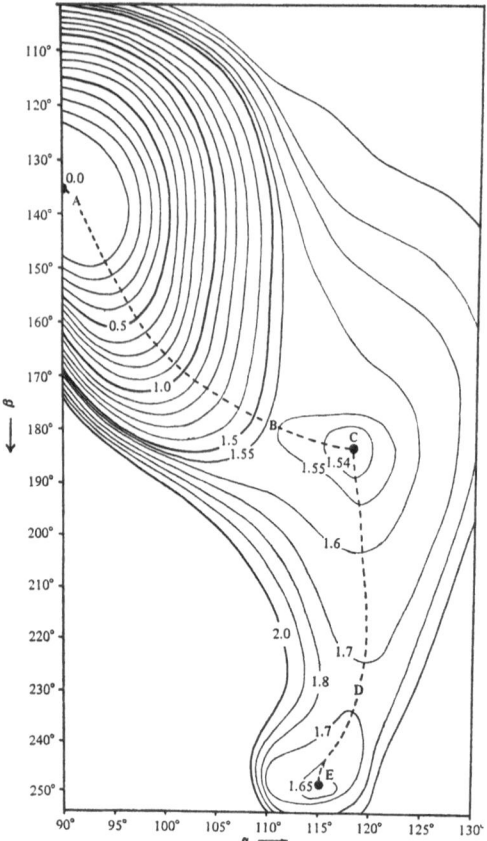

Abb. 22. Energiehyperfläche für den Grundzustand von *5* als Funktion der Winkel α und β aufgetragen; die Gebiete ohne isoenergetische Kurven sind solche mit rasch ansteigender Energie. Die angegebenen Energiewerte gelten relativ zum stabilsten Punkt, dem Cyclobutanon A

bond-Isomers der Abstand zwischen den Zentren C_1 und C_4 ist, um so berechtigter ist es, der betreffenden Struktur ein carbanionisches Zentrum an C_4 zuzuschreiben, das sich durch intramolekulare Charge Transfer-Wechselwirkung mit dem kationischen Acyliumsystem through space stabilisiert (s. Abschnitt 3.1). Infolge der Orientierung

— des „anionischen Orbitals" an C_4 nach innen (s. *6* in Abb. 25) wird die Überlappung zwischen C_4 und C_1 begünstigt;

— der O—C_1-Bindung nach außen (s. *6* in Abb. 25) wird die bindende Wechselwirkung zwischen C_4 und C_1 gefördert, die antibindende Wechselwirkung zwischen C_4 und O gemindert.

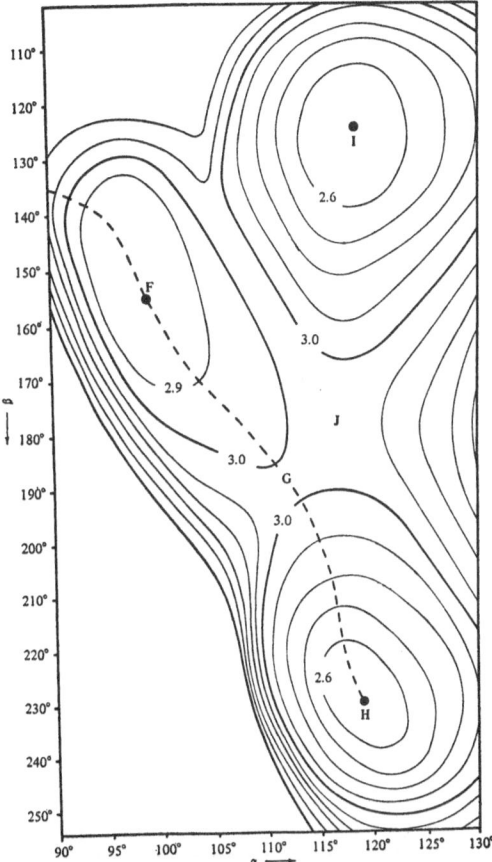

Abb. 23. Energiehyperfläche für den ersten Anregungszustand [75)] von *5* als Funktion der Winkel α und β aufgetragen; die Gebiete ohne isoenergetische Kurven sind Gebiete mit rasch ansteigender Energie. Die angegebenen Energiewerte gelten relativ zum stabilsten Punkt auf der Energiehyperfläche des Grundzustands in Abb. 22, dem Cyclobutanon A

Die bekannte Tendenz des Acylium-Kations zur Linearität steht den beiden zuletzt genannten Stabilisierungs-Effekten entgegen, wodurch die Abwinkelung des Dreizentren-Systems $O-C_1-C_2$ reduziert werden dürfte. Konformationen, die vor Erreichen des Maximums B relaxieren, bilden das ursprüngliche Edukt zurück (siehe die nodalen Eigenschaften von ψ_2).

Der hinter dem Maximum B liegende Konformations-Bereich ist dadurch charakterisiert, daß sich γ bei weiter ansteigendem Wert für β

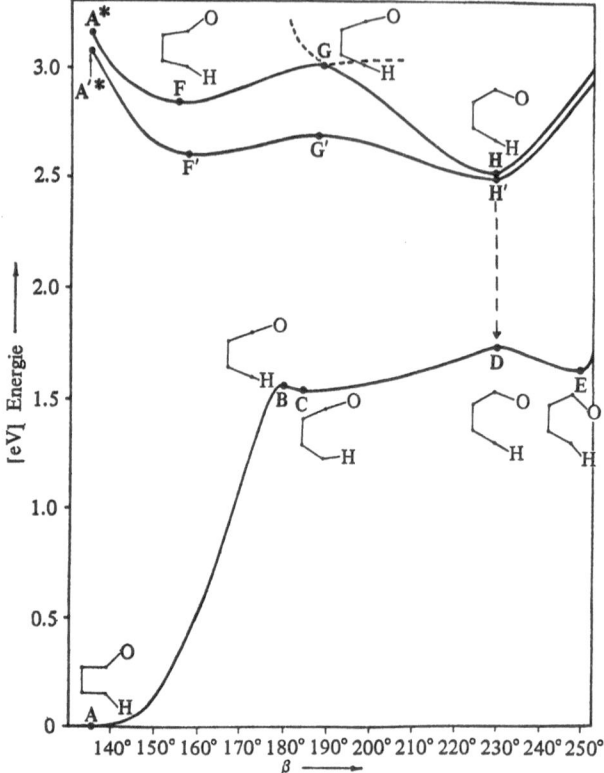

Abb. 24. Kurve A—E: Senkrechter Schnitt durch die Energiehyperfläche des Grund-zustands von *5* entlang des in Abb. 22 gestrichelt eingetragenen Reaktionspfads, als Funktion von β aufgetragen; Kurve A*—H: dasselbe für den ersten Anregungs-zustand [75] der Abb. 23. Die eingezeichneten Strukturen sind die für die ausgezeich-neten Punkte entlang der Reaktionsrouten berechneten Geometrien. Kurve A'*—H': analog zur Kurve A*—H, nur daß für die Berechnung der korrespondierenden Energiehyperfläche das O-Atom um 30° aus der Molekülebene herausgenommen wurde

unstetig ändert. Nachdem das stretch-Isomer nach Passieren von B den Charakter seines HOMO's relativ zum bond-Isomer gründlich gewandelt hat,

— invertiert das „anionische Orbital" an C_4 nach außen, um die through bond-Kopplung mit C_1 zu verstärken und die through space-Wechsel-wirkung mit C_1 zu schwächen (s. 7 in Abb. 25);

— schwingt das O-Atom nach innen, weil sich hierdurch die antibin-dende Wechselwirkung zwischen C_4 und C_1 erniedrigt und sich die

226

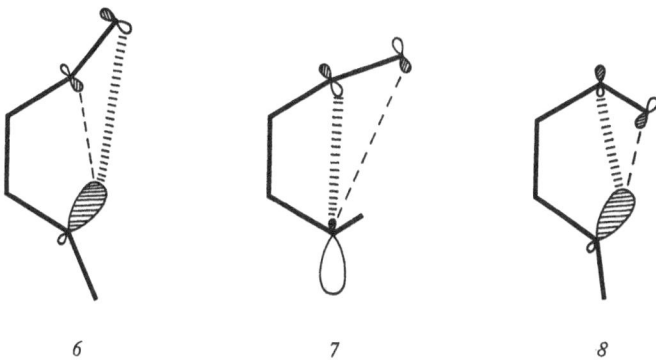

Abb. 25. Die jeweiligen HOMO's ausgewählter Konformationen auf der Grund-zustands-Hyperfläche (s. Abb. 24); 6: im Strukturbereich des bond-Isomers; 7: im Strukturbereich des stretch-Isomers (1,4-Alkyl/Acyl-Biradikal); 8: im Struktur-bereich des stretch-Isomers (Oxacarben)

bindende Wechselwirkung zwischen C_4 und O sowie die through bond-Kopplung zwischen C_1 und C_4 erhöht (s. 7 in Abb. 25); es wird wie-derum zu einem Kompromiß kommen, da das Acylium-Kation — für sich alleine genommen — die lineare Anordnung bevorzugt.

Dem Minimum C entspricht formal das 1,4-Alkyl/Acyl-Biradikal; sein oberstes, in der Rechnung zweifach besetztes Orbital ψ_3 ist im we-sentlichen auf das Zentrum C_4 konzentriert, und die Spezies kann als ein sp^3-hybridisiertes Carbanion verstanden werden, das stark durch through bond-Wechselwirkung mit dem Acylium-Kation stabilisiert wird (s. 7 in Abb. 25). Dem Minimum E entspricht das Oxacarben; diese Spezies läßt sich analog zur vorstehenden Überlegung formal als ein sp^3-hybridisiertes „Carbanion" auffassen, das durch through space-Wechselwirkung mit dem O-Atom stabilisiert wird (s. 8 in Abb. 25).

Der Übergang vom Biradikal zum Oxacarben, der mit einer Inver-sion an C_4 verknüpft ist, durchschreitet das Maximum D. In der hier vorliegenden Struktur stellt C_4 ein ebenes und somit weniger stabili-siertes Carbanion dar. Stark vereinfacht reduziert sich die Cyclisierung des Biradikals zum Oxacarben auf eine Inversion am carbanionischen Zentrum, und die Energiebarriere dieser Reaktion sollte dann wie eine Inversionsbarriere durch geeignete Substituenten an C_4 zu beeinflussen sein [77-79].

Der Reaktionspfad auf der Energiehyperfläche des untersten Anre-gungszustandes [75] für die coplanare Modell-Struktur 5 (s. Abb. 20) ist durch die Punkte A* bis H abgesteckt (s. Abb. 23 und 24). Im Minimum F weist das π^*,n-angeregte bond-Isomer gegenüber dem Cyclobutanon des Grundzustandes natürlich einen etwas längeren C_1—C_4-Abstand auf, da das bindende Orbital ψ_2 jetzt nur noch von einem Elektron besetzt ist. Aus dem gleichen Grund führt eine weitere Dehnung dieser Bindung zunächst zu einem wesentlich flacheren Energieanstieg als im Grundzustand, und anschließend bildet sich anstelle des flachen Minimums C das Hochplateau J aus.

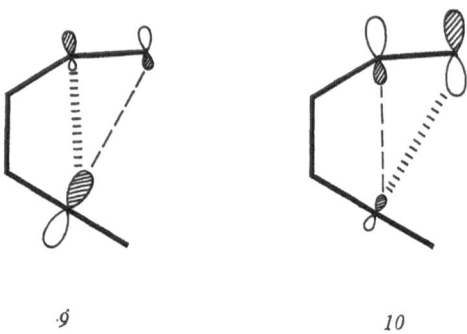

.9   10

Abb. 26. Mit jeweils einem Elektron besetzte Orbitale auf der Energiehyperfläche des untersten Anregungszustands [75]; 9: ψ_3 im π^*,n-angeregten Bereich; 10: ψ_2 im σ^*,n-angeregten Bereich

Der Übergang vom π^*,n-angeregten bond-Isomer zum π^*,n-angeregten stretch-Isomer ist — wie im Grundzustand — von einer Abwinkelung des Sauerstoffs nach innen auf C_4 hin begleitet ($\beta > 180°$), um die stabilisierenden Wechselwirkungen im einfach besetzten Orbital ψ_3 zu optimieren (s. 9 in Abb. 26); anders als im Grundzustand behält das System hierbei die ebene Anordnung am nun als Radikalzentrum anzusehenden Atom C_4 bei.

Die von ψ_3 induzierte Aufweitung des Winkels β bewirkt, daß sich das unbesetzte Orbital ψ_2, das im wesentlichen auf den Carbonylbereich konzentriert und hier antibindend ist, energetisch absenkt. Der Rechnung entsprechend kreuzt ψ_2 bei β ca. 190° das bis dahin besetzte, in seiner Energielage praktisch unbeeinflußte Orbital π^* (s. G in Abb. 24 [80] sowie Abb. 27) und übernimmt bei weiterem Absinken das dort befindliche Elektron; mit anderen Worten liegt jetzt das stretch-Isomer nicht mehr π^*,n-, sondern σ^*,n-angeregt vor und besitzt die Elektronen-Teilkonfiguration $\psi_3\psi_2$.

Dies hat zur Folge, daß sich die Aufweitung des Winkels β fortsetzt, die jetzt nicht nur von ψ_3, sondern zunächst auch von ψ_2 begünstigt wird; bis die anfangs schwache, im Verlauf der Einwinkelung aber stärker

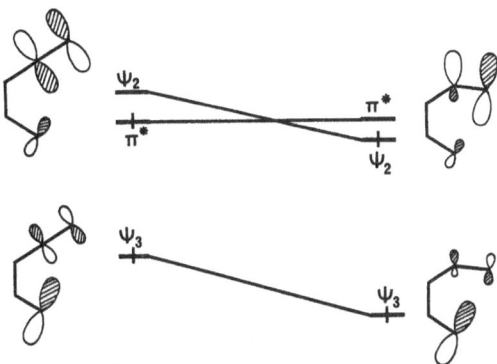

Abb. 27. Energiefolgen der Orbitale ψ_3, π^* und ψ_2 des zum Cyclobutanon gehörigen stretch-Isomers für $\beta <$ ca. 190° bzw. für $\beta >$ ca. 190°

werdende antibindende Wechselwirkung zwischen den Zentren C_4 und O (s. *10* in Abb. 26) die stabilisierende Bewegung im Minimum H auf der Energiehyperfläche mit σ^*,n-Charakter „stoppt". Ein weiteres stabilisierendes Einschwingen des O-Atoms nach C_4 hin ist von hier ab nur noch möglich, falls das Elektron aus dem C_4—O-antibindenden Orbital ψ_2 ins C_4—O-bindende Orbital ψ_3 überwechselt.

Dieser Orbitalwechsel („σ-gekoppelter [81] Übergang von H nach D) entspricht einer inneren Umwandlung des Moleküls vom untersten Anregungszustand[75] in den Grundzustand. Tritt dieser strahlungslose Übergang so rasch ein, daß eine thermische Äquilibrierung bei H nicht möglich ist[82], setzt das mobile Molekül die zuvor begonnene Bewegung auch nach Erreichen von D fort und gelangt so nach E. Daß sich Moleküle, die durch innere Umwandlung nach D gelangt sind, auf Reaktionswege nach E wie nach C hin verteilen, steht im Widerspruch zum stereochemischen Befund (s. Abschnitt 2.2).

Die Energiehyperfläche der Abb. 23 weist außer den Minima F und H noch das abseits der Reaktionsroute gelegene Minimum I auf. Dieser Strukturbereich ist in der Rechnung berücksichtigt worden, da für jeden α-Wert β im Bereich von 90 bis 250° variiert wurde; d.h. beim subtileren Struktur-Modell *5* (s. Abb. 20) sind nach Erreichen von Strukturen mit linearer Anordnung des Dreizentren-Systems O—C_1—C_2 im Hochplateau J auch solche Konformationen mit erfaßt worden, bei denen die O—C_1-Bindung eine Drehung um C_1 nach außen (unter Verkleinerung des Winkels β) erfährt. Für die hierbei auftretenden Konformationen gilt, daß ψ_2 — wie zuvor bei der gegenläufigen Bewegung — energetisch abgesenkt wird und nun noch deutlicher unter π^* zu liegen kommt, da beide Effekte, Reduktion der antibindenden Wechselwirkung für den Bereich $O-C_1$ wie für den Bereich $O-C_4$, in die gleiche Richtung wie die bindende through space-Wechselwirkung zwischen C_1 und C_4 wirken. Gleichzeitig wird die Energie des Orbitals ψ_3 angehoben. Da die Rechnung zeigt, daß sich die Orbitale ψ_2 und ψ_3 kreuzen, kann die Spezies im Minimum I demzufolge als ein σ^*,n-angeregtes bond-Isomer mit der Elektronen-Teilstruktur $\psi_2\,\psi_3$ betrachtet werden.

Zwei Argumente sprechen dafür, daß das elektronen-angeregte Edukt A* auf dem kürzesten Weg ins Minimum H und nicht unter Abzweigung nach I hin abreagiert. Die erste Überlegung ist elektronischer Art: Nach dem Überschreiten der (verhinderten) Orbital-Kreuzung schwingt die $O-C_1$-Bindung auf Grund der through bond-Kopplung nach innen und die Orbitale ψ_2 und π^* kehren sich in ihrer Reihenfolge um. Mit der Population von ψ_2 wird die Ablenkung, *deren Richtung bereits vorgegeben war*, weiter verstärkt. Die Rechnung reflektiert diese Einflüsse, denn der ermittelte Reaktionspfad verläuft erst gar nicht über das Hochplateau J, sondern läßt dieses links liegen. Das zweite Argument ist mechanistischer Art: Ein Molekül, das sich von A* entlang der Reaktionsroute verändert, schwenkt fortwährend die Bindung $O-C_1$ nach innen ein. Eine Abzweigung nach I wäre mit einer Umkehr der Drehbewegung um die Achse C_1 verbunden. Wegen der trägen Masse des O-Atoms bevorzugen wir denjenigen Prozeß, bei dem die Bewegungsrichtung beibehalten wird.

Ähnlich wie ein Elektron im Orbital ψ_2 des elektronen-angeregten stretch-Isomers das Dreizentren-System $O-C_1-C_2$ abwinkelt, wird beim relaxierten π^*,n-angeregten Cyclobutanon ein Elektron im π^*-Orbital der CO-Gruppe das Herausdrehen des Sauerstoff-Atoms aus der Molekülebene bewirken[86,87]. Um den tendenziellen Einfluß dieser Konformations-Änderung auf unser unrealistischeres, dafür transparenteres ebenes Modell aufzufinden, haben wir eine zweite Energiehyperfläche für den tiefsten EH-angeregten Zustand berechnet, wobei nun das O-Atom um 30° aus der Molekülebene herausragt. Die neu gewonnene Energiekurve A'* bis H' (s. Abb. 24) entspricht qualitativ dem durch die Punkte A* bis H charakterisierten Kurvenzug; die jeweilige Geometrie analoger Punkte ist praktisch gleich, sofern man von der Abwinkelung des O-Atoms aus der Molekülebene absieht.

Als Folge der $\sigma^*\pi^*$-Mischung im nichtplanaren System
— liegt die Kurve A'* bis H' energetisch tiefer als die Kurve A* bis H;
— wird vermieden, daß sich die Energiehyperflächen des „σ^*,n"- und „π^*,n"-angeregten stretch-Isomers kreuzen (s. *loc. cit.*[80]); sie „stoßen" sich vielmehr gegenseitig ab, wobei die untere Energiehyperfläche immer stärker σ^*,n- und die obere immer stärker π^*,n-Charakter annimmt. Anstelle des Kreuzungspunktes G beim ebenen Struktur-

Modell tritt wegen des nun gültigen Kreuzungsverbots das echte Maximum G' auf.

Die vorgebrachten EH-Ergebnisse, die natürlich nur grob-qualitativen Charakter haben, sprechen sich für die folgende diabatische [82] Photo-Isomerisierung von Cyclobutanonen zu Oxacarbenen aus (s. Abb. 24): Nach elektronischer Anregung des ketonischen Edukts A zu A* [88] findet auf der Energiehyperfläche des ersten Anregungszustandes Abreaktion durch das flache Minimum F' hindurch, über das sanfte Maximum G' hinweg, bis in das Minimum H' statt. Letzteres liegt genau über dem Maximum D der Grundzustands-Hyperfläche (s. Abb. 22 und 23 sowie 24); aus dieser guten strukturellen Übereinstimmung sollte wegen der quantitativ fragwürdigen Zuverlässigkeit der EH-Methode kein übertriebener Schluß gezogen werden. Durch einen strahlungslosen Übergang an dieser Stelle erreicht das Molekül in der Konformation von D den Grundzustand. Während des bisherigen Reaktionsverlaufs geschah eine kontinuierliche Vergrößerung des Winkels β — oder anders ausgedrückt — erfolgte ein stetiges Einschwenken der $O-C_1$-Bindung. Wie bereits erwähnt, wird das weiter relaxierende Molekül die einmal eingeschlagene Bewegungsrichtung beibehalten und zwanglos ins Minimum E des Oxacarbens gelangen, sofern es zuvor im Minimum bei H' nicht völlig „abgebremst" worden ist (s. *loc. cit.*[85]).

Das photochemische Primärprodukt Oxacarben reagiert als ein unter normalen Bedingungen instabiler Transient entweder zu typischen Folgeprodukten (Acetalen, Lactonen, Cycloaddukten; s. Abschnitt 2.1) weiter oder es invertiert über das Maximum D hinweg zum entsprechenden 1,4-Alkyl/Acyl-Biradikal. Derartige Biradikale sind bereits als Ringöffnungs-Produkte von Oxacarbenen erwogen worden (s. Abschnitt 2.3 sowie *loc. cit.*[61,89,90]).

Sie lassen sich abfangen [91], cyclisieren [89] in einer formal symmetrieverbotenen Reaktion zum Vierring-Keton, fragmentieren symmetrieerlaubt nach Grob [25a,b] zu Äthylen und Keten bzw. zu deren Derivaten oder verlieren CO (s. Abschnitt 1.2 sowie Tabelle 1). Da der Übergangszustand D bereits die erforderlichen sterischen [25a] und elektronischen [74] Voraussetzungen der Grob-Fragmentierung erfüllt, ist diese Reaktion bereits von hier aus denkbar. Wie detaillierte Betrachtungen der Grundzustands-Hyperfläche ferner erkennen lassen, ist vom Übergangszustand D aus auch eine Rückbildung des Vierring-Ketons unter Umgehen des Minimums C möglich. Nach der zur Hälfte erfolgten Inversion beim Übergang von E nach D halten wir das Abgleiten des Moleküls ins Minimum C für wahrscheinlich, aber nicht für zwingend; es mag von Fall zu Fall von den individuellen energetischen Verhältnissen abhängen.

Damit das Tetrahydrofuryliden fragmentiert oder sich zum Cyclobutanon umlagert, muß das Inversionsmaximum D auf jeden Fall erreicht werden. Für die kinetische Stabilität des Oxacarbens spielt die Höhe der Inversionsbarriere eine entscheidende Rolle. Inversionshemmende Substituenten an C_4 sollten die Fragmentierung bzw. die Rückbildung des Vierring-Ketons zugunsten einer bequemeren spektroskopischen Nachweisbarkeit (s. Abschnitt 2.3) sowie zugunsten der Bildung von Carben-Abfangprodukten (s. Abschnitt 2.1) zurückdrängen; inversionsfördernde Substituenten sollten den umgekehrten Effekt ausüben.

In diesem Sinn interpretieren wir z. B. die Bildung unterschiedlicher Typen von Reaktionsprodukten, je nachdem ob 2,2-Dimethyl-3-phenyl-cyclobutanon (s. Tabelle 1) oder 2,2-Bistrifluormethyl-3-phenyl-cyclobutanon (s. Abb. 4) bestrahlt werden. Für die erstere Verbindung ist eine höhere Inversionsbarriere zu erwarten als im letzteren Fall. Dementsprechend reagiert das methylierte Keton bei der Einwirkung von UV-Licht vornehmlich zum entsprechenden Oxacarben bzw. zu seinen Folgeverbindungen [27]; das fluorierte Keton unterliegt dagegen ganz der Grob-Fragmentierung (s. Abb. 4).

Zum Schluß sei noch bemerkt, daß bei expliziter Berücksichtigung der Elektronen/Elektronen-Wechselwirkung die EH-angeregte Kurve (s. Abb. 24) relativ zur derjenigen des EH-Grundzustandes abgesenkt würde und vielleicht sogar — etwa im Bereich des Minimums H, in dem die EH-berechnete HOMO/LUMO-Separation [53] klein ist — unter die Kurve des EH-Grundzustands zu liegen käme [92]. Dies könnte bedeuten, daß neben oder anstelle des closed shell-Biradikals C mit der Teilkonfiguration ψ_3^2 das open shell-Biradikal H mit der Teilkonfiguration $\psi_3 \psi_2$ als Minimum auf der Energiehyperfläche des wirklichen Grundzustands fungieren und auf Grund unserer Überlegungen und Berechnungen zwingend als Zwischenverbindung bei der Oxacarbenbildung auftreten würde. Gerade das Durchlaufen einer biradikalischen Zwischenverbindung haben wir jedoch experimentell ausgeschlossen (s. Abschnitt 2.2 und *loc. cit.* [41]) bzw. unwahrscheinlich gemacht (s. Abschnitt 2.3 und *loc. cit.* [56]), so daß unser einfaches Modell als erster Wegweiser für weiteres theoretisches Eindringen in das Problem der lichtinduzierten Cyclanon/Oxacarben-Isomerisierung zuverlässig erscheint.

Für anregende Diskussionen sind wir den Herren Professoren B. Ramsay und L. Salem verbunden. — Die Untersuchungen wurden von der Farbwerke Hoechst AG, der Deutschen Forschungsgemeinschaft, dem Fonds der Chemischen Industrie und der Stiftung Volkswagenwerk großzügig unterstützt und durch Forschungsmittel des Landes Hessen gefördert. Die Rechnungen wurden am Zentralen Recheninstitut der Johann Wolfgang Goethe-Universität durchgeführt. P. J. war zeitweilig Stipendiat der Schering AG. Den genannten Institutionen gebührt unser Dank.

4. Literaturangaben und Bemerkungen

1) Quinkert, G.: Pure Appl. Chem. *33*, 285 (1973).
2) Ciamician, G., Silber, P.: Ber. Deut. Chem. Ges. *36*, 1582 (1903) und nachfolgende Veröffentlichungen dieser Autoren in der gleichen Zeitschrift.
3) Carnap, R.: Philosophical foundations of physics. New York: Basic Books, Inc. 1966.
4) Dalton, J. C., Turro, N. J.: Ann. Rev. Phys. Chem. *21*, 499 (1970).
5) Turro, N. J., Dalton, J. C., Dawes, K., Schore, N.: Acc. Chem. Res. *5*, 92 (1972).
6) Quinkert, G., Heine, H.-G.: Tetrahedron Letters *1963*, 1659.
7) Quinkert, G., Heine, H.-G.: unveröffentlicht.
8) Quinkert, G., Wegemund, B., Blanke, E.: Tetrahedron Letters *1962*, 221.
9) Quinkert, G., Wegemund, B., Homburg, F., Cimbollek, G.: Chem. Ber. *97*, 958 (1964).
10) Quinkert, G., Moschel, A., Buhr, G.: Chem. Ber. *98*, 2742 (1965).
11) Norrish, R. G. W.: Trans. Faraday Soc. *33*, 1521 (1937).
12) Butenandt, A., Wolff, A.: Ber. Deut. Chem. Ges. *72*, 1121 (1939) sowie nachfolgende Veröffentlichungen dieses Arbeitskreises.
13) Quinkert, G., Blanke, E., Homburg, F.: Chem. Ber. *97*, 1799 (1964).
14) Quinkert, G.: Pure Appl. Chem. *9*, 607 (1964).
15) Calvert, J. G., Pitts, Jr., J. N.: Photochemistry, S. 390. New York: Wiley & Sons, Inc. 1966.
16) Barltrop, J. A., Coyle, J. D.: Chem. Commun. *1969*, 1081.
17) Closs, G. L., Doubleday, C. E.: J. Am. Chem. Soc. *94*, 9248 (1972); *95*, 2735 (1973).
18) Kaptein, R.: Privatmitteilung, Aug. 1973.
19) Lamola, A. A.: Energy transfer and organic photochemistry (ed. Lamola, A. A., Turro, N. J.). New York: Interscience Publ. 1969.
20) Wagner, P. J., Spoerke, R. W.: J. Am. Chem. Soc. *91*, 4437 (1969).
21) Quinkert, G., Cimbollek, G., Buhr, G.: Tetrahedron Letters *1966*, 4573.
22) Hostettler, H. U.: Tetrahedron Letters *1965*, 687.
23) Yates, P., Kilmurry, L.: Tetrahedron Letters *1964*, 1739.
24) Yates, P., Kilmurry, L.: J. Am. Chem. Soc. *88*, 1563 (1966).
25) Yates, P.: Pure Appl. Chem. *16*, 93 (1968).
25a) Grob, C. A.: Angew. Chem. *81*, 543 (1969); Angew. Chem. Intern. Ed. Engl. *8*, 535 (1969), dort weitere Literaturangaben aus diesem Arbeitskreis.
25b) Eschenmoser, A., Frey, A.: Helv. Chim. Acta *35*, 1660 (1952).
26) Ungezeichneter Artikel in Nachr. Chem. Techn. *1972*, 331.
27) Turro, N. J., Morton, D. R.: J. Am. Chem. Soc. *93*, 2569 (1971).
28) Turro, N. J., Cole, Jr., T.: Tetrahedron Letters *1969*, 3451.
29) Wagner, P. J., Stout, C. A., Searles, Jr., S., Hammond, G. S.: J. Am. Chem. Soc. *88*, 1242 (1966).
30) Morton, D. R., Lee-Ruff, E., Southam, R. M., Turro, N. J.: J. Am. Chem. Soc. *92*, 4349 (1970).
31) Kaplan, B. E., Turner, L. T.: Abstracts, 158th National Meeting of the American Chemical Society, New York, Sept. 1969.
32) Staab, H. A., Ipaktschi, J.: Tetrahedron Letters *1966*, 583.
33) Staab, H. A., Ipaktschi, J.: Chem. Ber. *101*, 1457 (1968).
34) Turro, N. J., Southam, R. M.: Tetrahedron Letters *1967*, 545.
35) Turro, N. J., McDaniel, D. M.: J. Am. Chem. Soc. *92*, 5727 (1970).
36) McDaniel, D. M., Turro, N. J.: Tetrahedron Letters *1972*, 3035.

[37] Morton, D. R.: Ph. D. Thesis, Columbia University, New York 1971.

[38] Erman, W. F.: J. Am. Chem. Soc. *89*, 3828 (1967).

[39] Scharf, H.-D., Küsters, W.: Chem. Ber. *104*, 3016 (1971).

[40] Hostettler, H. U.: Helv. Chim. Acta *49*, 2417 (1966).

[41] Quinkert, G., Jacobs, P., Stohrer, W.-D.: Angew. Chem. *86*, (1974); Angew. Chem. Intern. Ed. Engl. *13*, (1974) im Druck.

[42] Turro, N. J., Morton, D. R., Hedaya, E., Kent, M. E., D'Angelo, P., Schissel, P.: Tetrahedron Letters *1971*, 2535.

[43] Pacifici, J. C., Diebert, C.: J. Am. Chem. Soc. *91*, 4595 (1969).

[44] Turro, N. J., Lee-Ruff, E., Morton, D. R., Conia, J. M.: Tetrahedron Letters *1969*, 2991.

[45] Morton, D. R., Turro, N. J.: J. Am. Chem. Soc. *95*, 3947 (1973).

[46] Denis, J. M., Conia, J. M.: Tetrahedron Letters *1973*, 461.

[47] Smith, P. A. S.: Molecular rearrangements (ed. de Mayo, P.). New York: Interscience Publ. 1963.

[48] Hoffmann, R.: Acc. Chem. Res. *4*, 1 (1971).

[49] Hoffmann, R., Imamura, A., Hehre, W. J.: J. Am. Chem. Soc. *90*, 1499 (1968).

[50] Cookson, R. C., Edwards, A. G., Hudec, J., Kingsland, M.: Chem. Commun. *1965*, 98 erwähnen mit der Photo-Isomerisierung des 2,2,4,4-Tetramethylcyclobutan-1,3-dions, das sich nicht zu entsprechenden Oxacarben umlagert [30], zum konstitutionsisomeren Enoläther einen derartigen über ein 1,4-Biradikal formulierbaren Fall; s. ferner *loc. cit.* [50a].

[50a] Dowd, P., Sengupta, G., Sachdev, K.: J. Am. Chem. Soc. *1970*, 5726.

[51] Cram, D. J.: Steric effects in organic chemistry (ed. Newman, M. S.), S. 251. New York: Wiley & Sons, Inc. 1956.

[52] Miller, R. D., Dolce, D., Merritt, V. Y.: Manuscripts of Contributed Papers, IV. IUPAC-Symposium on Photochemistry, S. 172, Baden-Baden 1972.

[53] Gleiter, R., Hoffmann, R.: J. Am. Chem. Soc. *90*, 5457 (1968).

[54] Hoffmann, R., Zeiss, G. D., Van Dine, G. W.: J. Am. Chem. Soc. *90*, 1485 (1968).

[55] Rau, H.: Angew. Chem. *85*, 248 (1973).

[56] Quinkert, G., Kaiser, K. H., Stohrer, W.-D.: Angew. Chem. *86*, (1974); Angew. Chem. Intern. Ed. Engl. *13*, (1974) im Druck.

[57] Mauser, H.: Z. Naturforsch. *23b*, 1025 (1968).

[58] Quinkert, G.: Angew. Chem. *84*, 1157 (1972); Angew. Chem. Intern. Ed. Engl. *11*, 1072 (1972).

[59] Chapman, O. L., McIntosh, C. L., Barber, L. L.: Chem. Commun. *1971*, 1162.

[59a] Chapman, O. L., Mattes, K., McIntosh, C. L., Pacansky, J., Calder, G. V., Orr, G.: J. Am. Chem. Soc. *95*, 6134 (1973).

[60] Quinkert, G., Jürges, P.: unveröffentlicht.

[61] Hemminger, J. C., Rusbult, C. F., Lee, E. K. C.: J. Am. Chem. Soc. *93*, 1867 (1971).

[62] Zu Anwendungen der Störungstheorie in der Quantenchemie *loc. cit.* [63,64].

[63] Heilbronner, E., Bock, H.: Das HMO-Modell und seine Anwendung. Weinheim: Verlag Chemie 1968.

[64] Dewar, M. J. S.: The molecular orbital theory of organic chemistry. New York: McGraw-Hill 1969.

[65] Diese Elektronenkonfiguration entspricht dem linearen Anregungszustand des Acylradikals mit dem Einzelelektron in einem π-Orbital; energetisch nur wenig darunter liegt der gewinkelte Grundzustand des Acylradikals mit dem Einzelelektron in einem σ-Orbital. Zu Möglichkeiten, wie in einem Zustands-Korrelationsdiagramm das π^*,n-angeregte Keton direkt mit dem Grundzustand des Acylradikals korreliert, s. *loc. cit.* [66].

[66] Salem, L., Dauben, W. G., Turro, N. J.: J. Chim. Phys. *1973*, 694. — Salem, L.: J. Am. Chem. Soc., im Druck.

[67] Das Orbital ψ_3 wird durch das Orbital σ'^* der Bindung C_2—C_3 abgesenkt, ψ_2 wird durch das korrespondierende Orbital σ' angehoben. Dadurch wird die ursprüngliche Reihenfolge von ψ_2 und ψ_3 umgekehrt; denn mit wachsendem Abstand zwischen C_1 und C_4 nehmen die through space-Kopplung zwischen diesen Zentren und damit auch die Ausgangsseparation der Orbitale ψ_2 und ψ_3 ab.

[68] Hoffmann, R., Stohrer, W.-D.: Special Lectures at XXIIIrd International Congress of Pure and Applied Chemistry, Vol. 1, S. 157. London: Butterworth 1971.

[69] Stohrer, W.-D., Hoffmann, R.: J. Am. Chem. Soc. *94*, 1661 (1972).

[70] Stohrer, W.-D., Wiech, G., Quinkert, G.: Angew. Chem. *86*, (1974); Angew. Chem. Intern. Ed. Engl. *13*, (1974) im Druck.

[71] Die eingesetzten Parameter für die Rechnung waren die in *loc. cit.* [72,73] angegebenen, mit Ausnahme des H-Exponenten, für den der Wert 1,3 benutzt wurde; folgende Bindungslängen wurden verwendet: C_1—$C_2 = C_2$—$C_3 = C_3$—$C_4 = 1{,}54$ Å, C_1—$O = 1{,}22$ Å, C_2—$H = C_3$—$H = C_4$—$H = 1{,}09$ Å; die Ebenen H—C_2—H und H—C_3—H halbieren jeweils den Winkel α; die Winkel H—C_2—H und H—C_3—H wurden konstant zu $109{,}5°$ angenommen, der Winkel H—C_4—H wurde linear mit dem Winkel γ im Bereich von $109{,}5°$ bis $120°$ je nach Hybridisierung an C_4 variiert.

[72] Hoffmann, R.: J. Chem. Phys. *39*, 1397 (1963).

[73] Hoffmann, R.: J. Chem. Phys. *40*, 2745 (1964).

[74] Gleiter, R., Stohrer, W.-D., Hoffmann, R.: Helv. Chim. Acta *55*, 893 (1972).

[75] Wenn hier vom untersten Anregungszustand gesprochen wird, so steht lediglich die Elektronen-Konfiguration, nicht die Multiplizität zur Diskussion.

[76] Stohrer, W.-D., Wiech, G., Quinkert, G.: Angew. Chem. *86*, (1974); Angew. Chem. Intern. Ed. Engl. *13*, (1974) im Druck.

[77] Lehn, J. M.: Fortschr. Chem. Forsch. *15*, 311 (1970).

[78] Rauk, A., Allen, L. C., Mislow, K.: Angew. Chem. *82*, 453 (1970); Angew. Chem. Intern. Ed. Engl. *9*, 400 (1970).

[79] Müller, K.: Helv. Chim. Acta *53*, 1112 (1970).

[80] G charakterisiert den Schnittpunkt der Hyperflächen vom π^*,n- und σ^*,n-angeregten stretch-Isomer.

[81] Dekkers, A. W. J. D., Verhoeven, J. W., Speckamp, W. N.: Tetrahedron *29*, 1691 (1973); siehe dort weitere Literatur.

[82] Eine große Zahl photochemischer Reaktionen geht nicht adiabatisch auf der Energiehyperfläche eines angeregten Zustands vor sich, sondern erfolgt als diabatischer Prozeß, indem in einem bestimmten Strukturbereich dieser beiden Hyperflächen, der zwischen Edukt und Produkt liegt, ein vertikaler, strahlungsloser Übergang (innere Umwandlung bzw. Interkombination je nach Erhalt oder Wechsel des Spin-Zustands) stattfindet (s. *loc. cit.* [66,83,84]); zur Geschwindigkeit solcher Übergänge relativ zur zeitlichen Änderung der Kernkoordinaten s. *loc. cit.* [85].

[83] Förster, Th.: Pure Appl. Chem. *24*, 443 (1970); *34*, 225 (1973).

[84] Dougherty, R. C.: J. Am. Chem. Soc. *93*, 7187 (1971).

[85] Michl, J.: Mol. Photochem. *4*, 243 (1972).

[86] Moule, D. C.: Can. J. Phys. *47*, 1235 (1969).

[87] Condirston, D. A., Moule, D. C.: Theoret. Chim. Acta *29*, 133 (1973).

W.-D. Stohrer, P. Jacobs, K. H. Kaiser, G. Wiech und G. Quinkert

88) Es gibt gute Gründe anzunehmen, daß die Photo-Ringerweiterung beginnt, wenn A* im ersten angeregten Singulett-Zustand vorliegt und bevor die Boltzmann-Energieverteilung auf die Schwingungsniveaus geschehen ist. Für den Singulett-Charakter sprechen im Fall vieler Cyclobutanone erfolglose Energieübertragungen auf Ketontriplett-Quencher (s. *loc. cit.* 45); dort weitere Literaturangaben und Argumente); daß ^{1}A* auch schwingungsmäßig angeregt ist, wird durch Wellenlängenabhängigkeit der Fluoreszenz-Quantenausbeute gestützt (s. *loc. cit.* 61)).

89) Foster, A. M., Agosta, W. C.: J. Am. Chem. Soc. *94*, 5777 (1972); *95*, 133 (1973).
90) Carless, H. A. J., Metcalfe, J., Lee, E. K. C.: J. Am. Chem. Soc. *94*, 7221 (1972).
91) Dowd, P., Gold, A., Sachdev, K.: J. Am. Chem. Soc. *92*, 5724 (1970).
92) Buck, P., Gleiter, R., Köbrich, G.: Chem. Ber. *103*, 1431 (1970).
93) K. H. Kaiser, Dissertation 1973.

Manuskript eingegangen am 11. Oktober 1973

A. Schönberg

Preparative Organic Photochemistry

In cooperation with G. O. Schenck, O. A. Neumüller
Second, completely revised edition of "Präparative Organische Photochemie"
4 figures. 51 tables. XXIV. 608 pages. 1968
Cloth DM 148,—; US $ 57.00
ISBN 3-540-04325-X
Prices are subject to change without notice

This English edition of Schönberg's well-known monograph reflects the
remarkable progress that has been made since the appearance of the first
German edition in 1958. The book has been greatly enlarged, as a vast
number of new reactions had to be included. Nomenclature has been
standardized throughout, and the bibliography has undergone a thorough
revision. The typographical make-up and the presentation of chemical
formulae are much improved, so too are the detailed new indexes which make
this edition very easy to use. The aim of the book is unchanged. Theories
and reaction mechanisms are treated extensively in books already available.
This monograph is an exhaustive compilation of photochemical reactions
that are of preparative interest to the organic chemist. In every case sufficient
experimental details are given to make this book suitable as a manual
of photochemical laboratory techniques. It will be indispensable for any
worker in the field, and highly recommended for use in university courses.

Springer-Verlag
Berlin · Heidelberg · New York
München · Johannesburg · London · New Delhi · Paris
Rio de Janeiro · Sydney · Tokyo · Utrecht · Wien